Chalcogenide
Carbon Nanotubes and Graphene Composites

Abhay Kumar Singh
Department of Mechanical Engineering Science,
University of Johannesburg, South Africa

Tien-Chien Jen
Department of Mechanical Engineering Science,
University of Johannesburg, South Africa

 CRC Press
Taylor & Francis Group
Boca Raton London New York

CRC Press is an imprint of the
Taylor & Francis Group, an **informa** busine.

A SCIENCE PUBLISHERS BOOK

First edition published 2021
by CRC Press
6000 Broken Sound Parkway NW, Suite 300, Boca Raton, FL 33487-2742

and by CRC Press
2 Park Square, Milton Park, Abingdon, Oxon, OX14 4RN

© 2021 Taylor & Francis Group, LLC

CRC Press is an imprint of Taylor & Francis Group, LLC

Library of Congress Cataloging-in-Publication Data
Names: Singh, Abhay Kumar, 1976- author. | Jen, Tien-Chien, 1959- author.
Title: Chalcogenide : carbon nanotubes and graphene composites / Abhay
   Kumar Singh, Tien-Chien Jen.
Description: First edition. | Boca Raton : CRC Press; Taylor & Francis
   Group, 2021. | "CRC Press is an imprint of the Taylor & Francis Group,
   an informa business." | Includes bibliographical references and index.
Identifiers: LCCN 2020031837 | ISBN 9780367203146 (hardcover)
Subjects: LCSH: Chalcogenides. | Nanocomposites (Materials) | Carbon
   nanotubes. | Graphene.
Classification: LCC QD169.C5 S54 2021 | DDC 546/.72--dc23
LC record available at https://lccn.loc.gov/2020031837

ISBN: 978-0-367-20314-6 (hbk)
ISBN: 978-0-367-71076-7 (pbk)

Typeset in Times New Roman
by Shubham Creation

Preface

This book is intended for students, researchers and scientists working in the field of chalcogenide –nanocomposites or nanocomposite science from broad subjects such as physics, chemistry and nanotechnology and materials science. It would be mainly helpful for researchers, scientists, doctoral students as well as master and graduate levels university/college students, who wish to build a career in the area of chalcogenide-carbon nanotubes composites, chalcogenide-graphene composites, composite nanoscience and nanotechnology.

The prime goal of the book is to provide a clear idea on newly established chalcogenide-carbon nanotubes and chalcogenide-graphene composite materials in the glassy system and provide interpretations on their physio-chemical mechanisms as well as possible future applications in the area of nanoelectronics, optoelectronics, biomedical etc.

An effort has been made to present the topics in the book in a very simple manner. This current research book contains seven chapters, whose topics broadly deal in the beginning with an introduction of chalcogenide glasses to chalcogenide-carbon nanotubes, chalcogenide- graphene composites as well as their potential applications, including an interpretation of their deep physio-chemical mechanism at the nanoscale level.

The main features of the book are:

To review the scenarios of nanocomposite science for their wide range of applications in the distinct field of nanotechnology devices and developments.

To illustrate the scenarios of novel chalcogenide- carbon nanotubes and chalcogenide – graphene composite science for the glassy system and the vast potential utilities in the various technical areas for device fabrications and developments.

The detailed physio-chemical interpretations are addressed for various kinds of the chalcogenide glassy systems including preparation methods for the fabrication of the advanced nanomaterials functionalization ability to fabricate different kinds of composite materials.

To apprise the consequences of chalcogenide-carbon nanotubes composites materials for their prospective applications. It also points out the established concept of 'thermal quantum tunneling' with a demonstration of highly stiff multi-walled carbon nanotubes, which can disperse and diffuse (partially) in a low dimension chalcogenide glassy system.

To establish the consequences of chalcogenide-graphene composites for their wide range of future applications in various fields. It also acknowledges the concept of thermal quantum tunneling effect with the first successful diffusion (partial) and dispersion of the bilayer graphene in a low dimension chalcogenide alloy scheme.

To enlighten on the contradictory experimental findings of chalcogenide-carbon nanotubes and chalcogenide-graphene composites, in view of predictions of the theoretical concepts on these kinds of possible composites systems.

To notify on subjected composite systems breakthrough experimental findings as well as their interpretations for the possible wide range of applications in various fields in the future.

It discusses the imminent projections and tasks to demonstrate the newly established composite field and their potential nanoscale devices for the safety of the health of society and advanced technology.

Chapter 1

This chapter deals with chalcogenide glassy materials and their potential applications in several scientific and technological areas. Chalcogenide materials have attracted great attention due to their importance in nanoscience and nanotechnology. Their scientific and technological advances and adequate flexibility to make composites with the organic compounds were considered. it also briefly demonstrates the fundamentals of chalcogenide materials and their classifications as well as the emergent nano form of the materials. Therefore, crystalline structures of sulfur, selenium and tellurium are interpreted. The technologically significant polycrystalline chalcogenides such as polysulfides, polyselenides and polytellurides structures are also addressed. Since amorphous chalcogenide materials (or amorphous semiconductors) can have many technical applications, therefore, structural properties of these materials are also interpreted. Chalcogenides technical uses are always productive with their alloys along with other periodic table elements in the form of binary, ternary and multi-components alloys, hence a brief note on such compositions are also provided. Technologically sound chalcogenide glasses or amorphous semiconductors structures, types of bonding and related various theoretical models are addressed. These materials band structures are interpreted with the help of various models, while, the principles of defects are demonstrated from different models. A brief interpretation of the photoinduced effects in these materials is given. Moreover, ionic and electrical conductivities are also considered, and their basic concepts are discussed with the help of different theoretical models' descriptions. Subsequently, these materials electrical switching is one of the key properties, therefore, threshold switching and memory switching as well as various switching parameters are also addressed. A brief description on the potential utility of chalcogenide glasses (or amorphous semiconducting materials) is provided.

Chapter 2

The conceptual facts of chalcogenide glassy materials, are emphasized here and a basic description of glass formation and structural modifications are given. Additionally, a brief introduction on glass formation and infringements in chalcogenide materials are provided. Glass formation in such materials are described with the help of the kinds of glass formation criteria by interpreting the basic concepts. Structural characterizations of the glassy solids are important parameters for the scientific and technological applications as well as to define their physio-chemical properties. Hence, much attention is paid on interpreting the physical properties with the help of different criteria. Glassy materials are usually concerned with atomic ordering, therefore, an important description of the short- range and medium- range ordering in these systems are also incorporated. Moreover, in order to define various basic facts, the concept of rings and isolated molecules in chalcogenide glasses with a detailed description is provided. In terms of technological device applications, the disordered solids experimental physical parameters evaluations are significant according to well-established theoretical concepts, therefore, interpretation of the eutectoidal model for the stable electronic configuration as well as the glassy state with their experimental verification is provided. Additionally, the physicochemical analysis of the vitreous semiconducting systems interpretation is also described.

Chapter 3

The dimensionality of materials can offer distinct application compared to their conventional counterpart. Therefore, this chapter predominantly addresses carbon nanotubes emphasizing their historical facts and chronological developments including key physical properties. Specific attention is paid on the interpretation of zero, one, two- and three-dimensional carbon materials keeping in mind that carbon nanotubes application is largely influenced from their dimensions . A comprehensive description is provided on the basic concepts of the carbon nanotubes including their hybridizations. Moreover, graphite and different forms of carbon nanostructures as well as fullerene descriptions are also provided. A great deal of attention has also been paid to the technologically significant multi-walled carbon nanotubes (MWCNTs) and single-walled carbon nanotubes (SWCNTs) with their key physical properties interpretation, including a concrete description on their electronic properties, structural parameters, electronic structure, curvature effects and bundles of nanotubes. Since defective states are also technologically useful in many kinds of applications, therefore, an extensive description on distinct kinds of defects in carbon nanotubes are provided. The interpretations of various types of defects impact on carbon nanotubes physical properties are illustrated. Therefore, scientific and technologically important physical parameters such as mechanical properties, thermal properties, optical properties and electrical properties of the defective and non-defective carbon nanotubes are addressed. Thus, the detailed descriptions of the defect free and defects containing carbon nanotubes physical properties as well as types of the defects within these materials would be valuable to design high performance devices for the specified applications.

Chapter 4

Structurally well-ordered two-dimensional graphene sheets can offer robust physical properties in terms of their scientific/technical device performances. Therefore, this chapter contains a systematic chronological review of graphene developments with their various structural forms such as single and bilayer graphene. To fabricate these kinds of devices it is important to distinguish the structural differences between different forms of the graphene, therefore, a concrete description is provided on graphene, graphone, graphyne, graphdiyne, graphane. Since defects in a graphene sheet are generally useful to fabricate targeted devices with high performance, an important description on graphene defects is provided. Major graphene defects such as topological defects (Stone Wales defects, vacancies, ripples, ad-atom defects), line defects (dislocations and grain boundaries) are interpreted. Additionally, defects within the graphene also depend on their shape, size and number stacking layers, this indicates that with varying numbers of stacking layers the defects shape, size and characteristics are also changed, therefore, defects formation in the bilayer graphene are also provided. To get a better performance of the graphene based nano dimension devices, it is essential to know the deformation or cracks formation behavior of the materials. A description on cracks formation in graphene is also provided. In addition to this, general properties of graphene are also interpreted as well as brief discussion on the thermal, optical, mechanical and chemical properties of the both pure and defective graphene is given. Moreover, semiconducting properties of graphene including opening their band gaps for defective states are also described. Hence, the requirements of defective states of graphene is a debatable issue to improve the performances of graphene-based working systems.

Chapter 5

The Chalcogenide–Carbon Nanotubes (CNTs) composite system under the glassy system are dealt with in this chapter. Keeping in mind well established facts, composites can deliver more than individual counterparts in terms of their physical properties. Chalcogenide composites with different elements such as rare earth, polymer, oxide etc. materials have been intensively explored for more than five decades. Due to the fact that composite elements can influence the electrical, thermal optical properties of the materials, these are useful in various technical field applications such as optical fibers, optical non-linearity and optoelectronics based memory devices etc. Therefore, there is a significant emergence of composite materials for future potential utility in different scientific and technological applications. The novel chalcogenide –MWCNTs composites with the glassy system is introduced and their advantages in various forms are indicated. Therefore, in this chapter, a great deal of attention is paid on one of the growing potential research areas chalcogenide – nanotubes composites materials under a glassy configuration. Thus demonstrating different types of chalcogenide composite materials with carbon nanotubes . This chapter provides a detail description on the CNTs flexibility to make different kinds of nano composites including low dimension chalcogenide alloys. Additionally, their bond sharing ability in terms of covalent functionalization, non-covalent functionalization and alternative routes functionalization are also addressed in details. Subsequently, favorable conditions of chalcogenide alloys and carbon nanotubes for composite formation are interpreted. A brief overview on nano crystalline chalcogenides–CNTs and polycrystalline chalcogenides CNTs composites with their key advantages is also provided. In this order attention is drawn on deformation and diffusion (partial) of the MWCNTs in a chalcogenide glassy alloy, such composites exhibit a drastic enhancement in their physical properties (like optical and electrical properties) due to the structural modifications in MWCNTs. The scientific interpretation of the structural and configurational modifications in MWCNTs as well as chalcogenide alloys in terms of composite formation is also provided. The single step synthesized chalcogenide–MWCNTs glassy composite materials surface morphologies, Raman spectrum peaks structural interpretation are illustrated. Additionally, other physical characterizations such as UV/visible optical properties and infrared of the chalcogenide–MWCNTs composites are also described. Moreover, a brief overview on the mechanical and electrical properties of the composite materials for those synthesized from non- direct single methods are also provided. Thus, the evidence of different studies on two or more steps fabricated chalcogenide–CNTs composites materials have demonstrated a drastic enhancement in mechanical properties and electrical conductivity. However, available evidence on thermal properties with single step synthesized composites, have led to contradictory experimental findings which are also interpreted. The lower thermal stability and other inferior crystalline kinetic parameters experimental findings could be major challenges for theoreticians and experimentalists to reconsider a generalized concept for prospective chalcogenide–MWCNTs glassy composites. Though, this research area has been recently introduced and established, it requires intensive indepth investigations with these kinds of composite systems to make a conclusive view on it. In view of the chalcogenide alloys characteristics (or behavior) change with the selection of alloying established constitutes.

Chapter 6

The novel concept in the area of composite materials could provide important features of the materials. More precisely unexplored chalcogenide – graphene composite under the glassy system is one of the potential fields for innovation. Therefore, this chapter deals with newly invented chalcogenide- graphene glassy composite systems based on their experimental and theoretical conceptual interpretations by providing chalcogenide alloys structural properties and their composite formation ability with the inorganic/organic components. Therefore, the basic concept of building blocks in chalcogenide including clusters tetrahedral building blocks, inorganic–organic frameworks building from metal chalcogenide nanoclusters, properties of open-framework in chalcogenides are interpreted. Similarly, the basic properties of graphene are also addressed in terms of their composite formation abilities with various forms of the material, such as, graphene membranes, graphene energy, graphene sensors, graphene thermoacoustic devices, graphene magnets, graphene superconductors and graphene in biomedicine. The fullerene like structural similarities of the chalcogenides and graphenes are useful for their composite formation. This is described in the chapter by providing examples of the nanocrystalline chalcogenides-graphene composites, polycrystalline chalcogenides-graphene composites and amorphous chalcogenides-graphene composites systems. Therefore, this chapter demonstrates novel chalcogenide-garphene composites with the single layer as well as bilayer layers graphene systems under the glassy system. Considering the lower stiffness of the bilayer (or a few layer) graphene compared to single layer graphene, special attention is paid on low dimension chalcogenide composite with less stiff material. Hence, this work interprets innovative research with the chalcogenide -bilayer graphene composite systems under a glassy system by providing structural modifications, modifications in surface morphologies, modifications in Raman spectrum profiles, alternation in UV/Visible optical properties and PL property and FTIR transmission. Moreover, a brief overview on the unexplored topic of mechanical and electrical properties of such systems is also provided. Additionally, available experimental findings on thermal properties of the chalcogenide-graphene glassy composites are also interpreted by providing a description on the crystallization kinetic parameters, such as glass forming ability, crystallization activations energies and thermal stability. This introductory innovative research has established inferior crystallization kinetic parameters, while, according to theoretical predictions on such possible potential composites, this should improve these parameters impassively, but contradictory experimental findings were found and interpreted with the novel concept. Thus, intensive theoretical and experimental research attention is desired to resolve the issue for such composite materials, to explain experimental findings from a universal theoretical model. On the other hand, unexpected room temperature PL property and significant enhancement in IR transparency compared to parent chalcogenide glassy alloys are impressive features to fabricate efficient performing devices at nano levels.

Chapter 7

Innovative conceptual innovations in materials science are always associated with their existing technologies as well as future applications. Therefore, this chapter deals with the key technical application of the chalcognide composite materials in the form of nano crystalline chalcogenide systems-carbon nanotubes (CNTs), polycrystalline chalcogenide systems–CNTs and amorphous chalcogenide systems–CNTs. These composites are applicable for various distinct purposes such as optoelctronics, photovoltaic, nanoelectronics etc, with improved performances. Additionally, an overview on the possible applications of the distinct graphene composites including amorphous chalcogenide-graphene composite materials is provided. Moreover, a discussion on some possible future applications of recently developed new class of amorphous chalcogenide–MWCNTs and chalcogenide-graphene composite glassy system materials are described. The projected possible future applications are based on their unique IR transparency property, nonetheless, it is not limited to this, their range of application can be wider in various technological areas. However, future applications of these new class composite materials are still open for innovations to fabricate efficient performing devices in various technological areas.

Acknowledgement

The authors wish to acknowledge their colleagues, friends, and faculty members of various departments for their valuable suggestions and encouragement during the preparation of the contents of this book.

They are also grateful to the publisher of the book and the journals cited in the bibliography in respective chapters.

Abhay Kumar Singh
Tien-Chien Jen

Contents

Basic of the Chalcogenides

INTRODUCTION

Materials crystalline structures are the unacknowledged pillars of modern technology. Without knowing the form of the crystallinity of the materials, it will be difficult to predict for their potential use in distinct applications such as electronic industry, photonic industry, fiber optic communications, which widely depend on the materials/crystals forms such as semiconductors, superconductors, polarizers, transducers, radiation detectors, ultrasonic amplifiers, ferrites, magnetic garnets, solid state lasers, non-linear optics, piezo-electric, electro-optic, acousto-optic, photosensitive refractory of different grades and nanocrystalline films for microelectronics and computer industries. Hence the crystal growth is an interdisciplinary subject covering Physics, Chemistry, Material science, Chemical engineering, Metallurgy, Crystallography, Mineralogy, etc. In the recent years, a growing interest on various kinds of the crystal growth processes, particularly in view of the increasing demand has been made of materials for technological applications such as atomic arrays that are periodic in three dimensions with repeated distances, poly-crystalline material with a few uneven atomic structure and amorphous materials having short range random atomic arrangement. Each kind of materials have own technical advantages and disadvantages reason to grow perfect crystalline structure for many physical properties of solids that obscured or complicated by the effect of grain boundaries with the key advantages to anisotropy, uniformity of composition and the absence of boundaries between individual grains. While the polycrystalline materials useful to construct where partially non perfect crystalline are required. On the other hand, amorphous form of the materials also equally important for the various specific utility in which random atomic are required. Hence, to achieve high performance from the device, a well define form of the material is desired. Also, their crystallinity characterizations toward device fabrication have assumed great impetus due to their importance for both academic as well as applied research. Chalcogenides are also technologically an important class of materials that includes oxides, sulfides, selenides and tellurides. Three heaviest elements of the sub-group, namely sulfur, selenium, tellurium, and polonium, be collectively referred to as the "chalcogens," therefore, the term chalcogen be addressed only for these elements in practice. The chemically and technologically selenium and tellurium are to be considered more important to others. Although, according to the official guides to inorganic nomenclature, the term applies equally to all the elements for the 16 group of the periodic table, however, with being proper definition the oxygen and sulfur are also fall in this category. Although several text textbooks imply that of "chalcogens" oxygen is excluded from the chalcogens group based on the chemistry of oxygen.

As per available information's the term "chalcogen" was proposed around 1930 by Werner Fischer, when he worked in the group of Wilhelm Biltz at the University of Hannover, to explore 16 group elements. This nomenclature was quickly accepted among German chemists Heinrich Remy and recommended their official use in 1938 while being a member of the Committee of the International Union of Chemistry (later IUPAC) to Reform of the Nomenclature of Inorganic Chemistry. In this subsequent it was internationally accepted the elements oxygen, sulfur, selenium, and tellurium will be called chalcogens and their compounds chalcogenides. The term derives from the Greek terms $\chi\alpha\lambda\kappa'o\varsigma$ meaning copper and $\gamma\varepsilon\nu\nu'\omega$ meaning giving birth (cf. "hydrogen" similarly originating from $'\,\upsilon\delta\omega\rho$ meaning water; also "oxygen", etc.). The chemistry of soluble metal chalcogenide complex materials, either containing chalcogen–chalcogen bonds or only chalcogen–metal, has been studied extensively in beginning such as sulfur, afterward in the mid-1970s selenium and tellurium had also explored widely. Particularly, metal–sulfur systems have a long chemical history in all aspects, but from the 1960s due to growing interest in the related complex materials it was renewed for their significance in bioinorganic chemistry and hydrodesulphurization and other catalytic processes. In the early stage identification of the many possible coordination modes of the sulfide ligands has been summarized. A large number of synthetic molecular transition of the complexes with either terminal or bridging sulfide ligands and catalytic activity have been studied. In this order the coordination modes and structural types of soluble complex selenides and tellurides have been synthesized in solution or in the solid state. The excellent introduction to the synthetic and structural coordination chemistry of inorganic selenide and telluride ligands covering all facts as well as emphasis on compounds molecular nature has been reviewed with time to time. In the subsequent progress the metal containing chalcogenides complexes are become impressive due to their significance play an important role in the field of low-dimensional solids. Predominately, the origin of the resistivity anomalies observed in layered transition metal chalcogenides that stimulated the interest in low-dimensional inorganic materials has become a great area of the interest. The metal clustering and low-dimensional structures are frequently explored for the transition metal chalcogenides. In fact, in contrast to the ionic 3D-type oxides, these compounds tend to form covalent structures. Therefore, a reduced relative charge on the metal favors metal–metal bonding. In the metal-rich compounds the preferred coordination polyhedra occur for the non-metal (chalcogen) atoms. Linkage of the polyhedra takes place in such a way that they often end up with an ordered arrangement identical like isolated metal clusters. Usually, clusters are rarely isolated in the chalcogenide structures and they condense by sharing common vertices, edges, or faces. More unusually they may be connected via significant chemical bonding between the vertices. They can also form columns, in which the central metal atoms interact to give chains in the same direction. Specifically, in the layered chalcogenides enough d-electrons are available for the significant metal–metal (M–M) bonding in two dimensions. The interaction of M–M bonding could enhance the dimensionality of the material. However, in some cases, the cluster network is better in regard to 3D metal framework, in terms of metal packing arrangement. Hence the complexes in which metal clusters are coordinated by chalcogenide or polychalcogenide ligands occupy a special position among the so-called inorganic or high-valence clusters. The most characteristic being those of 4d- and 5d-metals of groups V-VII.

Additionally, the semiconducting nature and other fascinating structure related properties of transition metal dichalcogenides has triggered an intensive research on these materials. The accumulated knowledge on the properties of these materials including their derived nanostructures such as nanotubes, nanoribbons, nanoclusters, and fullerene-like nanoparticles envisions a multitude of potential applications. Nanostructures based on transition metal disulphides may also have applications in industry as catalysts and lubricants. They may also consider as the potential candidate for the electronic industries to make a verity of things such as transistors, logical elements, sensors, flexible electronics, energy technology (solar energy, hydrogen storage, batteries), nanotribology and advanced engineering materials (nanocomposites). The nanostructured materials unusual geometry and their promising physical properties are become a great interest of the area for the nanoscale chalcogenide research. Thus, different forms of the chalcogenide materials are the attractive due to a variety of reasons. They can be prepared in polycrystalline and amorphous form the variety of ways such as vapour-deposited thin films, melt - quenched glasses, melt quenched chalcogenide composites. They mostly form glasses continuously over wide composition ranges. Their physical properties also vary in a continuous behavior and physical characteristics often in a unique way that makes them useful for a number of actual and potential technological applications to these materials. Hence this work is predominately intended to classification of chalcogenides materials, with the brief description on nanostructured chalcogenides, crystalline, polycrystalline and amorphous structures of the key chalcogen materials such as S, Se and Te. A description on chalcogenide glasses including history and developments and their binary, ternary and multi-components compounds. Chalcogenide glasses structures has been addressed with a description on short-range, medium atomic arrangements as well as distinct morphologies and subsystem defects. Bond formation mechanism with the distinct interactions for the kind of chalcogenide systems and descriptive notes of the well stablished key theoretical models as well as network topological and critical chemical thresholds in amorphous semiconductors. A detailed description on the band structures of crystalline and amorphous semiconductors has been also addressed. In this study chemical view on the band structure of the amorphous semiconducting materials are also discussed. Subsequently, defects in amorphous semiconductors (including Street and Mott model for the charged dangling bond, Kastner-Adler-Fritzche Model) has been discussed. The photoinduced effects, ionic conductivity (described by the different models), electronic conductivity (by providing different models explanations), negative resistance and electrical switching are discussed in detailed. A brief section on the application of the chalcogenide glasses are also addressed.

CHALCOGENIDE MATERIALS AND CLASSIFICATION

The literal meaning of the word 'chalcogen' is 'ore' forming materials which represents a combination of the *Greek* word *khalkos,* meaning copper (ore or coin). The Latinized *'Greek'* word *'genes'* meaning is born or produced. Chalcogenides are Group 16 elements in the periodic table. Specifically, Sulfur (S), Selenium (Se), and Tellurium (Te) are the chalcogen elements. The S, Se, Te containing chalcogenes have technical application in electronics, photovoltaics and optoelectronics industries and are always in the form of chemical compounds with the other group of the elements. Chalcogenides chemical compounds consist of at least one chalcogen anion and one or more electropositive element. Chalcogenides are sometimes also referred as elements of the oxygen family including sulfides, selenides, tellurides and radioactive element Polonium (Po). Generally, element oxygen is treated separately from the chalcogen group elements due its distinct chemical behavior from sulfur, selenium, tellurium and polonium. Specifically, element 'S' containing materials show a high refractive index, nonlinearity, large kerr, good IR transmission beyond 1.5 μm, and good chemical resistance with a good response under the direct patterning exposure to near the band gap light. Collectively these adequate properties make them useful for the various technical applications like fiber Bragg gratings, fiber-optic communication, evanescent wave fiber sensors, etc. The element 'Se' containing chemical compositions have many applications such as solar cells, xerography, rectifiers, photographic exposure meters, anticancer agents, etc. It is also useful in eliminating the bubbles in the glass industry as well as removing undesirable tints in iron production. 'Se' has a high reactivity rate with suitable chemicals; this potential can be exploited to convert this element into functional materials. 'Se' based amorphous materials are used for imaging and biomedical applications. Crystalline 'Se' with other alloying elements such as Cu, In, Ga, etc. is widely used for photovoltaic and photo-detection applications. Tellurium 'Te' is a more metallic element in the chalcogen group. Te based materials are widely used for data storage devices due to their adequate transformation ability amorphous to crystalline and crystalline to amorphous. Se and Te are polymeric divalent materials with chain structures. They can form kinds of bonds with different bond strengths due to the cross-linking elements. The Se/Te cross-linked materials have a huge number of unbonded lone pair electrons which can easily be excited under optical and electrical fields. The crystalline phase change occurs when the amorphous phase cannot have lone pair excitation energy. Se/Te lone pair coating polymeric structures have a vibrational nature in which electronic transitions are possible owing to the motion of the chains. This property of such materials is mostly used for the phase change memory application. Due to the potential phase change memory property these materials have several high processing microelectronics utilities such as a microprocessor can contain transistors down to 15 nm, as a consequence well-known optical Moore's physical limits have been reached. The nanostructures of the materials make them front-liners for all kinds of industrial applications. Here the emphasis is on the technical utility of the inorganic and organic compounds formed by the two important chalcogenes Selenium (Se) and Tellurium (Te). Specifically, with a sound scientific concept on the host 'Se' chalcogen systems with additive organic compounds and their potential applications in the future in different areas such as imaging, biomedical, high processing electronics, sensors, memory devices and photovoltaics, etc. [1].

Chalcogen group elements (S, Se, Te) are accompanied with almost every group of the periodic table except noble gases and some of radioactive elements. The chalcogen-based alloys generally follow the traditional chemical valence trends such In_2Se_3, SeZn, etc. However, several chalcogen containing alloys are exceptions of this common behavior e.g., P_4S_3. The crystallographic structures of the key chalcogen group are in a directional manner under the dominance of the covalent directional bonding. In general chalcogen gives positive oxidation states with halides, nitrides and oxides [1].

NANOSTRUCTURED CHALCOGENIDES

Nanomaterials and relevant nanotechnology are considered key technologies for the 21st century. Materials like nano-glass ceramics are expected to play a major role since they offer improving certain properties which have been discovered recently.

Gustav Tammann had established the bases of crystallization in glasses in 1933, by adopting the two main concepts of nucleation and crystal growth. The nucleation and growth process depends on the matrix and crystalline nuclei composition [2]. The major theoretical studies concerning nucleation and crystallization of glasses were developed in the 60s by Jackson and Thakur et al. [2]. Despite the great potential application of nanocrystalline glass ceramics, the fundamentals of the crystallization mechanisms are not yet fully understood. Theories of crystallization are usually restricted to isochemical systems in which the crystalline phase has the same chemical composition as the glass matrix. Such a challenge can be realized by considering a large volume of alloys with varying concentrations of crystals with sizes in the 5–50 nm range with a narrow size distribution of the multicomponent systems, by governing the change in the chemical composition of the glass. As a consequence, their interphase can be formed during nucleation and crystal growth process, as an example Se–Zn–Te–In, etc. [3].

Thus, a nanomaterial is an object that has at least one dimension in the nanometer scale. Such materials can exhibit properties that are drastically different from its bulk, due to their increased surface-to-volume ratio and/or quantum confinement effects. The solution based metal chalcogenides are also receiving great attention due to their layered two-dimensional structures with well defined band gaps. The required band gap can be found in Transition Metal Dichacogenides (TMD) similar to graphite. Transition metal chalcogenides can be considered as a stack of triple-layers because each triple-layer consists of transition metal layer between two chalcogen layers. Atoms within a triple-layer can be chemically bonded together by strong ionic-covalent bonds, whereas triple-layers can be held together through weak van der Waals interactions, and can be mechanically extracted. Hence, two-dimensional chalcogenides (semiconductors) transition metal and their basic building materials can be useful for nanoelectronics [4]. The semiconducting nature and other fascinating structure-related properties of transition metal disulphides (dichalcogenides) have also become an intensive area of two-dimensional materials. The acquired knowledge on the properties of such materials and their nanostructures (such as, nanotubes, nanoribbons, nanoclusters and fullerene-like nanoparticles) can find a large number of potential applications. The transition metal disulphides (MoS_2, WS_2) nanostructures are already used as catalysts and lubricants in industrial applications. In the near future, they can have a profound impact on electronics (transistors, logical elements, sensors, flexible electronics), energy technology (solar energy, hydrogen storage, batteries), and advanced engineering materials (nanocomposites) [5].

CRYSTALLINE STRUCTURES OF THE CHALCOGENIDE MATERIALS

Chalcogen group elements can be found in nature in both free and combined states. The key elements S, Se and Te of this group structural property are attractive due to their large number of applications in various industries. For examples, many industries utilize sulfur, but emission of sulfur compounds is often seen to be more of a problem than the natural phenomenon. In the modern era it has been recognized that chalcogenide based materials applications are widely influenced from their metallic behavior. Usually the metallic character of chalcogen elements heightens with their increasing atomic numbers. The electronic and crystalline structures of the main chalcogen elements are briefly discussed here.

Sulfur

Sulfur (S) is solid at room temperature and 1 atm pressure. It is usually yellow, tasteless and almost odorless. It exists naturally in a variety of forms, including elemental sulfur, sulfides, sulfates and organosulfur compounds. Sulfur can be extracted by thewell-known Frasch process. Sulfur has the unique ability to form a wide range of allotropes compared to other chalcogen elements. The most common state is the solid S_8 ring; this is the most thermodynamically stable form at room temperature. Sulfur (S_α) has an orthorhombic crystalline structure at room temperature [6], as shown in Fig. 1.1(a). Sulfur also has a temperature dependence crystallographic structure like monoclinic sulfur (S_β), this is exhibited in Fig. 1.1(b) [7]. Sulfur crystallographic structure changes with increasing temperature; at 95.5°C it has monoclinic sulfur (S_β) and at 119°C, 160°C, liquid sulfur (S_λ), Liquid sulfur (S_μ), while at 445°C, sulfur vapor is formed. Sulfur exists in the gaseous form in five different forms (S, S_2, S_4, S_6 and S_8). The key physical parameters of sulfur are listed in the Table. 1.1.

Table 1.1   Important physical parameters of the chalcogen element Sulfur

Key physical parameters of Sulfur	
Atomic symbol	S
Atomic number	16
Electron configurations	1s 2s p 3s p4
Atomic weight (g/mol)	32.07
Structure	Orthorhombic
Phase at room temperature	Solid
Classification	Non-metal
Melting point (°C)	112
Boiling point (°C)	444.6
Ionization energy (kJ/mol)	1000
Ionic radius (pm)	184

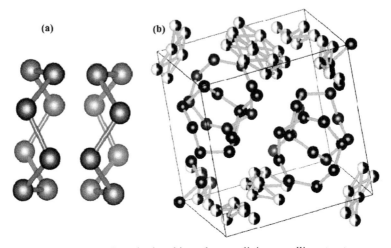

Figure 1.1(a, b) S_8 orthorhombic and monoclinic crystalline structures.

Selenium

Selenium was first discovered by Berzelius in 1818. Its meaning in Greek is 'moon', selene. Selenium is a red or gray crystalline or black amorphous solid structure. It can exist in multiple allotropes that are essentially different molecular forms of an element with varying physical properties. Selenium has a crystalline hexagonal structure of the metallic gray stable allotrope. The crystalline state Se can have several structural modifications. Under normal conditions only the trigonal phase is stable, while others phases are metastable. The trigonal Se crystal structure consists of rigid spiral chains which are weakly bonded in between. The structure of monoclinic Se consists of eight-membered rings [8]. Usually selenium is extracted from the electrolytic copper refining process. Selenium amphorous allotrope can also exist in the form of red powder. Selenium isotopes have the same atomic numbers but a different number of neutrons. It can have over 20 different isotopes; however, only five of them are stable. The five stable isotopes of the selenium are [74]Se, [76]Se, [77]Se, [78]Se, and [80]Se [9, 10]. The most stable crystallographic structure Se_8 is trigonal and the less stable α Se_8, β Se_8, γ Se_8 are monoclinic, and Se_6 is rohmohadral, all these structures are shown in Fig. 1.2(a, b, c, d, e).

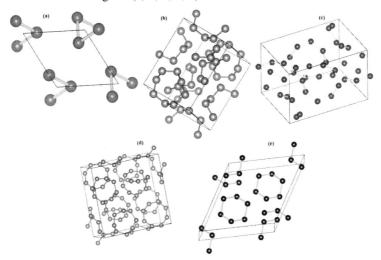

Figure 1.2(a, b, c, d, e) Se_8 trigonal, α Se_8 monoclinic, β Se_8 monoclinic, γ Se_8 monoclinic and Se_6 rohmohadral crystallographic structures.

The relevent details of crystallographic parameters have been reported by various investigators in the past [11–15]. Selenium often exists in soils and in plant tissues in the form of a bioaccumulated element. In a large dose this element is toxic; however, for many animals it is an essential micronutrient. The physical properties of selenium are similar to sulfur, however, because of its more metallic behavior, it is classified as nonmetal. The key physical properties of selenium are summarized in Table 1.2. Selenium can act as a semiconductor, therefore, it is often used in the manufacture of rectifiers (devices that convert alternating currents to direct currents). Selenium also has an adequate photoconductive property; this makes it useful for the light induced electrical conductivity derived devices. This means it has the ability to change light energy into electrical energy. Its adequate photoconductivity property makes it useful for various technical applications; such as photocells, photography, solar cells, production in plain-paper photocopiers, laser printers photographic toners, in the electronic industry as well as in the glass industry. Beside these it can also create a ruby-red colored glass. This element can also be used in the production of alloys and is an additive to stainless steel.

Table 1.2 Important physical parameters of the chalcogen element Selenium

Key physical parameters of Selenium	
Atomic symbol	Se
Atomic number	34
Electron configurations	$4s^2\ 3d^{10}\ 4p^4$
Atomic weight (g/mol)	78.96
Structure	Trigonal
Phase at room temperature	Solid
Classification	Metalloid
Melting point (°C)	220
Boiling point (°C)	685
Ionization energy (kJ/mol)	941, 2045 and 2973.7
Ionic radius (pm)	190

Tellurium

Tellurium was discovered in 1782, its name is derived from the Latin word 'tellus', literal meaning of this 'earth'. Tellurium is semi-metallic, lustrous, crystalline, brittle, silver white colored at room temperature and it is classified as a metalloid. It is usually available as a dark gray powder accompanied with metal and non-metal physical properties. Tellurium has eight isotopes. Tellurium six isotopes ^{120}Te, ^{122}Te, ^{123}Te, ^{124}Te, ^{125}Te, ^{126}Te [16, 17] are stable and remaining two isotopes ^{128}Te and ^{130}Te are slightly radioactive. The element tellurium stable trigonal crystalline structure is given in Fig. 1.3. Tellurium can form many compounds with sulfur and selenium. Like selenium, it also has a photoconductivity property. Tellurium is an extremely rare element, and commonly exists as a telluride of gold. It is often used in metallurgy with the combination of copper, lead and iron. Additionally, it is also used in solar panels and memory chips for computers. It is non-toxic or carcinogenic; however, when humans are overexposed to it, they develop a garlic-like smell on their breaths. The prime physical parameters of tellurium are listed in Table 1.3.

Figure 1.3 Te$_8$ trigonal crystallographic structure.

Table 1.3   Important physical parameters of the chalcogen element Telluium

Key physical parameters of tellurium	
Atomic symbol	Te
Atomic number	52
Electron configurations	$1s^2\,2s^2\,p^6\,3s^2\,p^6\,d^{10}\,4s^2\,p^6\,d^{10}\,5s^2\,p^4$
Atomic weight (g/mol)	127.6
Structure	Hexagonal
Phase at room temperature	Solid
Classification	Metalloid
Melting point (°C)	450
Boiling point (°C)	988
Ionization energy (kJ/mol)	869, 1790 and 2698
Ionic radius (pm)	140

POLYCRYSTALLINE STRUCTURES OF THE CHALCOGENIDE MATERIALS

Chalcogeon group elements can have a polycrystalline structure under a wide range of ions E^{2-}_n (E = S, Se, Te). Alkali metal polysulfide solutions have suitable cations chain-like dianions in the significantly stable range n = 2–6 [18, 19]. S_n^{2-} anions belongs to n = 7 and 8 have a crystallized structure in their bulk form [20, 21]. Polyselenides Se_n^{2-} and polytellurides Te_n^{2-} in the range 2–11, 8, 12 and 13 likewise have a solid state crystalline structure [22, 23]. Though sulfur and tellurium have a limited ability to form large homocyclic rings compared to selenium, they have more homopolyatomic anions than its lighter homologues. This is due to hypervalent bonding and weak $np^2 \rightarrow n\sigma^*$ ($n > 3$) interactions when moving down in the chalcogen group. Therefore, tellurium can not only form a classical bent $TeTe_2^{2-}$ units unbranched polytelluride chains but also linear $TeTe_2^{4-}$, T-shaped $TeTe_3^{4-}$ and square-planar $TeTe_4^{6-}$ units. Many polytellurides have a versatile adoptability and their $TeTe^{y-}_x$ building units under the weak secondary Te...Te bonding between neighboring tellurium atoms to make polymeric 1- to 3-dimensional networks. Schematic of the hypervalent bonding for square-planar $TeTe^{6-}_4$ unit and $5p^2 - 5\sigma^*$ secondary bonding between two Te atoms and the central atom of a $TeTe_2^{2-}$ unit are given in Fig. 1.4(a, b). General occupied np^2 lone pair orbitals and antibonding $\rightarrow n\sigma^*$ orbital energy difference is in a decreasing order with the increasing n. It allows to make intra- or intermolecular $np^2 \rightarrow n\sigma^*$ favorable bonds [24, 25].

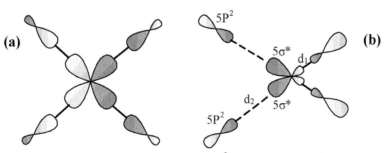

Figure 1.4   **(a)** Hyper bonding in a square-planar $TeTe_4^{6-}$ unit (3-centre, 4-electron), **(b)** $5p^2 - 5\sigma^*$ secondary bonding between two Te atoms with the central atom of a $TeTe_2^{2-}$ unit.

Polysulfides

Polysulfides have 2–8 members chain structures in a solid state form, in which five or more sulfur atoms can only be isolated as per $[NR_{4-x}H_x]^+$ (n = 6–8, R = alkyl or substituent) configuration. Crystal structures like A_2S_2, A_2S_3 and A_2S_5 are known as alkali metals A = K–Cs. The highly symmetrical cations have long-chain polysulfides ($n > 6$) which contain a helical all-trans conformation and all torsion angles are gauche with the same sign. In contrast, the packing forces can adopt an alternative chain in the presence of cations for the lower symmetry. In polysulfides, theoretically six possible configurations are possible for the heptasulfide anions. A trans–cis–cis–trans ($++--++$) conformation is adopted by the S_8^2 chain known as octasulfide. Polysulfides S–S–S bond angles 106–111° and S–S distance 201–208 pm are typical for S_n^2 chains [26, 27]. Their charge localization impacted on the first and last chain member, therefore, the shortest distances are usually associated to the terminal bonds. In polysulfides, short intermolecular secondary S....S bonds are often absent.

Polyselenides

The discrete polyselenide chains Se_n^2 have non-significant secondary intra- or intermolecular Se... Se contacts structurally defined as $n = 2–8$ [28]. Crystal structures of the binary alkali metal selenides such as, A_2Se_n in the range $n = 2–5$ (A_2Se_2 for A = Na, K, Rb, etc.), and longer Se_n^{2-} chains $n = 6–8$ only can have isolated encapsulated alkali metal cations or large non-coordinating organic monocations. Only well-known octaselenide Se_8, Se_7, Se_6 [29] usually have all-trans highly symmetrical counter cations. The terminal Se–Se bonds distances in the polyselenide chains are generally shorter than other existing interchain bonds. The heptaselenide anion average bond distances are between 228–233 pm and Se–Se–Se bond angles lie in the range 104–111°, with torsion angles in the all-trans chain between 66 and 84°. Moving down in Group 16, the hypervalent Se_n^2 anions reflect the structures with the value $n = 9–11$, that are not parallel to polysulfides. Existence of the Se_9 anion leads to additional bonds (length ~ 295 pm) stabilization possibly with the weak secondary Se_1...Se_6 interaction [28]. The weak secondary interaction in Se also causes to a significant lengthening of the opposite Se_5–Se_6 bond with a distance ~ 247 pm and a narrow Se_4–Se_5–Se_6 angle to 93°. The crystalline Se_9^{2-} and Se_{10}^{2-} anions internal bonding interactions cyclization are stabilized in all polyselenides Se_n^{2-} structures with $n > 8$. The cations packing factors involvement in structural directions can also play an important role in the determination of the extent coordination sphere distortion.

Polytellurides

Tellurium has a strong tendency to form intra- and intermolecular $np^2 - n\sigma^*$ bonding. The intra and intermolecular interaction leads to the distorted linear Te–Te... Te units in many polytellurides. Therefore, they have been classified as discrete or as polymeric chains, sheets or 3D frameworks often in an arbitrary structure. They can form isolated chains Te_{2n} under the strong Te–Te bonds (i.e., $d < 313$ pm) interaction, however, dianions with $n = 2–6, 8, 12$ and 13 can also exist. Similar to Se, Te also has the discrete bicyclic dianions at $n = 7, 8$ [30, 31]. In contrast to polysulfides and most of the polyselenides, a number of polytelluride anions also have total negative charges greater than 2 [32, 33]. The structural motif changes occur ($x/y = 5/3, 1/1, 2/3, 2/5, 1/3, 1/4, 1/6, 2/13, 1/7$ and $3/22$) in alkali metal polytellurides for those governed by the empirical A_xTe_y formula. The discrete anions can occur for $x/y \geq 2/3$, while the polymeric chain and lamellar anionic network form for the $x/y \leq 2/5$ [34, 35]. The isolated polytelluride anions can only exist in binary alkali metal A_xTe_y with $x/y \geq 2/3$. The hypervalent bonding leading to linear $TeTe^{4-}_2$, T-shaped $TeTe^{4-}_3$ and square-planar $TeTe^{6-}_4$ building units can be in polymeric anions with $x/y \leq 2/5$. The electron count of the heavy main group elements consist of finite length linear chains with an electron-rich multicentre bonding [36, 37]. Such as Te_5 linear units cross each other at right angles and their bonding have significant s–p mixing with the relative unimportant p-bonding, in which five p_x orbitals split into bonding (x_2), non-bonding (x_1) and anti bonding orbitals. Similarly, in Te_6 two multicenter bonds corresponding to 2×6 electrons can exist. Additional, s and p_z lone pair electrons of the six Te atoms (6×4) with four in-plane p_y lone pair electrons of the two central Te atoms (2×2) can form the Te^4_6 configuration. However, the weak secondary interactions could exist between the Te_8 rings. Hence, polytellurides can form a verity of structures under the different circumstances which is governed by well-defined rules.

AMORPHOUS STRUCTURES OF THE CHALCOGENIDE MATERIALS

In practice, all solids are found in a disordered state in nature. Solids are grown at finite temperatures, because of this they contain defects. Even, crystalline solids e.g., Si and Ge are disordered, while it is believed that the crystalline state of these materials is free from defects and impurities. In amorphous materials, homogeneous disorder occurs due to the large number of atoms having the same average (or bulk) properties such as specific heat, electrical and thermal conductivities, optical properties and density, etc. Materials can be classified as amorphous or vitreous for those possessing homogeneous disorder, the schematic of the amorphous solid is given in Fig. 1.5.

Figure 1.5   Amorphous structures of solids.

Thus, amorphous solids or non-crystalline materials can be described for those not possessing long-range periodical ordering in their atomic organization. These solids exhibit lack of long-range periodic arrangement of atoms. Amorphous materials can be obtained by quick freezing of the liquid state into a solid state. Due to a sudden phase change such materials undergo physiological changes. The key properties of amorphous or non-crystalline solids are primarily influenced by electronic configuration and chemical bonding between the adjacent atoms compared to crystalline solids. In crystalline solids these properties are essentially determined by the long-range periodic order of their constituent atoms. While in amorphous materials, the freedom from the constraint of atomic periodicity permits a wide range of material compositions depending on their preparation. The non-crystalline or amorphous materials can be electrically insulating, semiconducting or metallic in nature. They can have widespread applications in the microelectronics industry [38, 39]. Particularly, amorphous semiconductors have attracted much attention in the area of Phase Change Memory (PCM) applications due to their electrical and optical properties [40, 41]. They can also be considered as potential candidates for many commercial applications [42, 43].

The cohesive energy between atoms in crystals and amorphous materials are the same. Therefore, amorphous semiconductors can be classified based on their chemical bonding into two broad categories; namely the covalently bonded amorphous semiconductors and the semiconducting oxide glasses [44]. Further covalent semiconductors can be classified into two major groups [45, 46]; (i) the tetrahedrally bonded amorphous solids such as amorphous silicon (a–Si), germanium (a–Ge), etc., (ii) the chalcogen semiconductor contain one or more chalcogen elements, (sulfur, selenium and tellurium) in a combination with elements from III, IV, or V group of the periodic table.

CHALCOGENIDE GLASSES

Chalcogenide glassy semiconductors have been in existence for more than about 50 years [47]. Chalcogenide glasses contain one or more chalcogen elements, S, Se or Te of VIth group of periodic table. Chalcogen elements predominantly form covalent atomic structures and their bonds have a pronounced tendency to link together to form chains and rings. Usually chalcogenide glasses have much lower mechanical strength and thermal stability as compared to existing oxide glasses, but they have higher thermal expansion, refractive index, larger range of infrared transparency and higher order of optical non-linearity.

It is believed that chalcogenide glasses were discovered in the 1950's [48] when Frerichs investigated the As_2S_3 glass and published an article titled "New optical glasses in infrared up to 12 μm". In year 1955 Goryunova and Kolomiets [49] discovered $TlAsSe_2$ chalcogenide glasses which possess semiconductor properties. It was the first discovery of semiconducting glass and opened a new field in semiconductor physics named as 'amorphous semiconductors'. Ovshinsky and his co-workers in 1968 discovered that some chalcogenide glasses exhibited memory and switching effects [50]. After this discovery it became clear that the electric pulses could switch the phases of chalcogenide glasses back and forth between amorphous and crystalline state. Around the same period in the 1970's, Sir N.F. Mott and E.A. Davis developed the theory on the electronic processes in non-crystalline chalcogenide glasses [51]. In this order Kawamura [51] discovered xerography in 1983. Chalcogen based solar cells applications were developed by Ciureanu and Middehoek [52] in 1992 and by Robert et al. in 1998 [53]. Infrared optic applications were studied by Quiroga et al. in 1996 [54] and Leng et al. in 2000 [55]. The switching device applications were introduced by Bicerono and Ovshinsky (1985) [56] and Ovshinsky 1994 [57]. In 2001, Boolchand et al. discovered an intermediate phase in chalcogenide glasses [58]. The photo-induced ductility (2002), thermo-stimulated inter-diffusion (2003), multistate switching effect (2004) have been studied by different investigators [59]. The bond constraint theory for phase change memory of chalcogenide glasses was studied in 2008 by Paesler et al. [60]. In this order in 2009, 2010 Singh and Singh. introduced Se–Zn–In and Se–Zn–Te–In two new semiconducting series [61, 62]. Furthermore, in 2013 Singh first introduced advanced challenging single step synthesis of chalcogenide-multi-walled carbon nanotubes and chalcogenide – graphene composite glasses with a new Se–Zn–Sb composition. Proceeding with this newly discovered inorganic and organic materials composite glass area, in the same year (2013) Singh also reported the microscopic study with a new Se–Te–Ge amorphous semiconducting alloy [63, 64].

While describing key developments in the field of chalcogenide glasses several review books have been published on this topic e.g., "The Chemistry of Glasses" by A. Paul in 1982, "The Physics of Amorphous Solids" by R. Zallen in 1983 and "Physics of Amorphous Materials" by S.R. Elliott in 1983. However, the first book was mainly dedicated to chalcogenide glassy materials entitled "Chalcogenide Semiconducting Glasses", published in 1983 by Borisova. In this order, Vinogradova published her monograph "Glass formation and Phase Equilibrium in Chalcogenide Systems" in 1984. Further, Andriesh published a book on specific applications of chalcogenide glasses entitled "Glassy Semiconductors in Photo-electric Systems for Optical Recording of Information". Popescu provided a large and detailed account on the physical and technological aspect of chalcogenide systems in his book "Non-Crystalline Chalcogenides". The compendium of monographs on the subject of photo-induced processes in chalcogenide glasses entitled "Photo-induced Metastability in Amorphous Semiconductors" was compiled by Colobov in 2003. Robert Fairman and Boris Ushkov-2004 described physical the properties in the book "Semiconducting Chalcogenide Glass I: Glass formation, structure, and simulated transformations in Chalcogenide Glass". Furthermore, Zakery and Elliott published "Optical Nonlinearities in Chalcogenide Glasses and their Applications" in 2007.

The work on chalcogenide glasses is well accepted and their thermal, electrical and optical properties widely depend on the alloying concentration. In these glasses a chemical threshold occurs at a particular concentration of the alloy. Most of the work in chalcogenide glasses has been reported on binary, ternary and multicomponent systems [65–70]. More recently the field of chalcogenide glasses has mainly focused on more metallic chalcogen alloys and composites in order to obtain more stable and harder chalcogenide glasses.

Binary Chalcogenides

Amorphous chalcogenide can form a binary alloy together with Se–S, Se–Te and S–Te [71–73] compounds. They can also form many binary compounds with other group alloying element from a periodic table. Most extensively studied binary compounds with indium, antimony, copper, germanium, etc. have attracted significant interest due to their extensive technical applications [74, 75]. Specifically Se–based binary compounds have attracted much attention to sulfur and tellurium owing to their versatile technical utility [76]. The VI-III family compounds can form the layered structures with strong covalent bonds. The VI-III group Se–In compounds layered hexagonal symmetry structure consists of tetrahedrally or pentagonally coordinated Se and In are given in Fig. 1.6.

Figure 1.6　Layered In_2Se_3 structure.

Ternary Chalcogenides

Ternary chalcogenide glasses have been broadly studied for more than three decades. Ternary chalcogenides can be prepared by introducing a suitable additive element in any well-known or new binary matrix. In the recent years, Zn containing ternary chalcogenide glasses attracted large attention due to their higher melting point, metallic nature and advanced scientific interest [77]. The crystalline state of zinc has a hexagonal close-packed crystal structure with an average coordination number of four. However, in amorphous structures the metallic Zn can dissolve in Se chains and make homopolar and heteropolar bonds. The additional third element concentration in binary alloy can affect the chemical equilibrium of existing bonds, therefore, newly formed ternary glass stochiometrics can be heavily cross-linked and make homopolar and heteropolar bonds within the alloying. As an example Se–Zn–In ternary chalcogenide glasses can form Se–In heteronuclear bonds with strong fixed metallic Zn–In, Zn–Se bonds. Due to incorporation of third additive element indium concentration as -cost of selenium amount, the whole ternary Se–Zn–In matrix become heavily cross -linked, therefore, steric hindrance of the system increases.

In such glasses a chemical boundary occurs at a threshold of indium concentration. Incorporation of the indium content beyond the threshold concentration leads to drastic change in physical properties of these glasses [78–80].

Multicomponent Chalcogenides

There is a great interest on the study of multi-component chalcogenide glasses for sophisticated device technology as well as from the point of view of basic physics. Among the chalcogen group elements Se rich binary and ternary chalcogenide glasses can have high resistivity, greater hardness, lower aging effect, enhanced electrical and optical properties with good working performance. Due to the technical application limitations of the Se binary and ternary glasses, investigators have preferred to make multicomponent glasses. Thus adding more than two components into a selenium matrix to produce considerable changes in the properties of the complex glasses are known as multicomponent glasses. Particularly, metal and semi-metal containing multi-component amorphous semiconductors have been identified as promising materials for investigations of Ge–Bi–Se–Te, Al–(Ge–Se–Y), Ge–As–Se–Te, Cd(Zn)–Ge(As), $GeSe_2–Sb_2Se_3–PbSe$, $Cu_2ZnSnSe_4$, Se–Zn–Te–In, etc. [81–87].

STRUCTURE

As described above, non-crystalline substances do not possess long-range ordering of atoms in their periodic arrangement. To define non-cystalline solids, some necessary and sufficient parameters need to be followed [88]. To describe the simplest ideal single crystal structure, it is essential to know the structure of an elementary cell or a short-range order of the arrangement of atoms. Therefore, it is necessary to add at least one defective subsystem to define any real single crystal. In the case of describing the polycrystal structure, both short-range ordering and addition of defects are desired. Along with materials morphology such as crystal size distribution, crystal texture, formation of spherulites are also important parameters. Therefore to consider any substance as a non-crystalline solid the following four parameters should be defined:
  (i)   Short-range order of atomic arrangement
  (ii) Medium-range order of atomic arrangement
  (iii) Morphology
  (iv) Defect subsystem

  The different kinds of the solids structural parameters are listed in Table 1.4. Here necessary structural parameters are mentioned in terms of + and –, this is used to describe the increasing growth complexity. To described structural parameters of an ideal single crystal, only one parameter is sufficient, while in the case of non-crystalline solids all four characteristic parameters are required. Therefore, non-crystalline solids require indepth understanding of structural parameters compared to crystalline solids. This is necessary to define short- and medium-range ordering of the atomic arrangement, morphology and defect subsystem for the description of a non-crystalline solid structure [88].

Table 1.4   Key structural parameters for the different solids [88].

Structure	Solid states			
	Ideal single crystal	*Real single crystal*	*Polycrystal*	*Non-crystalline solid*
Short-range order	+	+	+	+
Medium-range order	–	–	–	+
Morphology	–	–	+	+
Defect subsystem	–	+	+	+

BONDING

The individual chalcogen elements bonding is classically demonstrated as each di-coordinated atoms contain two lone pair orbitals. In their s–p separation, mainly ns^2 and np^2 ($n = 3$–5) of two lone pair orbitals are involved. Usually ns^2-type orbital is not involved in the stereochemistry due to repulsion between the adjacent occupied np^2 lone pair orbitals. Hence, these systems always try to minimize coulombic repulsion through adopting E–E–E–E torsion angles of about $90 \pm 20°$. This is shown in Fig. 1.7.

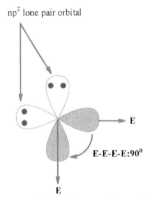

Figure 1.7   Repulsion of the np^2 lone pair orbitals on adjacent E atoms leads to a preference for an orthogonal arrangement.

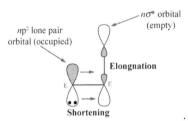

Figure 1.8   Relief of the repulsion of coplanar np^2 lone pair orbitals by np^2-$n\sigma^*$ interactions.

np^2-nr^* Interactions

Due to steric reasons it is not possible to adopt torsion angles in a well-defined manner, such as sulfur (S_7). However chalcogen systems try to minimize the occupation of the neighboring np^2 lone pair orbital with low torsion angles. This can be due to an interaction with empty orbitals in the vicinity. Generally, the empty E–E σ^* orbitals have the right orientation and suitable orbital energies as shown in Fig. 1.8. This kind of np^2-$n\sigma^*$ interaction can lead to bond lengths alternation. Such bond lengths alternation allows the extreme structural effect in their interactions to obtain neutral and cationic polychalcogen compounds.

π and π^*–π^* Bonding

From the well-known Huckel description, π bonding in chalcogen can occur according to the ($4n + 2$) π electron rule (E_4^{2+}; S_6^{2+}) or through an interaction of partially depleted np^2 lone pair orbitals, as exhibited in Fig. 1.9. Such depletion could proceed either through a np^2–$n\sigma^*$ interaction or through oxidation and cation formation. This leads to the oxidation of neutral chalcogens, which can provide polychalcogen cations. The electron density removal could then occur from the occupied np^2 lone pair orbitals [89, 90]. The partially occupied molecules antibonding π^* orbital interactions is represented in Fig. 1.10. Under these circumstances the interaction of two individual and partially occupied π^* orbitals may be too weak to have a significant bonding. In a more detailed study, this was found for I_4^{2+} dication [91] and designated as π–π bond, as an example Te_6^{4+} exo and endo π^*–π^* bonds structure of E_8^{2+} are given in Fig. 1.11. Usually in solid chalcogenide alloys such bonding occurs in a complex form and it can be explained with the help of well- established network theories.

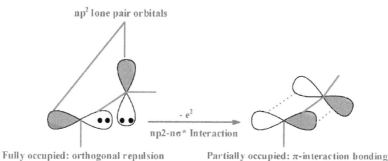

Figure 1.9 Partial p bonding of the depleted np^2 lone pair orbitals.

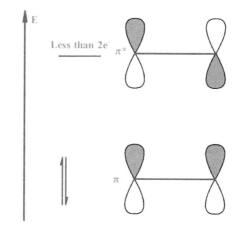

Figure 1.10 View of the depleted adjacent np^2 lone pair orbitals.

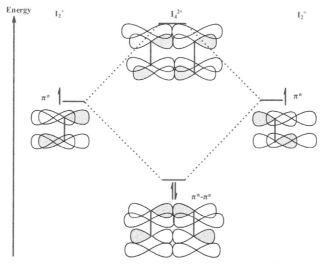

Figure 1.11 Prototypical π^*–π^* bonding in I_4^{2+}.

Continuous Random Network (CRN) Model in an Ideal Form

The Continuous Random Network (CRN) model is based on the assumption that atoms coordination numbers $Z \leq 4$ should be associated with the direct covalent bonds in a linked network. The local atomic structure in non-crystalline semiconductors is not completely random; therefore, the basic principles of the CRN model can be described as:

(i) The coordination of the constituent atoms should satisfy 8–N rule [92], here N represents the column in the periodic table to which the element belongs.

(ii) The nearest neighbor distances (bond-lengths) are allowed to vary <1% only when the bond angles are allowed to be noticeably dispersed ($\pm$ 10%) [93]

(iii) This CRN model does not permit any dangling (unsatisfied) bonds or over-coordinated atoms.

Therefore the structure of glass and crystal differs from a significant spread in bond-angles for a glassy network. Under this action the structural unit is allowed to rotate and hence leads to the absence of long-range ordering. Based on this concept Zachariasen constructed A_2B_3 glass structural network, this construction is shown in Fig. 1.12.

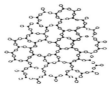

Figure 1.12 A_2B_3 glassy network.

Random Covalent Network (RCN) and Chemically Ordered Network (CON) Models

To extend the versatility of an ideal CRN model, investigators presented another version of this model considering both atomic and chemical local networking in amorphous semiconductors. It is known as the Random Covalent Network (RCN) and Chemically Ordered Network (CON) model [94]. Typically the word 'Random' describes the type of bonds statistical distribution. As an example a binary alloy system can be expressed in terms of $A_{1-x}B_x$, here A and B are two different species and x is a normalized concentration variable. Let us suppose a and b are the columns in the periodic table corresponding to elements A and B atoms. Further, let us assume this system satisfies the (8–N) rule for normal covalent bonding with A and B atoms coordination. Then Y_a and Y_b can be described as $Y_a = (8-a)$ and $Y_b = (8-b)$, respectively [93]. Thus, the types of the bonds distribution can be determined by using the local co-ordinations Y_a and Y_b and the fractional concentrations of A and B atoms $(1-x)$ and x, respectively. In a general way, if we neglect the relative bond energy then it can be expressed in the form A–A, A–B, B–B for all compositions other than $x = 0$ and $x = 1$.

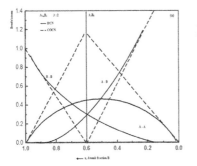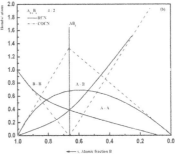

Figure 1.13 Bond counting statistics using; **(a)** Random Covalent Network (RCN) model for 3:2 network, **(b)** Chemically Ordered Covalent Network (COCN) model for 4:2 network.

On the other hand, the Chemically Ordered Network (CON) model based on relative bond energies for the amorphous semiconductors, in which heteropolar bonds are favored over homopolar bonds. This model predicts a chemically ordered glass composition, which can have only heteropolar bonds at the critical value X_c.

$$X_c = \frac{Y_b}{Y_b - Y_b} \tag{1.1}$$

A rich compositions $(0 < X < X_c)$ may contain A–A and A–B bonds and B rich compositions $(1 > X > X_c)$ can have A–B and B–B type bonds. Statistical schematics of these two types of bond models for 3:2 and 4:2 for the simplest binary alloys are given in Fig. 1.13(a, b). Hence with necessary and sufficient conditions, the RCN and CON approaches are appropriate only for covalently bonded systems and not for random close-packed structures [95].

Network Topological Thresholds for the Amorphous Semiconductors

To explain the random closed packet bonds structures of the amorphous semiconductors, the CRN model was further extended by adopting the network topological thresholds concept [96, 97]. Hence amorphous semiconducting systems exhibit distinguishable changes in their physical properties at certain specific composition. Such a particular composition of an alloy is known as mechanical threshold (or rigidity percolation threshold) and the chemical threshold [98, 99]. The increasing coordination of the atoms in a glassy network leads to a gradual transformation flexible (floppy) and weakly cross-linked chains containing a rigid three dimensional network. Further, when the fraction of cross-linking reaches a threshold value, the transition occurs in the percolation rigidity and a good glassy physical property could arise near the threshold [100, 96]. In these systems the degree of cross linking is established by the mean coordination number $\bar{r}$. If total number of atoms is N and atoms with coordination r $(r = 2, 3, 4)$. Then it can be expressed by the following relationship:

$$N = \Sigma_{r=2}^{4} n_r \tag{1.2}$$

The mean coordination for a system can be obtained as

$$\bar{r} = \frac{\Sigma_{r=2}^{4} r n_r}{N} = \frac{\Sigma_{r=2}^{4} r n_r}{N = \Sigma_{r=2}^{4} n_r} \tag{1.3}$$

For the covalent network, the bond length and angles are usually fixed. Kirkwood (or Keating potential) has been used to describe the equilibrium structure for small shifts, correlated with the relationship and presented in Fig. 1.14.

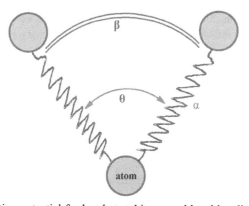

Figure 1.14 The Keating potential for bond-stretching α and bond-bending β terms [101]

$$V = \frac{\infty}{2}(\Delta l)^2 + \frac{\beta l}{2}\Delta\theta^2 \tag{1.4}$$

Here l is the bond length, Δl is the change in the bond length, and $\Delta\theta$ is the change in the bond angle, β is the bond bending force and bond stretching term. Other potential terms are much smaller and can be neglected.

Further, when a single constraint is associated with an individual bond under equal bond shearing conditions, then such constraints can be associated with each r coordinated atom. If an atom's nearest neighbor coordination is r, which is between the two bonds, then an angle should to be specified. For the other remaining $(r-2)$ bonds, the constraints of the two bond angles have to be also specified. Hence the total number of bond angle constraints on an atom can be considered as $[1 + 2(r-2)]$ or $(2r-3)]$.

Thus the total number of constraints:

$$\Sigma^4_{r=2} n_r \left[\frac{r}{2} + (2r-3) \right] \tag{1.5}$$

In the case of zero frequency mode function, this function can be written as

$$f = \frac{\left[3N - \Sigma^4_{r=2} n_r \left[\frac{r}{2} + (2r-3) \right] \right]}{3N} \tag{1.6}$$

$$= 1 - \frac{1}{3N}\left[\Sigma^4_{r=2}\frac{rn_r}{2} + 2\Sigma^4_{r=2}rn_r - 3\Sigma^4_{r=2}n_r \right] \tag{1.7}$$

$$= 1 - \frac{1}{3N}\left[\frac{\bar{r}n}{2} + 2\bar{r}N - 3N \right] \tag{1.8}$$

Finally we will get

$$f = 2 - \frac{5}{2}\bar{r} \tag{1.9}$$

Here for the Selenium chain value of $\bar{r} = 2$, this means $f = \frac{1}{3}$, reflecting the fact that one-third of the entire mode is floppy, therefore, indicating that a glass is in a floppy mode. When the coordination value is higher (>2), then the network is cross-linked, therefore, f drops to zero at $\bar{r}_C = 2.4$. Hence a phase transition occurs from a floppy to rigid mode in the glassy network. This significant transformation is also known as stiffness transition. It can be verified from various experimental characterizations, like Raman scattering [102], vibrational density of states [103], Brillouin scattering [104], Lamb-Mossbauer factors [105], resistivity [106], etc.

In the case of halogens (F, Cl, Br, I), the above-described constraint counting is not valid due to their one-fold coordination atoms, because the number of angular forces $(2r-3)$ become equal to -1 when $r = 1$, instead of zero. Therefore, investigators have introduced a compact concept to address this issue [107, 108].

With this concept, a valuable correction is required in zero frequency mode when n_1 one fold coordinated atoms, and can be expressed as:

$$f = \frac{\left[3N - \Sigma^4_{r=2} n_r \left[\frac{r}{2} + (2r-3) \right] - n_1 \right]}{3N} \tag{1.10}$$

This gives us

$$f = 2 - \frac{5}{2}\vec{r} - \frac{n_1}{3N} \qquad (1.11)$$

If the stiffness transition takes place at a lower mean coordination, then r_C can be obtained as:

$$\vec{r}_C = 2.40 - 0.4\frac{n_1}{N} \qquad (1.12)$$

Here $\frac{n_1}{N}$ is the simple correction term fraction for one fold coordinated atoms. Therefore, the transition can take place at a lower average coordination, because one fold coordinated atoms play no role in the network connectivity. In several systems the stiffness transition has been found to span over a range of compositions/average coordination numbers [109]. In these systems, there is an intermediate region between the floppy and the stressed rigid phase, existing in this range. The current intermediate region and floppy and rigid phases can be expressed as $\vec{r}_{c1} \leq \vec{r} \leq \vec{r}_{c2}$, the corresponding schematic is represented in Fig. 1.15(a, b, c).

(a)   **(b)**   **(c)**

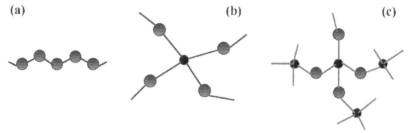

Figure 1.15   **(a)** underconstrained, **(b)** optimally constrained, and **(c)** overconstrained glassy network.

Critical Chemical Threshold in Amorphous Semiconductors (X_c)

In these kinds of systems, chemical threshold can be defined using the CON model through the critical composition (X_c), at which chemical ordering is maximized due to the preference of heteropolar bonding. Thus composition at which the highest chemical stability occurs is called Chemical Threshold (CT) of the glassy system. As an example, the most extensively studied Ge_xSe_{1-x} glassy system, when the Ge concentration increases, the network is more cross-linked and forms heteropolar bonds of Ge–Se. At a particular value of x, the network has only heteropolar bonds, which reflects that the system is fully chemically ordered. This stage is known as a chemical threshold of the system. The divalent- Se and tetravalent- Ge fulfill the chemical composition as:

$$2(1 - x_c) = 4\,x_c \qquad (1.13)$$

$$x_c = 0.33 \text{ or } \langle r \rangle = 2.67$$

Similarly, other binary composition systems like $As_x Se_{1-x}$, In_xSe_{1-x}, etc. also compensate the threshold chemical composition rule [110–112].

BAND STRUCTURES

Understanding the energy band structure is an essential property to explore the electronic band structure of amorphous semiconductors. This can be directly related to DC and AC conductivity, electrical switching and other physical amorphous semiconductors. To understand the band structure of amorphous semiconductors several models have been proposed by investigators based on the assumption of the local structural environment. A slight change in interatomic distance and valence angle is allowed for an amorphous solid in the absence of long-range ordering. Therefore, the overall density of energy states in non-crystalline semiconductors is almost similar to the corresponding crystalline material. Thus, the concept of valence and conduction band and the band gap are equally valid for these materials with a remarkable difference in electron energy states, which is extended within the band gap due to the non-existence of long range periodicity.

The electronic state of the crystalline as well as amorphous material is given in Fig. 1.16(a). This electronic state can be schematically interpreted as, the electronic states of the crystalline materials are distributed discontinuously at the band edges whereas in glassy semiconductors the distribution of states continues into the forbidden gap [113]. A special fluctuation in potential occurs due to configuration disorder in amorphous materials. This leads to the formation of localized states and above and below the band tail to the normal band [114, 115]. The localized states of electrons can be correlated to their diffusion near the region at zero temperature. A sharp boundary of the mobility edge also exists between the extended and localized state, because these two states cannot coexist in the same energy state. Hence in amorphous semiconductors the conduction band edge (E_c) and the valance band edge (E_v) are separated from the localized and extended state, and their mobility gap is the band gap (E_g) [116]. The energy band gap separation in such materials is given in Fig. 1.16(b).

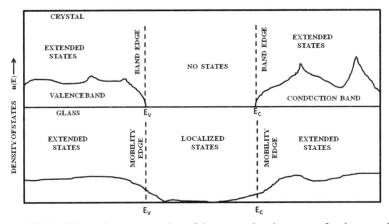

Figure 1.16(a) Schematic representation of the energy band structure for the crystalline and amorphous semiconductors.

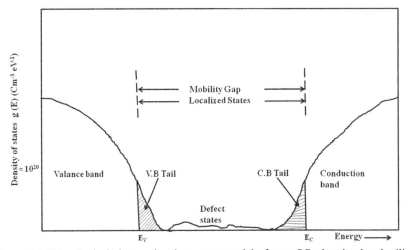

Figure 1.16(b) Typical electron band structure model of an a–SC, showing band tailing and states within the mobility gap. E_v and E_c are the mobility edges.

Cohen–Fritzsche–Ovshinsky (CFO) Model

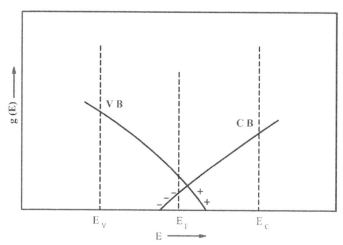

Figure 1.17(a)   The CFO model. Positive and negative signs indicate
ionization of impurities due to overlap of bands.

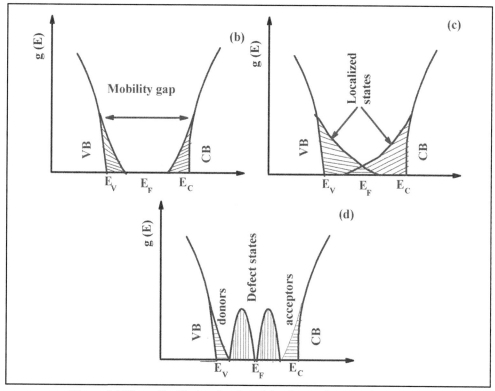

Figure 1.17(b, c, d)   Various band models of amorphous semiconductors,
the Mott model, the CFO model and the Davis–Mott Model.

Cohen, Fritzche and Ovshinsky (CFO) proposed a model to explain the properties of the amorphous semiconductors, as shown in Fig. 1.17(a). They assumed [117] that the extensive tailing of band edges, is a result of both compositional and topological disorders. Therefore, conduction and valence bands extend far into the gap and overlap to each other, and an appreciable density of the state exists in the middle of the gap. Therefore, the redistribution of charge electrons transformation takes place from the higher region of the Valence Band (VB) tail into the lower region of the Conduction Band (CB) tail. The involvement of localized states makes the negatively charged filled conduction band and empty positively charged valance band states. Their self-compensation and pinning the Fermi level near the mid gap gives the overlapping tails. The formation of negatively and positively charges filled and unfilled states are given in Fig. 1.17(b, c, d).

Davis–Mott (DM) Model

Considering the drawbacks of the CFO model, Davis and Mott introduced another model that there is no broad tailing in the density of state [118, 119], rather, tails are localized and extended to a few tenths of an electron-volt into the forbidden gap. Additionally, this model also proposed the existence of a band of compensated levels near the middle of the gap originating from the defects in the random networks. They also suggested that the central band splits into donor and acceptor bands, which pin the Fermi level (E_p), as shown in Fig. 1.17(d). Furthermore, Marshall and Owen proposed the concept of having donor and acceptor like bands in the upper and lower halves of the mobility gap respectively in 1971 [120]. They also suggested that these bands adjust themselves by a self compensation mechanism and keep their concentrations equal, thereby fixing the E_F near mid-gap.

Chemical View of Band Structure

Formation of the band gap in amorphous semiconductors can also be understood by adopting a chemical approach [121–123]. According to this approach, the nature of density of states in a solid mainly depends on the nature of chemical bonds and coordination number of the atoms and valance states [124]. The relevance of this model can be understood by considering tetrahedrally bonded Ge and chalcogenide Se as a specific example of two main classes of semiconductors. The four-fold coordinated Ge has highly directional sp^3 hybridized orbitals. These sp^3 hybridized orbitals can split into σ and σ^* bonding/anti-bonding states, as shown in Fig. 1.18(a, b) [121].

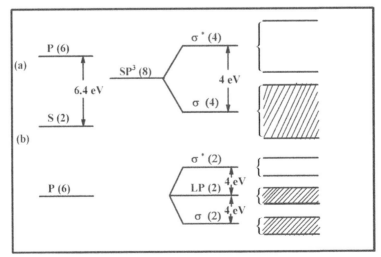

Figure 1.18(a, b) Bonding in **(a)** Ge and **(b)** Se

The chalcogen 'Se' atom contains two s and four p electrons in its outermost shell. As the figure shows, it is obvious that out of the three p states, two states can be utilized for the bonding.

This leaves one p-state unoccupied, which is activated by two paired electrons of the opposite spin. This is known as non-bonding electron pair or lone-pair electrons. Such non-bonded electrons do not participate in the bonding. Therefore, the bonding is actually due to the remaining two p-states (orbitals) each occupied by single electron. On the other hand, selenium is usually present in 2-fold coordination. The conduction band in Se also initiate from σ^* states like Ge. With a remarkable difference the highest occupied VB is formed from non-bonding states instead of the σ states as shown in Fig. 1.18(b). The unshared or non-bonding electrons states near the original p-state energy acts as the valence band [121, 125]. Hence non-bonding electrons constitute the highest VB band and reveal the conduction properties of chalcogenides. Such materials are also called lone-pair semiconductors [121].

DEFECTS IN AMORPHOUS CHALCOGENIDES

The models described above, such as RCN and CON models are based on only structural network in glassy semiconductors and explains the ideal glassy network when they do not have any defects. In reality, all kind of glassy solids possess a wide range of defects and their electrical, optical or in general any transport property is largely controlled by the existing bonding defects. Various types of defects are present in a glassy amorphous, such as, important charge defects [126]. Considering the charge defect Street and Mott were the first to propose a model for chalcogenide glassy semiconductors based on Andoreson proposition [124]. In this order Kastner et al., also proposed another model which successfully interpreted various properties of the amorphous semiconductors [127].

Street and Mott Model for the Charged Dangling Bond

Street and Mott proposed a model which elucidates the nature of defects in amorphous semiconductors [128]. This model assumes that the localized gap states are constituted by dangling bonds with concentrations of 10^{18}–10^{19} cm^{-3}, having a doubly occupied role out of the gap states. This could originate from a dangling bond; as a consequence, it creates a charged dangling bond. Such original charged dangling bonds can have their counterparts in positively charged states. They are designated as empty (donor) orbitals. Thus the newly formed empty state and existing filled state make two equivalent charged defects states. Street and Mott in 1975 demonstrated these defects as D° (neutral dangling bond), D$^+$ and D$^-$ are positively and negatively charged centers respectively. Further, they also established the term D indicates defects in an amorphous lattice. Using the Anderson's proposition, Street and Mott suggested that formation of energetically favorable D$^+$ and D$^-$ (charged pairs) from the 2D° states. These energetically favorable charged pairs raises due to the coulombic interactions between each other and it can also enhance the local relaxations. This can decrease the positive Hubbard energy U into an effective negative Hubbard energy U_{eff}, this as shown in Fig. 1.19 [129].

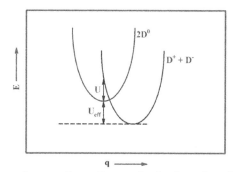

Figure 1.19 Configuration coordinate diagram for the formation of a D$^+$–D$^-$ pair [129].

Since dangling bonds (D°) are point defects at which the normal coordination (e.g., two for Se) cannot be satisfied due to the constraints of local topography. This is because neutral dangling bonds contain an unpaired electron. In these kinds of dangling bonds a strong electron-lattice distortion has a profound effect on the energy levels of electrons. This allows the electron occupation of a dangling bond to change under the spontaneous transformation into a pair of charged defect states D^+ (overcoordinated) and D^- (undercoordinated) according to the following reaction mechanism;

$$2D° = D^+ + D^- \tag{1.14}$$

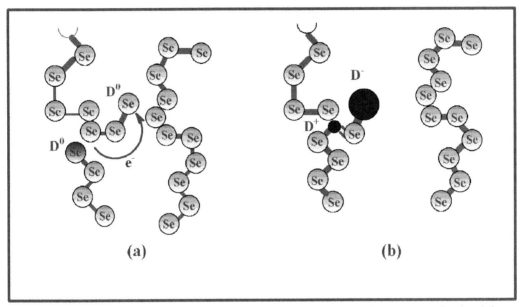

Figure 1.20(a, b) Schematic representation of the process of formation of charged coordination defects from two neutral dangling bonds ((**a**)→(**b**)) in a–S.

This reaction is considered exothermic due to its effective negative correlation energy. The two neutrally charged chain-end dangling bonds (D°) transformation into charged defect sites is given in Fig. 1.20(a, b) [128]. Therefore, this model successfully explains many experimental results. However, this model could not explain the assumption of high density of dangling bonds, such as the origin of negative effective energy U_{eff} [130]. Nor does it concretely describe why a large negative effective energy U_{eff} arises in chalcogenide glasses and not in the tetrahedrally bonded amorphous semiconductors [131].

Kastner–Adler–Fritzche Model (KAF)

The KAF model is considered as a valence-alternation model owing to it being based on the process of spin pairing at defects in amorphous materials. This model introduced a useful subscript and superscript denotation for the coordination and charged state of the defect sites in amorphous semiconductors. As an example, a doubly coordinated neutral chalcogen atom can be symbolized as C_2^0. Therefore, this model essentially emphasizes on the behavior of charged defects and the dominant contribution to the negative chemical correlation energy in origin. Additionally, this model has also suggested that specific interactions between the non-bonding orbitals which arise due to unusual bonding configurations. This is known as Valence Alternation Pairs (VAP's), requiring relatively small energy for their formation.

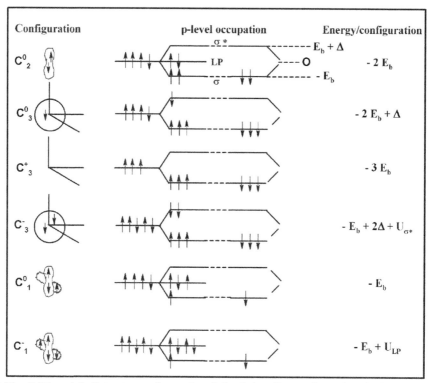

Figure 1.21   KAF model of structure and energies of simple bonding configurations in chalcogens. Straight lines represent bonding (σ) orbitals, lobes represent the lone-pair orbitals, and large circles represent the anti-bonding (σ^*) orbitals. Here arrows indicate the spin state of the electrons. Energies are given considering the non-bonding state energy as zero.

Figure 1.21 shows the KAF model; the structure and energy creation for several simple bonding configurations in amorphous chalcogenides. According to this model it is obvious the non-bonding energy can be taken as zero and assigned the σ-orbital bonding energy $-E_b$ and σ^*-antibonding orbital energy $+E_b+\Delta$. Here Δ is the excess energy of the anti-bonding orbital compared to the bonding orbital. Further, C_2^0 is the normally bonded chalcogen atom with the energy $-2E_b$. The three-fold coordinated atom C_3^0 is the lowest bonding energy configuration of the neutral defect. When the non-bonding electrons are moved from σ-bonding to σ^*-antibonding orbitals. Then three fold coordinated C_3^0 and the singly coordinated C_1^0 chalcogen atoms have to be obtained to possess energies $-2E_b+\Delta$ and $-E_b$, respectively. As C_3^0 has a lower energy configuration, its transformation can take place through an unstable state and finally transforms into (C_3^+ to C_3^-).

$$3C_3^0 \rightarrow C_3^+ + C_3^- \tag{1.15}$$

This reaction could occur at the cost of positive correlation energy of U_{σ^*}, cause, this formation can take place between two electrons which exist in the σ^* state. Hence C_3^- transforms into C_2^0, while the existing C_2^0 converts into a C_1^- center:

$$C_3^- + C_2^0 \rightarrow C_2^0 + C_1^- \tag{1.16}$$

Thus, due to the transformation of two σ and σ^* electrons system, this can lower its energy into the non-bonding states. One of the singly coordinated and two fold coordinated chalcogens, leading to the following reaction:

$$2C_2^0 \rightarrow C_3^+ + C_1^- \tag{1.17}$$

Net formation of coulombically charged defects pair is known as Valence Alternation Pairs (VAP) with the creation of additional lone pair on one of the chalcogen atoms C_1^-. Here $2\Delta \geq U_{\text{non-bounding}}$, where $U_{\text{non-bounding}}$ is the correlation energy if the electron is placed in a single localized non-bonding state orbital [123]. Such VAPs pin the Fermi energy at the mid-gap. Thus, the Davis–Mott and Kastner models were also used extensively to describe defects in chalcogenides.

PHOTOINDUCED EFFECTS IN AMORPHOUS CHALCOGENIDES

Amorphous semiconductors are metastable in nature which exhibit a wide variety of changes in their physical properties, specifically when light is irradiated on their band gap. The chalcogen element selenium has exhibited a peculiar phenomenon when light illuminated on it with the comparable optical band gap energy [132, 133]. Some important well described observations in chalcogenides are given in Fig. 1.22. It describes that when a suitable wave length light photon impinging on such materials the electron-hole pairs are created, which can contribute in these ways; (i) the electrical response of the material (e.g., photoconductivity), (ii) recombine, either radiatively (giving rise to photoluminescence (PL)) or (ii) non-radiatively. The amorphous semiconductors and insulators materials usually exhibit a wide variety of photo-induced phenomena due to their freedom and flexibility of the relaxation of crystallographic constraints. The presence of structural disorder in these materials can lead to localization of electron and hole states at the band edges in the vicinity of the gap. This is likely to induce metastable changes.

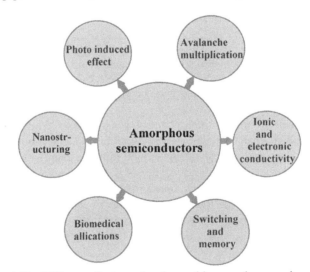

Figure 1.22   Different effects can be observed in amorphous semiconductors.

In amorphous semiconductors the optical absorption in general involves either lattice or electronic contributions. Such lattice absorption is the result of interaction of light with optical phonons, it can be observed in the infrared region. In which the electronic absorption involvement of the electronic states energy bands are well defined, these carriers may be intra- or inter-band in behavior. The optical energy band gap can be calculated at which absorption increases sharply as well as considering the nature of the transition (viz., direct or indirect) [134]. In direct energy band gap transitions involve the interaction of an electromagnetic wave with an electron in the valence band; this could be raised across the fundamental gap to the conduction band. However, indirect transitions also involve simultaneous interaction with lattice vibrations. Thus, the wave vector of an electron can change in the optical transition, as a consequence, phonons can enhance the momentum change.

If we neglect the exciton formations (electron–hole interactions), then formation of the absorption coefficient 'α' can be expressed as a function of phonon energy $h\omega$ which depends on the energy $N(E)$ of their bands containing the initial and final states. As a simple parabolic bands $N(E) \propto E^{1/2}$ can be correlated for direct transitions as:

$$\alpha n_0 h v_0 \sim (h v - E_0)^n \qquad (1.18)$$

Here $n = 1/2$ or $3/2$ depending on the transition is allowed or forbidden in the quantum mechanical sense, E_0 is the optical gap with no refractive index.

In the case of indirect energy band gap the equitation can be expressed as:

$$\alpha n_0 \hbar \omega = \frac{(\hbar\omega - E_0 + hv_{ph})^n}{\left(\exp\left(\dfrac{hv_{ph}}{KT}\right) - 1\right)} + \frac{(\hbar\omega - E_0 + hv_{ph})^n}{\left(1 - \exp\left(\dfrac{-hv_{ph}}{KT}\right)\right)} \qquad (1.19)$$

The two terms on the right side of this equation represent the contributions of phonons which are involved in the transitions of phonon absorption and emission respectively. This include the coefficients of proportionality and temperature dependency. Note that for the allowed transitions $n = 2$ and for the forbidden transitions $n = 3$.

IONIC CONDUCTIVITY

There has been a great deal of interest on the ionic conductivity of chalcogenide glasses due to high polarizability of sulfur and selenium. Therefore, chalcogenide glasses can have higher conductivity than their oxide counterparts. The alkali (mainly lithium) conducting chalcogenide glasses have also been considered as potential candidates for the development of solid state batteries. Similarly silver-doped chalcogenide films can also be considered for the development of ionic memories for data storage. Additionally, the ionic conductivity of metal (silver) doped chalcogenide membranes can also be useful for the sensors development to the environmental control. Hence, important research and innovations have been emphasized on the structure/property relationship in order to gain insight into the ion dynamics of chalcogenide glasses.

Several models have been used to describe ion transport in glasses or chalcogenide glasses. Specifically, several models have been used for oxide and chalcogenide glasses. The ionic conductivity in the glasses is a thermally activated phenomenon. Below the vitreous transition temperature T_g, the conductivity follows the Arrhenius law:

$$\sigma = \sigma_0 \exp\left(\frac{-E_\sigma}{k_B T}\right)$$

where σ_0 is the pre-exponential factor E_σ is the activation energy of conductivity, k_B is the Boltzmann factor and T the temperature. Above T_g, the ion motion is more frequent due to the cooperative motion of the macromolecular chains constituting the glass. Therefore, ionic conductivity can be described by the free volume law, considering a general law that accounts for many other properties of liquid and overcooled liquids above T_g:

$$\sigma = \sigma_0 \exp\left(\frac{-E_\sigma}{k_B}\right)(T - T_0)$$

where T_0 is the vitreous temperature. Thus, ionic conductivity can be expressed with the relationship $\sigma = nq\mu$, where n is the concentration in mobile carriers of charge q and μ is the mobility of these carriers. However, it is very difficult to measure the mobility and the number of mobile carrier separately. Therefore, it is not well known which factor is the main contributor to the large variation of σ with the modifier content.

For model description based on the involvement of independent thermally activated jumps of carriers between equivalent sites, the mobility can be expressed as

$$\mu = \left(\frac{\alpha \lambda^2 v_0 q}{k_B T} \right) \exp\left(\frac{-E_m}{k_B T} \right)$$

where α is the degree of freedom, λ the average jump distance, v_0 the attempt frequency of the ions and E_m the migration energy. The number of charge carriers is also thermally

$$n = n_0 \exp\left(\frac{-E_C}{k_B T} \right)$$

activated, where n_0 is the total number of modifer cations and E_C the energy of creation of mobile carriers.

Thus, the conductivity through the activation energy is the sum of the two terms; the creation energy of mobile carriers E_C (directly linked to n) and the migration energy of the ions E_m (directly linked to μ). Since the pre exponential factor does not alter much with the change in the composition of the chalcogenide glasses [135], therefore several models have been proposed by investigators to explain such conductivity changes [135].

Anderson and Stuart Model

The Anderson and Stuart model was proposed in 1954, which describes conductivity of ion conducting for silicate glasses [136]. This model can be applied for chalcogenide glasses with adjustment of a few parameters. According to this model, the activation energy of conductivity is the result of two contributions: an electrostatic one, which accounts for the M^+ .....O^- binding energy E_b and other one is the stain energy E_s which arises from the elastic bending of the glass when the ion passes through.

$$E_\sigma = E_b + E_s \tag{1.20}$$

$$= \frac{1}{\gamma\left(\dfrac{ZZ_0 e^2}{r + r_0} - \dfrac{ZZ_0 e^2}{\dfrac{\lambda}{2}} \right)} + 4\pi G R_D (r + r_d) \tag{1.21}$$

where γ is a covalence parameter which is arbitrarily taken to be equal to the relative permittivity ϵ_r of the glass; Z_0 and r_0 are the charge and the radius of the oxygen cation, λ is the jump distance, r is the mobile cation radius, r_d is the doorways radius and G is the elastic modulus. Anderson and Stuart calculated the activation energy for several compositions of ion conducting silicate glasses as an example (Li^+, Na^+, Ag^+), $E_b > E_s$. The variation of E_b rules over the variation of E_s with composition. With this in mind, the covalence parameter which scales with the relative permittivity of the glass, were predicted to have the lower activation energy of conductivity (a factor of 2) for chalcogenide glasses.

Weak Electrolyte (W.E.) Theory

Ravaine and Souquet proposed a model in 1977. It was based on data obtained by thermodynamic activity measurements on a series of sodium silicate glasses [137]. Later Reggiani et al.[138] and Levasseur et al. [139] studied this model's utility for silver phosphate and lithium borates glasses. They demonstrated that the data of the glasses acted as weak electrolytes. With this in mind, Ravaine and Souquet demonstrated that glass acts as an electrolyte with the glassy network being the solvent and the modifier being the solute. In any weak electrolyte, the solute would be weakly disassociated leading to only a small fraction of the cations mobility at time t.

Since the original model was based on the macroscopic properties descriptions, Souquet et al. later proposed their microscopic version [140]. They assumed that ionic displacement arises due to the migration of cationic pairs formed by a partial dissociation. This would correspond to the creation of a Frenkel defect in an ionic crystal. As a consequence an ion would leave a 'normal' position close to the non-bridging oxygen (eventually chalcogen) and jump in an already occupied position, finally creating a cationic pair. The energy requirement for such a jump limits the concentration of the cationic pairs. Therefore, Souquet et al. formulated the expressions for variation in ionic conductivity as a function of temperature. Below the T_g,

$$\sigma = n \left(\frac{e^2 \lambda^2 v}{6 k_B T} \right) \exp \left(\frac{\frac{\Delta S_f}{2} + \Delta S_m}{k_B T} \right) \exp \left(-\frac{\frac{\Delta H_f}{2} + \Delta H_m}{k_B T} \right).$$

This expression fitted with experimental data described for several alkali disilicate glasses with permissible mobility of effective charge carrier calculation close to 10^{-4} cm^{2}s^{-1}V^{-1} at room temperature. They also comprised a ratio between the number of effective charge carriers and the total number of alkali cations around 10^{-8} to 10^{-10}. This is comparable to the concentration of intrinsic defects in an ionic crystal or dissociated species from the weak electrolyte solution [140]. The weak electrolyte theory usually deals with the very fast ion conductive chalcogenide glasses preparation [141]. This model predicts that the conductivity would rise by increasing the dielectric constant of the glassy network, by replacing oxygen by a more polarizable ion. Hence, the growth in conductivity by increasing the dielectric constant of the medium was predicted by both the Anderson and Stuart model and Weak Electrolyte Theory.

Dynamic Structure Model

Owing to strong computer simulation progress in the early 1990s it was possible to produce simulated (virtual) glasses; whose physical behavior could be compared to real materials. Considering this, Maass et al. developed the dynamic structure model based on the concept of site memory effect [142, 143]. This model basically adopted the idea that mobile ions can create well defined sites in specific pathways [144]. Consider an $\bar{A}$ cation occupying an $\bar{A}$ site and leaving the same site behind when it moves through the glassy matrix. Such empty sites act as a 'stepping stone' for the other $\bar{A}$ cations to move through the glass. Nevertheless, with time 't' goes on, the empty $\bar{A}$ site will relax and lose the memory of the passing $\bar{A}$ cation. It can create a 'less good' $\bar{C}$ site. This newly created site would be energetically less favorable for an cation to enter a 'wrong' $\bar{C}$ site. The jump probability of $\bar{A}$ cation to a vacant nearest-neighbor $\bar{A}$ site can be described with the relationship

$$\omega_{A\bar{A}} = V_{A\bar{A}} \exp \left(\frac{-E_{A\bar{A}}}{k_B T} \right).$$

Similarly, its jump probability to a neighboring $\bar{C}$ site can be expressed as

$$\omega_{A\bar{C}} = V_{A\bar{C}} \exp \left(\frac{-E_{A\bar{C}}}{k_B T} \right).$$

In such an environment, $\bar{C}$ site must distort in order to accommodate the arrival of A cation. Therefore, appropriate distortion elastic energy can be expressed as $\Delta E_{A\bar{C}}$. This can be written in the equation form $\Delta E_{A\bar{C}} = E_{A\bar{A}} + \Delta E_{AC}$, when mismatch energy $E > 0$.

In a similar manner, glasses containing two kinds of cations A and B can be considered with the presence of two types of sites in which $\bar{A}$ is favorable to A, and $\bar{B}$ is favorable to B, along with sites $\bar{C}$ but it cannot be adjusted to cations A or cations B. As per Monte Carlo simulations, the emergence of fluctuating pathways facilitated the migration of particular ions in the glass. The usual occurrence of the mixed cation effect in glassy materials can also be explained by the dynamic structure model. Thus this model is also able to explain the dependence of conductivity on the modifier content in single alkali glasses following a simple power-law relation [142–145].

Coupling Model

This model was introduced by Ngai [146], based on general concepts dealing with all types of relaxations in complex systems including electrical excitations. This model's major assumption can be demonstrated as, the primitive relaxation of a microscopic unit at short time, $t \ll t_c$, can be independent of other relaxing units, while, after the critical time t_c, whereas interactions between the relaxing units cannot be neglected, therefore, relaxation can occur in a cooperative way. Considering this assumption, the time dependent relaxation rate $W(t)$ can expressed as:

$$W(t) = W_0 \text{ for } t \ll t_c \tag{1.22}$$

$$W(t) = W_0 \left(\frac{t}{t_0} \right)^{-n}, 0 < n < 1, \text{ for } t > t_c \tag{1.23}$$

Further, Ngai and his coworkers assumed that time t_c is independent of temperature. With this in mind, they were able to demonstrate the short-time relaxation rate is thermally activated and it can be correlated as:

$$W_0 = v \exp \left(\frac{-E_0}{k_B T} \right).$$

The associated normalized relaxation function $\phi(t)$ can be calculated as:
Using the relationship

$$\frac{d\phi(t)}{dt} = -W(t)\phi(t)$$

we can get

$$\phi(t) = \exp(-W_0 t) \quad \text{for } t \ll t_0 \tag{1.24}$$

$$\phi(t) = \exp \left(-\frac{t}{\approx \tau_\sigma} \right)^{1-n} \quad \text{for } t \gg t_c \tag{1.25}$$

Here

$$\tau_\sigma = \left(\frac{t_c^{-n}}{W_0} \right)^{\frac{1}{(1-n)}}$$

In case of short time intervals Debye relaxation can be observed, while, for longer periods the Kolrausch-Williams-Watts-type relaxation can take place. Using this relationship, the conductivity with frequency can be calculated by a plot. The coupling model predicts a constant conductivity for $t > \tau_\sigma$ and $\tau_\sigma < t_c$. Therefore, a plateau at low and high frequencies and a dispersive behavior can be obtained, according to $\sigma = A\omega^n$ ($0 < n < 1$) for $\tau_\sigma < t < t_c$.

Counter–Ion Model

Dieterich and co-workers proposed the Counter–Ion Model (CM) in the 1990s [147, 148]. According to this model, the mobile ions move among immobile counterions for these are placed randomly at the centers of cubic lattice cells. Such site energies encountered from the mobile ions can be attributed to the Coulomb fields of the immobile counterions. This model has been explained by the typical conductivity spectrum $\sigma(\omega)$ in four different systems; i) at high frequency when every hop contributes to conductivity, therefore, a plateau may exist, ii) at the lower frequency, the dispersive region occurs ($\sigma(\omega)$, $\alpha\omega''$), this comprises two schemes, iii) an additional arrangement occurs only to correlate dipolar reorientations which are able to escape out of Coulombic trap ($n \sim 0.5$–0.6), iv) final dc plateau occurs when long-range diffusion becomes possible.

Frequency-dependent Conductivity Models

This model is based on the evaluation of the conductivity with frequency, in which the complex impedance is usually analyzed in the frequency range a few mHz up to a few kHz. According to this model a dc plateau can exist at a low frequency, as a consequence a dispersive part can appear which exhibits a power law dependency with frequency, $A\omega^n$ with $n \sim 0.5$. This also is also known as Jonscher's universal law [146]. However, in the wider frequency domain, this model can have a more complete outcome with the exponent n increasing value with a growing frequency up to unity [149, 150, 151–153]. Thus, this model was developed to account for ion transport in glasses.

Jump Relaxation Model

Funke and Riess presented a Jump Relaxation Model (JRM) [154]. This model deals with the dispersion in the ion conductivity which arises due to strong forward-backward jump correlations during the motion of an ion. According to this model when an ion jumps from its previously relaxed position into a new site, then it is no longer in an equilibrium condition. The equilibrium can be recovered from the relaxation of the system in one of the two following ways: the surrounding relaxation is due to the motion of other ions or the ion itself can perform a backward jump and go back to its original position. The first or second process event probability depends on the change of the time. The longer ion time stays in new position, may result in a smaller probability occurrence of a backward jump. Therefore, the physical treatment of the system can lead to the time dependent relaxation rate $W(t)$ in terms of the frequency-dependent conductivity [147]. This model predicts high and low frequency plateaus separated by a dispersive region at intermediate frequencies.

This concept also combines the idea developed for the dynamic structural model. However, basically JRM considered that all sites can be equivalent and the new Unified Site Relaxation Model (USRM) [155] introduced the presence of two types of sites. An $\overline{A}$ site adapted to the A cation and a less favorable $\overline{C}$ site (as described in the dynamic structure model). The probability for a backward jump can be larger for an ion $\overline{A}$ jump to a $\overline{C}$ site rather than to jump to an $\overline{A}$ site. Therefore, a superlinear frequency dependence of conductivity is possible in the case of the USRM framework, while, it was not possible in JRM.

Further, the mismatch and relaxation [155, 156] and MIGRATION concept [157, 158] steps were also included in the development of JMR model. For the CMR concept, the neighboring ions, rearranged in the mismatch induced dipole field can have the same dynamic properties as the 'central' hopping one. This assumption provided a new formulation for the relaxation rate $W(t)$. As a consequence, the conductivity spectrum would be revealed as a continuously increasing slope. At the end, the MIGRATION concept could also explain the time-dependent shielding of the dipole-field associated with the mismatch of the central ion [159].

ELECTRONIC CONDUCTIVITY

The physical properties of electrical conduction of amorphous semiconductors based on the Davis and Mott model have been extensively studied and well understood. According to this model there are three different types of electrical conduction in amorphous semiconductors; (i) extended state conduction, (ii) conduction in band tail and (iii) conduction in localized states near the Fermi level.

Extended State Conduction

When the objects temperature is high enough the charge carriers in the extended states gain enough energy to be excited beyond the mobility gap. Thus conductivity for this region can be expressed as [160]

$$\sigma = \sigma_0 \exp\left(-\frac{(E_C - E_F)}{kT}\right) \tag{1.26}$$

Here the value of pre-exponential factor is given as $\sigma_0 = eN(E_C)kT\mu_C$, where, $N(E_C)$ reflects the density of state at the mobility edge, μ_C is mobility. $(E_C - E_F)$ is activation energy of the electrical conductivity. In the case of extended state conduction, the carrier mobility is in the order of $2\ cm^{-1}V^{-1}s^{-1}$ [118].

Conduction through the Band Tail

Electrical conduction in such materials can occur from the band tail under a thermally activated process through hopping. In this process two charge carriers move from one localized state to another with an exchange of energy with a phonon. The hopping conductivity at the tail state can be expressed by the following relationship [161]:

$$\sigma_{hop} = \sigma_{0hop} \frac{kT}{\Delta E} C_1 \exp-\left(\frac{(E_A - E_F + W)}{kT}\right) \tag{1.27}$$

Here 'W' is the hopping energy:

$$\sigma_{hop} = \frac{1}{6} v_{hop} e^2 R^2 N(E_C) \tag{1.28}$$

and

$$C_1 = 1 - \exp\left(-\frac{\Delta E}{kT}\right)\left(1 + \frac{\Delta E}{kT}\right) \tag{1.29}$$

where v_{hop} is the hopping frequency, $\Delta E = (E_C - E_A)$ and R is the covered distance.

Conduction through Localized State at the Fermi Energy

Charge carriers can move between the localized states near Fermi energy (E_F) when E_F lies in a band of localized states. Through a phonon-assist tunneling process analogous to impurity conduction of heavily doped and compensated semiconductors at low temperatures. This region conduction can be described by the following relationship [162]:

$$\sigma = \sigma_0 \exp\left(-\frac{W}{kT}\right) \tag{1.30}$$

Here

$$\sigma_0 = \frac{1}{6}e^2 R^2 v_{ph} N(E_F)\exp(-2\alpha R)$$

The term $\left(-\dfrac{W}{kT}\right)$ represents the probability of finding a phonon with excitation energy equal to energy difference 'W' between the states, v_{ph} is the frequency in the range 10^{12}–10^{13} s^{-1} [161]. Here R is the jumping distance at a high temperature equal to the inter-atomic spacing. The whole quantity represents the rate of fall-off of the wave function at a site.

Negative Resistance and Electrical Switching in Amorphous Semiconductors

Normally semiconducting materials show the linear proportionality behavior in between the applied Voltage (V) and the Induced current (I), following the classic Ohmic law. This common behavior in many passive and active devices consists of well-established linear system characteristics. However, certain materials do not follow the well-established linearity when the electric field is sufficiently high (10^6 V/cm), therefore, a nonlinear electrical behavior can appear [46, 163]. Due to high electric fields, the insulating materials would undergo a destructive breakdown, as a consequence, amorphous semiconductors exhibit the effect of switching and negative resistance phenomenon. Therefore, because of the electrical switching effect, amorphous semiconductors materials can have a high conducting 'ON' state from a low conducting 'OFF' state when a suitable electric field is applied. Generally, negative resistance can be classified into six possible types, as shown in Fig. 1.23(a–f).

Voltage Controlled Negative Resistance (VCNR)

Figure 1.23(a) s shows that VCNR has a low resistance with the Ohmic behavior up to a critical voltage V_t. Beyond the critical V_t, it crosses the negative resistance region and reaches into the high resistance region. During the reduction of voltage, the I–V retraced characteristic material shows that no memory was observed for VCNR, whereas memory was observed when the high resistance state is reached. It is retained after the reduction of the field (Fig. 1.23(b)) [164].

Current Controlled Negative Resistance (CCNR)

Figure 1.23(c) illustrates the CCNR behavior, it can be seen that with increasing voltage, the current initially also gets greater up to a threshold voltage V_t. Later this voltage decreases with increasing current and passing a negative resistance zone corresponding to a low resistance state. When the current reduces down to zero, the corresponding I–V characteristics retraced and showed the CCNR memory with the retained low resistance state (Fig. 1.23(d)) [164].

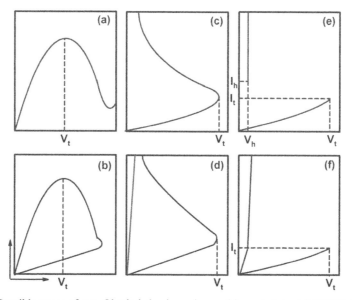

Figure 1.23 Possible types of non-Ohmic behaviour observed in materials. **(a)** VCNR; **(b)** VCNR with memory **(c)** CCNR; **(d)** CCNR with memory; **(e)** threshold switching; **(f)** memory switching.

Electrical Switching

In amorphous semiconductors, electrical switching is one of the most commercially utilized phenomenon. The electrical switching in amorphous semiconductors was discovered by Ovshinsky [46]. This important discovery gained popularity due to phase change memory applicability in such materials [164]. The electrical switching in amorphous semiconducting materials can be classified into two types; reversible (threshold switching) and irreversible (memory switching) depending on the way of the removal electric field.

Threshold Switching

In threshold switching, with increasing current amorphous semiconducting material switches from the 'OFF' state to the 'ON' state beyond the threshold voltage. Furthermore with reduced current, the voltage increases up to a certain threshold, and afterwards it reaches zero under the reverse threshold process, as shown in Fig. 1.23(e).

Memory Switching

In the memory switching system, once the system is switched on, it will remain in the low resistance ON state even though the current is reduced down to zero. Thus memory devices have a permanently ON state and their current can fall down up to zero without affecting the behavior of the ON state. However, the OFF state can be restored by passing a strong pulse of current as demonstrated in Fig. 1.23(f).

ELECTRICAL SWITCHING PARAMETERS IN AMORPHOUS SEMICONDUCTORS

The electrical parameters that are involved in the switching process are described as follows:

Electrical Switching Voltage

The threshold of the memory materials which switches the high conducting ON state from the low conducting OFF state is called threshold or switching voltage (V_{th}). The threshold or switching voltage are dependent on the specimen thickness, electrode separation, resistivity of the material, ambient temperature and nature of excitation. Therefore, specifications and conditions are important to define the material V_{th}.

Switching Time

When the voltage drops across the specimen exceeding the switching voltage (V_{th}), the time taken by the specimen changes from its high resistance state to a low resistance state is known as switching time (ts).

Delay Time

When the applied electric field surpasses the switching voltage (V_{th}), the amorphous semiconductors remains in an OFF state for a short period of time. The time gap between the applied switching voltage pulse and the actual initiation of switching is called the Delay time (t_D). The delay time typically is in the order of the 10 s and decreases exponentially when voltage is over the switching voltage [164–167].

Lock-on Time

During the memory switching, a device switches to a low resistance state after the delay time t_D. In the entire process, it is essential to maintain the material low resistance state for the shortest period of time to keep the memory state. This minimum time period setting is known as Lock-on time (t_{lo}). When plus duration is less than ($t_D + t_{lo}$) then the memory switch can be reverted back to the high resistance state. If the lock -on time is greater than 100 (micro second), then it will depend on the applied voltage [166–169].

Recovery Time

In a threshold switch specimen, after switching, the high resistance state comes back with the reduced current below I_h. Later a definite time is required for the specimen to regain its original high resistance state. This time is called the recovery time (t_r) which is typically in the order of 1 µs. During the recovery time, the specimen switches on a second time with lesser voltage (V_{th}) [164, 168].

Set Time

The required high resistance OFF state to low resistance ON state setting of the device in a small interval of time is known as the set time. The set time is equal to the sum of the delay and lock-on time ($t_D + t_{lo}$).

Reset Time

This is the time required to bring back the device in the initial high resistance OFF state by applying appropriate voltage/current pulses.

SWITCHING MECHANISMS IN AMORPHOUS SEMICONDUCTORS

There are several mechanisms to describe the electrical switching in amorphous semiconductors. The important electrical switching in amorphous semiconductors will be briefly described here.

Purely Electronic

This concept was proposed by Adler and Peterson explaining the threshold switching based on the field induced filling of charged defects [170, 171]. In amorphous semiconductors, the special charge defect states (traps) C_3^+ and C_1^- are known as Valence Alternation Pairs (VAP). The sufficiently large electric field application leads to the excitation of the charge carriers. Such charge carriers can fill the existence charge defects states in the system. The filling of the entire traps could improve the lifespan of excited charge carriers. This process provides a desired sudden increase of the charge carries to govern the specimen thickness. The higher order of lifetime increases the carrier's conductivity of the material, as a consequence a voltage drop occurs and the specimen switches to an ON state. This process is reversible, therefore, threshold switching is also referred to as reversible switching.

Space Charge Mechanism

This model was introduced by Henisch et al., and Lucas, providing an explanation of double-injection space charge in an electronic process [172, 173]. According to this model, electron and holes are injected through the applied field at the electrodes. When this process starts, these charge carriers instantly recombine and neutralize with the positively and negatively charged traps. As a consequence, a negative and positive space charge is created around the anode and cathode, respectively. When the electrical field is sufficiently strong then more charges are injected, therefore, space charge regions spread rapidly and overlap in the central area. Since in the overlapping region, the material is neutral, it leads to an increase in conductivity. Under a continuously increased electric field, the injection of charge carriers also increases at the electrode, therefore, electrons and holes are accelerated sufficiently and move cross the neutral region. In a combination, both effects increase the rate of space charge by overlapping with the increased rate of defects filling. This leads to a sudden increase in conductivity in amorphous semiconductors and has been characterized as an ON state.

Electro-thermal

In amorphous semiconductors, electrical switching contains both electronic and thermal effects with the necessary and sufficient satisfying switching condition by itself. Thus to explain any qualitative analysis of switching in such materials both thermal and electronic effects from the developed coupled response must be considered. Hence the electronic process initiates the switching, in which a current channel forms within the specimen due to the temperature rising to several hundred degrees along the conducting channel before the current shows any substantial change [174]. As a consequence, a thermally induced amorphous to crystalline phase transition occurs in the conducting channel, the specimen is then latched to an ON state [45]. The electronic process includes the thermal influence that can be considered responsible for the memory switching phenomenon. After the switching initiation due to Joule heating of a larger current flow through the specimen which is sufficient to thermally induce the amorphous to crystalline state phase transition, leading to memory switching. Otherwise, the specimen will show threshold behavior. There are several factors which decide whether an amorphous semiconductor will exhibit memory or threshold type electrical switching:

The Thermal Stability

Such specimens undergo a simple devitrification that can exhibit memory behavior even at lower ON state currents.

Thermal Diffusivity

The conducting filament temperature rise can be determined by the rate of dissipation of heat away to the bulk of the material. This in turn is decided by the thermal diffusivity of the amorphous semiconductor. The lower thermal diffusivity of amorphous semiconductors memory switching is preferable compared to those with higher thermal diffusivity.

Network Topological Effects

The network connectivity, rigidity and network topological thresholds, etc., can play a crucial role in the electrical switching in amorphous semiconductors.

APPLICATIONS OF THE CHALCOGENIDE GLASSES

In 1950 the A_2S_3 amorphous semiconductor as a commercial utility was first demonstrated as a passive optical material in the mid-IR range. At that time, widespread supplications of amorphous semiconductors were initiated. Some examples include infrared detectors, moldable infrared optics such as lenses and infrared optical fibers. The key advantage of these materials is that they can transmit across a wide range of the infrared electromagnetic spectrum. Amorphous semiconductors with the properties of high refractive index, low phonon energy, high nonlinearity also make them ideal candidates for lasers, planar optics, photonic integrated circuits and other rare earth ions doped active device applications. Apart from these, several amorphous semiconducting materials exhibited photon-induced refraction and electron-induced permittivity modification [46]. A few of such materialse also showed thermally driven amorphous crystalline phase changes. This property makes them useful for the rewritable optical disks and non-volatile memory PC-RAM devices. Amorphous semiconductors are not restricted only to memory devices applications, but are also useful for cognitive computing and reconfiguring logic circuits.

In the past few decades sulfide, selenide and selenidetelluride based amorphous semiconductors were extensively exploited for various commercial applications [175, 176]. Specifically, for the infrared optics including energy management, thermal fault detection, electronic circuit detection, temperature monitoring and night vision [177]. Amorphous semiconductors are applicable for the blackbody radiation at room temperature, specifically, in the human body IR thermal imaging range 8–12 p.m. These materials are also considered suitable for the active electronic device components in photocopying and switching applications. They are also applicable in Far-IR under suitable optimization in terms of acousto-optic figure devices in the range 10.6 μm [178]. The desirable metal photodissolution effect in such materials makes them useful for image creation and storage. These materials also have the ability of negligible image degradation. Amorphous semiconductors are used in the production of inorganic resists for VLSI lithography [179]. These materials with wide IR window and their high resolution Fourier transform infrared spectrometry allows remote sensing of gases and liquids with good absorption fundamental vibrational modes [180]. Due to the wide range of utility of these materials, their use as the active element in all-optical switching devices has been possible.

The surface relief structural in these materials makes them useful for the infrared diffractive optics and small-scale integrated optics [181]. This property also enables the necessary requirement of the planar devices for those operated by diffraction rather than by reflection or refraction with the lightweight components such as mirrors, lenses and filters. These materials photoinduced phase transition amorphous crystalline, and vice versa make them useful for optical mass memory applications. Hence a focused laser writes information can be induced by a localized phase transition, and data reading can be achieved through the utilization of the difference in reflectance between the amorphous and crystalline phases [182].

Amorphous semiconductors are well defined for chemical-sensing applications. These materials can be used as fiber-optic chemical-sensor systems for quantitative remote detection and identification, as well as detecting chemicals in mixtures. Amorphous semiconductors can also be used for the Attenuated Total Reflectance (ATR), diffuse reflectance and absorption spectroscopy. These numerous systems have studied oils, freon, soaps, paints, polymer-curing reactions, glucose/water, benzene and derivatives, chlorinated hydrocarbons, alcohols, carboxylic acids, aqueous acids, perfumes and pharmaceutical products.

Thus amorphous semiconductors are potential optoelectronics materials for various kind of ions based commercial applications. The wide variety of optical applications are presented in Fig. 1.24. Amorphous semiconductors have the potential to be the basis for future optical computers, much as silicon is the basis for today's microprocessors and computer memories [183]. These materials can also be used as [184] grating materials. They can also be used for the high-speed optical switching [185], demultiplexing signals up to 50 Gbit/s with the potential to exceed 100 Gbit/s operations. Amorphous semiconducting fibers can also be applied in various fields due to their high band gap, long wavelength multiphonon edge and low optical attenuation [186]. They have potential chemical stability in air for long core-clad fibers. They are also likely to permit new applications that are unachievable with current infrared materials [187–190].

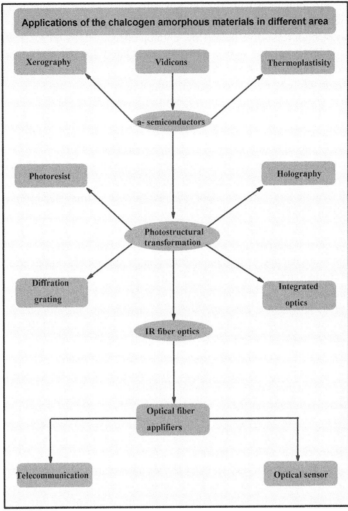

Figure 1.24　Applications of amorphous semiconductors in different fields.

CONCLUSIONS

In brief, The fundamentals of chalcogenide materials and their classifications including their emergent nano form have been discussed. Specific emphasis has been paid to provide a separate characterization on crystalline structures of sulfur, selenium and tellurium. Considering the technological significance the key features of chalcogenide based polycrystalline materials structures have been described; such as, polysulfides, polyselenides and polytellurides. As pointed out here, amorphous glassy form of chalcohenide materials are also known as amorphous semiconductors, therefore, these materials are also extensively used in semiconducting industries to fabricate types of optoelectronics devices. Therefore, considering their emergence in term of a wide range of utility the structural properties of these class of materials have been also interpreted. Taking into account the basic properties of key chalcogen elements (S, Se, Te), investigators used their alloy forms by adding suitable periodic table elements in a specific stoichiometry. Such chalcogenide materials can offer improved performance than their individual element based technical devices, therefore, the key properties of binary, ternary and multicomponent glassy alloys have also been discussed. In reporting such materials, their structure, bonding types, including kinds of theoretical approaches and those providing a correlation with their physical properties, have described to a large extent. Thus, this class of semiconducting materials band structures description has been provided with the help of the different existing models based on the fundaments of the defects in amorphous semiconductors. Considering the defects governing properties and their correlation with the photo-induced phenomenon, photoinduced effects in amorphous chalcogenides has also been briefly interpreted. Moreover, various aspects of the amorphous semiconductors such as ionic and electrical conductivities have to be taken into consideration and their basic concepts have been described by providing an explanation from different existing theoretical models. Since electrical switching in amorphous semiconductors is one of the electrical properties, the significance of the different kinds switching including their key parameters have also been discussed; such as, threshold switching and memory switching. Considering the key advantages and their existing application in various technological areas, a brief description was also provided. Thus, in this introductory work a concrete view has been provided on chalcogenide materials from earlier to the current era by demonstrating their various key aspects based on the available scientific/technological information's.

■ References

1. Ahluwalia, G.K. (2017) Applications of Chalcogenides: S, Se, and Te. Springer International Publishing AG, Gewerbestrasse 11, 6330 Cham, Switzerland.

2. de Pablos-Martın, A., Duran, A., Pascual, M.J. (2012) Nanocrystallisation inoxyfluoride systems: mechanisms of crystallisation and photonic properties. Int. Mater. Rev., 57, 165–186.

3. Singh, A.K. (2012) Amorphous and nanophase microstructures of bulk Se based chalcogenide alloys. Optoelectronics Lett., 8, 0165–0167.

4. Sorkin, V., Pan, H., Shi, H., Quek, S.Y., Zhang, Y.W. (2014) Nanoscale transition metal dichalcogenides: Structures, properties, and applications. Crit. Rev. Solid State Mater. Sci., 39, 319–367.

5. Pershin, Y.V., Ventra, M.D. (2011) Memory effects in complex materials and nanoscale systems. Adv. Phys., 60(2), 145–227.

6. Diéguez, O., Marzari, N. (2009) A first-principles characterization of the structure and electronic structure of -S and Rh-S chalcogenides. Phys. Rev. B, 80, 214115/1–214115/6.

7. Templeton, L.K., Templeton, D.H., Zalkin, A. (1976) Crystal structure of monoclinic sulphur locality: synthetic. Inorg. Chem., 15, 1999–2001.

8. Golosovsky, I.V., Smirnov, O.P., Delaplane, R.G., Wannberg, A., Kibalin, Y.A., Naberezhnov, A.A. ,et al. 2006. Atomic motion in Se nanoparticles embedded into a porous glass matrix. Eur. Phys. J. B, 54, 211–216.

9. Minaev, V.S., Timoshenkov, S.P. Kalugina, V.V. (2005) Structural and phase transformations in condensed selenium. J. Optoelectron. Adv. Mater., 7, 1717–1741.

10. Degtyareva, O., Gregoryanz, E., Mao, H.K., Hemley, R.J. (2005) Crystal structure of sulfur and selenium at pressures up to 160GPa. High Pressure Res., 25, 17–33.

11. Cherin, P., Unger, P. (1967) The crystal structure of trigonal selenium. Inorg. Chem., 6, 1589–1591.

12. Cherin, P., Unger, P. (1972) Refinement of the crystal structure of alpha-monoclinic Se. Acta Crystallographica B, 28, 313–317.

13. Marsh, R.E., Pauling, L., McCullough, J.D. (1953) The crystal structure of beta-selenium. Acta Crystallographica, 6, 71–75.

14. Foss, O., Janickis, V. (1980) Crystal structure of gamma-monoclinic selenium. Dalton Transactions, 1980, 624–627.

15. Miyamoto, Y. (1980) Structure and phase transformation of rhombohedral selenium composed of Se_6 molecules. Jpn. J. Appl. Phys., 19, 1813–1819.

16. Devillanova, F.A. (2006) Handbook of Chalcogen Chemistry: New Perspectives in Sulfur, Selenium and Tellurium. RSC Publishing, Cambridge CB4 0WF, UK.

17. Laitinen, R.S., Oilunkaniemi, R. (2015) Catenated Compounds: Group 16 Elements Selenium and Tellurium, Reference Module in Chemistry, Molecular Sciences and Chemical Engineering, Elsevier.

18. Licht, S., Hodes, G., Manassen, J. (1986) Numerical analysis of aqueous polysulfide solutions and its application to cadmium chalcogenide/polysulfide photoelectrochemical solar cells. Inorg. Chem., 25, 2486–2489.

19. Dubois, P., Lelieur, J.P., Lepoutre, G. (1988). Identification and characterization of lithium polysulfides in solution in liquid ammonia. Inorg. Chem., 27, 73–80.

20. Muller, C., Bottcher, P. (1995). Darstellung und kristallstruktur von bis(cyclohexyIammonium) heptasulfid • Cyclohexylamin. Zeitschrift für Naturforschung, 50b, 1623–1626.

21. Schliephake, A., Falius, H. (1988). Darstellung und Kristallstruktur des Bis(triethylammonium) octasulfids, $[HN(C_2H_5)_3]_2S_8$. Zeitschrift für Naturforschung, 43b, 21–24.

22. Kanatzidis, M.G., Huang, S.P. (1994) Coordination chemistry of heavy polychalcogenide ligands. Coord. Chem. Rev., 130, 509–621.

23. Smith, D.M., Ibers, J.A. (2000) Syntheses and solid-state structural chemistry of polytelluride anions. Coord. Chem. Rev., 200–202, 187–205.

24. Kanatzidis, M.G., Sutorik, A.C. (1995) The application of polychalcogenide salts to the exploratory synthesis of solid state multinary chalcogenides at intermediate temperatures. Prog. Inorg. Chem., 43, 151–265.

25. Sheldrick, W.S., Wachhold, M. (1997) Solventothermal synthesis of solid-state chalcogenidometalates. Angew. Chem. Int. Ed. Engl., 36(3), 206–224.

26. Muller, C., Bottcher, P. (1994). Darstellung und kristallstruktur des bis(trimethylammonium)hexasulflds $[HN(CH_3)_3]_2S_6$. Zeitschrift für Naturforschung, 49b, 489–493.

27. Muller, C., Bottcher, P. (1993) Darstellung und kristallstruktur des bis(diisobutylammonium)-heptasulfids $[H_2N(i-C_4H_9)_2]_2S_7$. Zeitschrift für Naturforschung, 48b, 1732–1736.

28. Muller, V., Grebe, C., Muller, U., Dehnicke, K. (1993) Synthese und kristallstruktur des nonaselenids $[Sr(15-Krone-5)_2]Se_9$. ZAAC, 619, 416–420.

29. Staffel, R., Muller, U., Ahle, A., Dehnicke, K. (1991) Die Kristallstruktur von $[Na(12-Krone-4)_2]^+$ $I_2Se_8{}^{2-}$ (Se_6, Se_7). Zeitschrift für Naturforschung, 46b, 1287–1292.

30. Kaiber, F., Petter, W., Hulliger, F. (1983) The structure type of Re_2Te_5, a new [M6X14] cluster compound. J. Solid State Chem., 46, 112–120.

31. Schreiner, B., Dehnicke, K., Maczek, K., Fenske, D. (1993) $[K(15-Krone-5)_2]_2Te_8$ — ein bicyclisches polytellurid. ZAAC, 619, 1414–1418.

32. Sheldrick, W.S., Wachhold, M. (1995) Discrete crown-shaped Te_8 rings in Cs_3 Te_{22}. Angew. Chem. Int. Ed. Engl., 34, 450–451.

33. Bottcher, P., Keller, R. (1985) The crystal structure of NaTe and its relationship to tellurium-rich tellurides. Journal of the Less Common Metals, 109, 311.

34. Bottcher, P. (1988) Tellurium-rich tellurides. Angew. Chem. Int. Ed. Engl., 27, 759–772.

35. Sheldrick, W.S. (2000) Network self-assembly patterns in main group metal chalcogenide-based materials. J. Chem. Soc., Dalton Trans., 18, 3041–3052.

36. Papoian, G., Hoffmann, R. (2001) The first intermolecular transition metal-catalyzed [5+2] cycloadditions with simple, unactivated, vinylcyclopropanes. J. Am. Chem. Soc., 123, 179–180.

37. Ienco, A., Proserpio, D.M. Hoffmann, R. (2004) Main group element nets to a T. Inorg. Chem., 43, 2526–2540.

38. Lucovsky, G., Parsons, G.N. (2004) U.S. Patent No., 6, 787, 861.

39. Wilk, G.D., Wallace, R.M., Anthony, J.M. (2001) High-κ gate dielectrics: Current status and materials properties considerations. J. Appl. Phys., 89, 5243–5275.

40. Kiss, L.F., Bakonyi, I., Lovas, A., Baran, M., Kadlecova, J. (2001) Magnetic properties of amorphous $Ni_{81.5-x}Fe_xB_{18.5}$ alloys (x=1,2,3): A further key to understand the magnetism of amorphous $Ni_{81.5}B_{18.5}$. Journal of Physics Review B, 64, 064417/1–064417/7.

41. Egami, T. (1977) Low-field magnetic properties of amorphous alloys. J. Am. Chem. Soc., 60, 128–133.

42. Singh, J., Shimakawa, K. (2003) Advances in Amorphous Semiconductors. CRC Press.

43. Hamakawa, Y. (1984) Amorphous Semiconductors Technologies & Devices. Elsevier Science Ltd.

44. Adler, D. (1971) Amorphous semiconductors. Journal of CRC Critical Reviews in Solid State Sciences, 2-3, 317–465. DOI: 10.1080/10408437108243545.

45. Madan, A., Shaw, M.P. (1988) The Physics and Applications of Amorphous Semiconductors. Academic Press, Inc. London.

46. Tauc, J. (ed.) (1974) Amorphous and Liquid Semiconductors. Plenum Press, New York.

47. Ovshinsky, S.R. (1968) Reversible electrical switching phenomena in disordered structures. Phys. Rev. Lett., 21, 1450–1453.

48. Frerichs, R. (1950) New optical glasses transparent in the infrared up to 12m. Phys. Rev., 78, 643.

49. Goryunova, N.A., Kolomiets, B.T. (1955) Properties and structure of ternary semiconductor systems. I. Electric properties and structure of some materials in the Tl-Sb-Se System. Zh. Tekh. Fiz., 25, 984–994.

50. Stocker, H.J. (1969) Bulk and thin film switching and memory effects in semiconducting chalcogenide glasses. Appl. Phys. Lett., 15, 55–57.

51. Mott, N.F., Davis, E.A. (1979) Electronic Processes in the Non-Crystalline Materials, 2nd Ed. Oxford University Press, Oxford.

52. Kawamura, T., Yamamoto, N., Nakayama, Y. (1983) Electrophtographic application of amorphous semiconductors- amorphous semiconductor technologies and devices. JARECT Hamakawa, Y. North Holland OHMSHA, 6, 325–336.

53. Ciureanu, P., Middelhoek, S. (1992) Thin Film Resistive Sensors. CRC Press, London.

54. Robert, E.J., Kasap, S.O., Rowlands, J., Polischuk, B. (1998) Metallic electric contacts to stabilized amorphous selenium for use in X-ray image detectors. J. Non-Cryst. Solids, 227–240, 1359–1362.

55. Quiroga, I., Corredor, C., Bellido, F., Vazquez, J., Villares, P., Garay, R.J. (1996) Infrared studies of Ge-Sb-Se glassy semiconductor. J. Non-Cryst. Solids, 196, 183–186.

56. Lenz, G., Zimmermann, J.K., Katsufuji, T. (2000) Large kerr effect in bulk Se-based chalcogenide glasses. Optical Letter, 25, 254–256.

57. Bicerono, J., Ovshinsky, S.R. (1985) Chemical bond approach to the structure of chalcogenide glasses with reversible switching properties. J. Non-Cryst. Solids, 74, 75–84.

58. Ovshinsky, S.R. (1994) An history of phase change technology. Memories Optics Systems, 127, 65.

59. Boolchand, P., Georgiev, D.G., Goodmana, B. (2001) Discovery of the intermediate phase in chalcogenide glasses. J. Optoelectron. Adv. Mater., 3, 703–720.

60. Popescu, M. (2006) Chalcogenides–Past, present, future. J. Non-Cryst. Solids, 352, 887–891.

61. Singh, A.K., Singh, K. (2009) Crystallization kinetics and thermal stability of $Se_{98-x}Zn_2In_x$ chalcogenide glasses. Philosophie Magazine, 89, 1457–1472.

62. Singh, A.K., Singh, K. (2010) Observation of meyer neldel rule and crystallization rate constant stability for $Se_{93-x}Zn_2Te_5In_x$ chalcogenide glasses. Eur. Phys. J. Appl. Phys., 51, 30301/1–30301/5.

63. Singh, A.K. (2013) SeZnSb alloy and its nano tubes, graphene composites properties. AIP Advances, 3, 042124/1–042124/11.

64. Singh, A.K. (2013) Microscopic study on the Se-Te-Ge Alloy and its composite with carbon nanotubes and graphene. J. Adv. Microsc. Res., 7, 1–7.

65. Mehta, N., Kumar, A. (2007) Comparative analysis of calorimetric studies In $Se_{90}M_{10}$ (M=In, Te, Sb) Chalcogenide Glasses. J. Therm. Anal. Calorim., 87, 345–150.

66. Song, S.M., Choi, S.Y., Yong-Keun, L. (1997) Crystallization property effects in $Ge_{30}Se_{60}Te_{l0}$ glass. J. Non-Cryst. Solids, 217, 79–82.

67. Usuki, T., Uemura, O., Konno, S., Kameda, Y., Sakurai, M. (2001) Structural and physical properties of Ag-As-Te glasses. J. Non-Cryst. Solids, 293, 799–805.

68. Bhanu Prashanth, S.B., Asokan, S. (2008) Composition dependent electrical switching in $Ge_xSe_{35-x}Te_{65}$ ($18 \leq x \leq 25$) glasses–the influence of network rigidity and thermal properties. Solid State Commun., 147, 452–456.

69. Othman, A.A., Aly, K.A., Abousehly, A.M. (2006) Crystallization kinetics in new $Sb_{14}As_{29}Se_{52}Te_5$ amorphous glass. Solid State Commun., 138, 184–189.

70. Vassilev, V., Parvanov, S., Vasileva, T.H., Aljihmani, L., Vachkov, V., Evtimova, T.V. (2007) Glass formation in the As_2Te_3–As_2Se_3–SnTe system. Mater. Lett., 61, 3676–3578.

71. Gill, W.D., Street, G.B. (1973) Drift mobility in amorphous selenium-sulfur alloys. J. Non-Cryst. Solids, 13, 120–130.

72. Sarrach, J., de Neufville, J.P., Haworth, W.L. (1976) Studies of amorphous Ge-Se-Te alloys (I): Preparation and calorimetric observations. J. Non-Cryst. Solids, 22, 245–267.

73. Hawes, L. (1963) Sulphur–selenium and sulphur–tellurium cyclic interchalcogen compounds. Nature, 198, 1267–1270.

74. Abu EL-Fadl, A., Hafiz, M.M., Wakaad, M.M., Aashour, A.S. (2007) Calorimetric studies of the crystallization process in $Cu_{10}Se_{90}$ and $Cu_{20}Se_{80}$ chalcogenide glasses. Physica B, 398, 118–125.

75. Alnajjar, A.A. (2009) The role of thermal treatment on the optical properties of $Ge_{0.15}Se_{0.85}$ system. Renewable Energy, 34, 71–74.

76. Singh, A.K., Singh, K. (2007) Correlative study of optical, electrical and thermal transport properties of $Se_{100-x}In_x$ chalcogenide glasses. J. Optoelectron. Adv. Mater., 9, 3756–3759.

77. Boo, B.H., Cho, H., Kang, D.E. (2007) Ab initio and DFT investigation of structures and energies of low-lying isomers of Zn_xSe_x ($x = 1$–4) clusters. Journal of Molecular Structure: THEOCHEM, 806, 77–83.

78. Singh, A.K., Singh, K. (2009) Composition dependence UV-Visible and MID-FTIR properties of $Se_{98-x}Zn_2In_x$ ($x = 0$, 2, 4, 6 and 10) chalcogenide glasses. J. Mod. Opt., 56, 471–476.

79. Singh, A.K., Singh, K. (2011) Localized structural growths and kinetics of $Se_{98-x}Zn_2In_x$ ($0 \leq x \leq 10$) Amorphous Alloys. Physica Scripta, 83, 025605/1–025605/5.

80. Singh, A.K. (2011) Effect of indium additive on the heat capacity of Se-Zn chalcogenide glasses. Eur. Phys. J. AP, 55, 11103/1– 11103/4 .

81. Thingamajig, B., Ganesan, R., Asha Bhat, N., Sangunni, K.S., Gopal, E.S.R. (2000) Determination of thermal diffusion length in bismuth doped chalcogenide glasses, by photoacoustic technique. J. Optoelectron. Adv. Mater., 2, 91–94.

82. Petkov, P. (2002) Multicomponent germanium chalcogenide glasses. J. Optoelectron. Adv. Mater., 4, 747–750.

83. Tikhomirov, V.K., Furniss, D., Seddon, A.B., Savage, J.A., Mason, P.D., Orchard, D.A., et al. (2004) Glass formation in the Te-enriched part of the quaternary Ge-As-Se-Te system and its implication for mid-infrared optical fibres. Infrared Physics & Technology, 45, 115–123.

84. Vassilev, V. (2006) Multicomponent Cd (Zn)-containing Ge(As)-chalcogenide glasses. Journal of Chemical Technology and Metallurgy, 41, 257–276.

85. Vassilev, V., Tomova, K., Parvanova, V., Parvanov, S. (2007) New chalcogenide glasses in the $GeSe_2$–Sb_2Se_3–PbSe system. Mater. Chem. Phys., 103, 312–317.

86. Wibowo, R.A., Kim, W.S., Lee, E.S., Munir, B., Kim, K.H. (2007) Single step preparation of quaternary $Cu_2ZnSnSe_4$ thin films by RF magnetron sputtering from binary chalcogenide targets. J. Phys. Chem. Solids, 68, 1908–1913.

87. Singh, A.K., Singh, K. (2010) Observation of meyer neldel rule and crystallization rate constant stability of $Se_{93-x}Zn_2Te_5In_x$ chalcogenide glasses. Eur. Phys. J. AP, 51, 30301/1–30301/5.

88. Popov, A. (2004) Semiconducting Chalcogenide Glass I. Elsevier Academic Press.

89. Webpage, http://webbook.nist.gov/chemistry/.

90. Cameron, T.S., Deeth, R.J., Dionne, I., Du, H., Jenkins, H.D.B., Krossing, I., et al. (2000) Bonding, structure, and energetics of gaseous E_8^{2+} and of solid $E_8(AsF_6)_2$ (E = S, Se). Inorg. Chem., 39, 5614–5631.

91. Brownridge, S., Krossing, I., Passmore, J., Jenkins, H.D.B., Roobottom, H.K. (2000) Recent advances in the understanding of the syntheses, structures, bonding and energetics of the homopolyatomic cations of groups 16 and 17. Coord. Chem. Rev., 197, 397–481.

92. Mott, N.F. (1969) Conduction in non-crystalline materials. Philosophical Magazine, 19, 835–852.

93. Zallen, R. (1983) Physics of Amorphous Solids York. Wiley-VCH Verlag GmbH & Co., KGaA, Weinheim, Germany.

94. Lucovsky, G., Hayes, T.M. (1979) Amorphous Semiconductors. edited by M.H. Brodsky. Berlin, Springer-Verlag, p. 36.

95. Elliot, S.R. (1984) Physics of Amorphous Materials. Longman Inc., New York.

96. Phillips, J.C. (1979) Topology of covalent non-crystalline solids I: Short-range order in chalcogenide alloys. J. Non-Cryst. Solids, 34, 153–181.

97. Tanaka, K. (1988) Layer structures in chalcogenide glasses. J. Non-Cryst. Solids, 103, 149–150.

98. Senapati, U., Varshneya, A.K. (1995) Configurational arrangements in chalcogenide glasses: A new perspective on Phillips' constraint theory. J. Non-Cryst. Solids, 185, 289–296.

99. Mullmann, R., Mosel, B.D., Eckert, H. (1999) Physical and chemical threshold behavior in chalcogenide networks: 119Sn Mössbauer spectroscopy of Ge(Sn)-As-Se glasses. Phys. Chem. Chem. Phys., 1, 2543–2550.

100. Thorpe, M.F. (1983) Continuous deformations in random networks. J. Non-Cryst. Solids, 57, 355–370.

101. Wooten, F., Weaire, D. (1974) Solid State. Academic, New York 40, 1.

102. Feng X., Bresser, W., Boolchand, P. (1997) Direct evidence for stiffness threshold in chalcogenide glasses. Physical Review Letters, 78, 4422–4425.

103. Kamitakahara, W.A., Cappelletti, R.L., Boolchand, P., Halfpap, B., Gompf, F., Neumann, D.A., et al. (1991) Vibrational densities of states and network rigidity in chalcogenide glasses. Phys. Rev. B, 44, 94–100.

104. Sreeram, A.N., Varshneya, A.K., Swiler, D.R. (1991) Molar volume and elastic properties of multicomponent chalcogenide glasses. J. Non-Cryst. Solids, 128, 294–309.

105. Boolchand, P., Bresser, W.J., Zhang, M., Wu, Y., Wells, J., Enzweiler, R.N. (1995) Lamb-Mössbauer factors as a local probe of floppy modes in network glasses. J. Non-Cryst. Solids, 182, 143–154.

106. Asokan S., Prasad, M.Y.N., Parthasarathy, G. (1989) Mechanical and chemical thresholds in IV-VI chalcogenide glasses. Phys. Rev. Lett., 62, 808–810.

107. Boolchand P., Zhang, M., Goodman, B. (1996) Influence of one-fold-coordinated atoms on mechanical properties of covalent networks. Phys. Rev. B, 53, 11488–11494.

108. Angus, J.C., Jansen, F.J. (1988) Dense "diamondlike" hydrocarbons as random covalent networks. J. Vac. Sci. Technol. A, 6, 1778–1782.

109. Boolchand, P., Bresser, W.J. (2001) Mobile silver ions and glass formation in solid electrolytes. Nature, 410, 1070–1073.

110. White, R.M. (1974) Random network model for amorphous alloys. J. Non-Cryst. Solids, 16, 387–398.

111. Azoulay, R., Thibierge, H., Brenac, A. (1975) Devitrification characteristics of Ge_xSe_{1-x} glasses. J. Non-Cryst. Solids, 18, 33–53.

112. Narayanan, R.A. (2000) Amorphous semiconductors. PhD thesis, IISc.

113. Zallen, R. (1983) The Physics of Amorphous Solids. John Wiley & Sons, New York.

114. Anderson, P.W. (1958) Absence of diffusion in certain random lattices. Phys. Rev., 109, 1492–1505.

115. Mott, N.F. (1970) Conduction in non-crystalline systems. Philos. Mag., 22, 7–29.

116. Adler, D., Schwartz, B., Martin, C.S. (1985) Physical Properties of Amorphous Materials. Plenum Press, New York.

117. Cohen, M.H., Fritzche, H., Ovshinsky, S.R. (1969) Simple band model for amorphous semiconducting alloys. Phys. Rev. Lett., 22, 1065–1068.

118. Mott, N.F., Davis, E.A. (1979) Electronic Processes in Non Crystalline Materials. Clarendon, Oxford.

119. Davis, E.A., Mott, N.F. (1970) Conduction in non-crystalline systems V. Conductivity, optical absorption and photoconductivity in amorphous semiconductors. Philosophical Magazine, 22, 903–922.

120. Marshall, J.M., Owen, A.E. (1971) Drift mobility studies in vitreous arsenic triselenide. Philosophical Magazine, 24, 1281–1305.

121. Kastner, M. (1972) Bonding bands, lone-pair bands, and impurity states in chalcogenide semiconductors. Phys. Rev. Lett., 28, 355–357.

122. Adler, D. (1980) Chemistry and physics of amorphous semiconductors. J. Chem. Educ., 57, 560–563.

123. Kastner, M., Adler, D., Fritzsche, H. (1976) Valence-alternation model for localized gap states in lone-pair semiconductors. Phys. Rev. Lett., 37, 1504–1507.

124. Street, R.A., Mott, N.F. (1975) States in the gap in glassy semiconductors. Phys. Rev. Lett., 35, 1293–1296.

125. Mooser, E., Pearson, W.B. (1960) Progress in Semiconductors. Heywood and Company Ltd., London.

126. Adler, D. (1980) Defects in amorphous semiconductors. J. Non-Cryst. Solids, 35-36, 819–824.

127. Anderson, P.W. (1975) Model for the electronic structure of amorphous semiconductors. Physical Review Letters, 34, 953–955.

128. Elliott, S.R. (1991) Material Science and Technology: A Comprehensive Treatment. VCH Publishers Inc, New York.

129. Rao, K.J. (2002) Structural Chemistry of Glasses. Elsevier Science.

130. Kastner, M. (1978) Defect chemistry and states in the gap of lone-pair semiconductors. J. Non-Cryst. Solids, 31, 223–240.

131. Kastner, M., Fritzche, H. (1978) Defect chemistry of lone-pair semiconductors. Philosophical Magazine B, 37, 199–225.

132. Lu, Q., Gao, F., Komarneni, S. (2006) Cellulose-directed growth of selenium nanobelts in solution. Chem. Mater., 18, 159–163.

133. Gates, B., Yin, Y., Xia, Y. (2000) J. Am. Chem. Soc., 122, 12582

134. Xiong, S., Xi, B., Wang, W., Wang, C., Fei, L., Zhou, H., et al. (2006) The fabrication and characterization of single-crystalline selenium nanoneedles. Cryst. Growth Des., 6, 1711–1716.

135. Pradel, A., Ribes, M. (1989) Lithium chalcogenide conductive glasses. Mater. Chem. Phys., 23, 121–142.

136. Anderson, O.L., Stuart, D.A. (1954) Calculation of activation energy of ionic conductivity in silica glasses by classical methods. J. Am. Ceram. Soc., 37, 573–580.

137. Ravaine, D., Souquet, J.L. (1977) A thermodynamic approach to Ionic conductivity in oxide glasses: I. Phys. Chem. Glasses, 18, 27–31.

138. Reggiani, J.C., Malugani, J.P., Bernard, J. (1978) Étude des systèmes vitreux $AgPO_3$—Ag X (X = I, Br, Cl) par calorimetrie de dissolution. Corrélation entre l'activité thermodynamique de Ag X et la conductivité ionique du verre. J. Chim. Phys., 75, 849–854.

139. Levasseur, A., Brethous, J.C., Kbala, M., Hagenmuller, P. (1981) Synthesis and characterization of new amorphous solid electrolytes. Solid State Ionics, 5, 651–654.

140. Souquet, J.L., Nascimento, M.L., Rodrigues, A.C. (2010) Charge carrier concentration and mobility in alkali silicates. J. Chem. Phys., 132, 034704/1–034704/7.

141. Barrau, B., Latour, J.M., Ravaine, D., Ribes, M. (1979) Silicates Industriels, 12, 275.

142. Maass, P., Bunde, A., Ingram, M.D. (1992) Ion transport anomalies in glasses. Phys. Rev. Lett., 68, 3064–3067.

143. Malugani, J.P., Saida, A., Wasniewski, A., Robert, G. (1979) Comptes Rendus Hebdomadaires des Seances de L'Academie des Sciences Serie C, 289, 69–71.

144. Ingram, M.D., Robertson, A.H.J. (1997) Ion transport in glassy electrolytes. Solid State Ionics, 94, 49–54.

145. Bunde, A., Ingram, M.D., Russ, S. (2004) A new interpretation of the dynamic structure model of ion transport in molten and solid glasses. Phys. Chem. Chem. Phys., 6, 3663–3668.

146. Ngai, K.L. (1992) Structural relaxation and conductivity relaxation in glassy ionics. Journal de Physique IV, 2, 61–73.

147. Funke, K. (1993) Jump relaxation in solid electrolytes. Prog. Solid State Chem., 22, 111–195.

148. Knodler, D., Pendzig, P., Dieterich, W. (1996) Ion dynamics in structurally disordered materials: Effects of random Coulombic traps. Solid State Ionics, 86–88, 29–39.

149. Durand, B., Taillades, G., Pradel, A., Ribes, M., Badot, J.C., Belhadjtahar, N. (1994) Frequency dependence of conductivity in superionic conducting chalcogenide glasses. J. Non-Cryst. Solids, 172, 1306–1314.

150. Cutroni, M., Mandanici, A., Piccolo, A., Fanggao, C., Saunders, G.A., Mustarelli, P. (1996) Ionic conduction in silver phosphate glasses doped with silver sulphide. Philosophical Magazine B, 73, 349–365.

151. Burns, A., Chryssikos, G.D., Tombari, E., Cole, R.H., Risen, W.M. (1989) Dielectric spectra of ionic conducting ylasses to 2 GHz. Phys. Chem. Glasses, 1989, 30, 264–270.

152. Cramer, C., Funke, K. (1992) Observation of two relaxation processes in an ion conducting glass yields new structural information. Berichte der Bunsen-Gesellschaft–Phys. Chem. Chem. Phys, 96, 1725–1727.

153. Pradel, A., Taillades, G., Cramer, C., Ribes, M. (1998) Ion dynamics in superionic chalcogenide glasses studied in large frequency and temperature ranges. Solid State Ionics, 105, 139–148.

154. Funke, K., Riess, I. (1984) Debye-hückel-type relaxation processes in solid ionic conductors. Zeitschrift für Physikalische Chemie Neue Folge, 140, 217–232.

155. Funke, K., Banhatti, R.D., Bruckner, S., Cramer, C., Krieger, C., Mandanici, A., et al. (2002) Ionic motion in materials with disordered structures: conductivity spectra and the concept of mismatch and relaxation. Phys. Chem. Chem. Phys., 4, 3155–3167.

156. Funke, K., Wilmer, D. (2000) Concept of mismatch and relaxation derived from conductivity spectra of solid electrolytes. Solid State Ionics, 136, 1329–1333.

157. Funke, K., Banhatti, R.D., Cramer, C. (2005) Correlated ionic hopping processes in crystalline and glassy electrolytes resulting in MIGRATION-type and nearly-constant-loss-type conductivities. Phys. Chem. Chem. Phys., 7, 157–165.

158. Funke, K., Banhatti, R.D. (2006) Ionic motion in materials with disordered structures. Solid State Ionics, 177, 1551–1557.

159. Funke, K., Banhatti, R.D. (2008) Translational and localised ionic motion in materials with disordered structures. Solid State Sci., 10, 790–803.

160. Marshall J.M., Owen, A.E. (1971) Drift mobility studies in vitreous arsenic triselenide. Philosophical Magazine, 24, 1281–1305.

161. Nagels, P. Edited by Brodsky, M.H. (1985) 2nd Edition, Springer-Verlag, Berlin, P 125.

162. Nagels, P. Edited by Brodsky, M.H. (1985) 2nd Edition, Springer-Verlag, Berlin, P 126.

163. Peck, W.F., Jr., Dewald, J.F. (1964) The hall effect in semiconducting glasses. J. Electrochem. Soc., 111, 561–563.

164. Fritzsche H., Edited by Tauc, J. (1974) Amorphous and Liquid Semiconductors. Plenum Press, New York.

165. Wuttig, M. (2005) Towards a universal memory? Nat. Mater., 4, 265–266.

166. Ovshinsky, S.R., Fritzsche, H. (1973) Amorphous semiconductors for switching, memory, and imaging applications. IEEE Transactions on Electron Devices, 20, 91–105.

167. Bunton, G.V., Quillam, R.M. (1973) Switching and memory effects in amorphous chalcogenide thin films. IEEE Transactions on Electron Devices, 20, 140–144.

168. Owen, A.E., Robertson, J.M. (1973) Electronic conduction and switching in chalcogenide glasses. IEEE Transactions on Electron Devices, 20, 105–122.

169. Warren, A.C. (1973) Reversible thermal breakdown as a switching mechanism in chalcogenide glasses. IEEE Transactions on Electron Devices, 20, 123–131.

170. Adler, D. (1977) Amorphous-semiconductor devices. Scientific American, 236, 36–49.

171. Petersen, K.E., Adler, D. (1976) On state of amorphous threshold switches. J. Appl. Phys., 47, 256–263.

172. Henisch, H.K., Fagen, E.A., Ovshinsky, S.R. (1970) A qualitative theory of electrical switching processes in monostable amorphous structures. J. Non-Cryst. Solids, 4, 538–547.

173. Lucas, I. (1971) Interpretation of the switching effect in amorphous semiconductors as a recombination instability. J. Non-Cryst. Solids, 6, 136–144.

174. Warren, A.C. (1969) On state of chalcogenide glass switches. Electron letter, 5, 461–462.

175. Hilton, A.R., Jones, C.E., Brau, M. (1966) Non oxide chalcogenide glass. Glass forming region and variation in physical properties. Phys. Chem. Glasses, 7, 105–112.

176. Savage, J.A. (1985) Infrared Optical Materials and their Antireflection Coatings. Adam Hilger, Bristol.

177. Johnson, R.B. (1988) Proc. Soc. Photo-optical instrumentation engineers. SPIE., 915, 106–115.

178. Seddon, A.B. (1995) Chalcogenide glasses: A review for their preparation, properties and applications, J. Non-Cryst. Solids, 184, 44–50.

179. Andriesh, A.M., Ponomar, V.V., Smiruov, V.L., Mironos, A.V. (1986) Applications of chalcogenide glasses in integrated and fiber optics. Quantum Electron., 16, 721–736.

180. Nishii, J., Morimoto, S., Inagawa, I., Iizuka, R., Yamashita, T., Yamagishi, T. (1992) Recent advances and trends in chalcogenide glass fiber technology: A review. J. Non-Cryst. Solids, 140, 199–2008.

181. Ewen, P.J.S., Zekak, A., Slinger, C.W., Dale, G., Pain, D.A., Owen, A.E. (1993) Diffractive infrared optical elements in chalcogenide glasses. J. Non-Cryst. Solids, 164–166, 1247–1250.

182. Ovshinsky, S.R. (1992) Optically induced phase changes in amorphous materials. J. Non-Cryst. Solids, 141, 200–203.

183. Sanghera, J.S., Florea, C.M., Shaw, L.B., Pureza, P., Nguyen, V.Q., Bashkansky, M., et al. (2008) Nonlinear properties of chalcogenide glasses and fibers. J. Non-Cryst. Solids, 354, 462–467.

184. Ganeev, R.A., Ryasnyanski, A.I., Usmanov, T. (2003) Effect of nonlinear refraction and two-photon absorption on the optical limiting in amorphous chalcogenide films. Phys. Solid State, 45, 207–213.

185. Ganeev, R.A., Ryasnyansky, A., Kodirov, M.K., Usmanov, T. (2002) Twophoton absorption and nonlinear refraction of amorphous chalcogenide films. J. Opt. A: Pure Appl. Opt., 4, 446–451.

186. Nasu, H., Kubodera, K.I., Kobayashi, M., Nakamura, M., Kamiya, K. (1990) Thirdharmonic generation from some chalcogenide glasses. J. Am. Ceram. Soc., 73, 1794–1796.

187. Shiryaev, V.S., Churbanov, M.F. (2017) Recent advances in preparation of high-purity chalcogenide glasses for mid-IR photonics. J. Non-Cryst. Solids, 475, 1–9.

188. Andriesh, A. (2005) Chalcogenide glasses as multifunctional photonic materials. J. Optoelectron. Adv. Mater., 7, 2931–2939.

189. Lezal, D. (2003) Chalcogenide glasses-survey and progress. J. Optoelectron. Adv. Mater., 5, 23–34.

190. Sanghera, J.S., Aggarwal, (1990) Active and passive applications of chalco-genide glass fibers a review. J. Non-Cryst. Solids, 256–257, 6–16.

Chapter 2

Glass Formation and Structural Modification in Glasses

INTRODUCTION

Word glass is derived from a latin term glaseum, this literal meaning a lustrous and transparent material. The glassy substances are also defined as vitreous. In term of crystallization it is early civilisations that considered as luster and durability, this is one of the most important characteristics of glasses. In the current age, too luster, transparency and durability of the glasses are exploited in the applications. But it can be emphasized that these properties of luster, transparency and durability are neither sufficient nor necessary to describe the glass forming systems. However, the presence of glasses in our surroundings is so common that can be noticed for their existence. The ancient Egyptians considered glasses as precious materials as evidenced by the glass beads found in the tombs of ancient Pharaohs. Humans were produced glasses from the melting of raw materials since thousands of years. As per known knowledge the first crude manmade glasses were used to produce beads or to shape into tools requiring sharp edges. Subsequently, methods of the production with the controlled shapes were developed. In this order the ancient Roman historian Pliny suggested that Phoenician merchants had made the first glass in the region of Syria around 5000 B.C. However, archaeological evidence suggests that the first true glass was made in Mesopotamia in western Asia. The history of the glass is also revealed that around 4000 B.C. ago it has been used to cover the colour glazes of the copper containing compounds. In between 14th and 16th centuries B.C. the glass vessels recovered in Egypt were made and draw out threads, wind them around to sand or clay core which was itself held on an iron rod and king-melt the glass threads. They were used many chemical agents to colour the glasses. The patterns used to be made by assembling pieces of coloured glass and re-melting them. Syria and Palestine were emerged as the major glass making centuries after 1000 B.C. using the same techniques. Syrians and Palestinians made core-formed vessels, but in different shapes and sizes. Predominately in they were developed the new glass blowing technique. Macedonia and Greece had also recognized centers of glass making around 400 B.C. Particularly, Greeks were developed the sandwiching technique that useful to trapped gold layers in between clear glass parts. Around same time the Mosaic forming technique was also developed to impart special colour effects. In the early nineteenth century scientific study of glass began from the Faraday and others. At present the glass science is a well-developed subject and many new materials can be produced in the amorphous form with special properties and considerable technological applications. A glassy material can be defined as "an amorphous solid with the complete lack of the long-range periodicity in atomic structure that exhibits a region of glass transformation behaviour. The American Society for testing materials has also defined the glass as "an inorganic product" of fusion, which should cool under a rigid condition without crystallizing".

Glasses can be defined as a non-crystalline material that obtained by the melt-quench process. The term non-crystalline solids and glass transition indicate that a glass cannot be classified either in the category of crystalline materials such as quartz, sapphire, etc. or in the category of liquid. Their atomic arrangements are also different from those of crystalline materials possessing lacks long-range regularity. In present days non-crystalline materials cannot distinguished only from the melt-quenched glass of the same composition, but also obtainable by using various techniques, like, melt quenching method, ion implantation, chemical vapour deposition or sol-gel process etc. Hence, most of glass scientists regard the term "glass" as covering all non-crystalline solids that show a glass transition regardless of their preparation method. Moreover, the glass is an amorphous solid, amorphous solid is a substance whose constituent particles do not regular, orderly arranged, it is also recognized as super cooled liquids or pseudo solids. Thus, the amorphous solids are non-crystalline solids in which the atoms and molecules are not organized in a periodic lattice pattern as crystalline solids. The atomic arrangements of amorphous solids and liquids are in close proximity to each other, but their physiochemical properties are entirely different. Such as a solid material has both a well-defined volume and shape, while liquid has a well-defined volume with undefined shape. The liquid shape depends on the shape of the container. By means a solid exhibits resistance to shear stress while a liquid does not. At the atomic level such macroscopic distinctions arise due to the nature of the atomic motion. Since atoms of a solid are not stationary but they oscillate rapidly about fixed points that can be viewed as a time-averaged center of gravity of the rapidly jiggling atom. The spatial arrangement of these fixed points constitutes the solid's durable atomic scale structure, whereas a liquid possesses no enduring arrangement of atoms with mobile and continues wandering characteristics throughout the material.

Traditionally glass transformation behavior has been discussed on the basis of either enthalpy or volume versus temperature. Usually enthalpy and volume behave in a similar fashion of the choice of the arbitrary parameter. However, another case can also envision a small volume of a liquid at a temperature well above the melting temperature of that substance. When cool the liquid the atomic structure of the melt will gradually change and will be a characteristic of the exact temperature at which the melt is held. Whereas the cooling at any temperature below the melting temperature of the crystal would normally result in the conversion of the material to the crystalline state with the formation of a long range periodic atomic arrangement. In this situation the enthalpy will decrease abruptly to the value appropriate for the crystal. If to be continue cooling of the crystal could result in a further decrease in enthalpy due to the heat capacity of the crystal. When liquid will be cooled below the melting temperature of the crystal without crystallization a supercooled liquid may obtained. Usually, structure of the liquid continually rearrange as the temperature decreases without any abrupt decrease in enthalpy. Moreover, as the liquid is cooled further the viscosity increases. This increase in viscosity eventually becomes so excessive and restrict to the atoms can no longer completely rearrange to equilibrium liquid structure in the allowed time during the experiment. This process could allow to structure begin to lag, if enough time allowed to reach equilibrium. As the consequence, the enthalpy begins to deviate from the equilibrium and make a curve with the gradual decreasing slope. Eventually obtained the heat capacity of the frozen liquid. Therefore, viscosity becomes so great and structure of the liquid become fixed with no longer temperature dependent. The temperature region lying between the limits where the enthalpy is that of the equilibrium liquids, the existed frozen solid is known as glass transformation or transition region. This process obtained frozen liquid is called a glass. Furthermore, since the temperature where the enthalpy departs from the equilibrium is controlled by the viscosity of the liquid. This could be directly connected to it control by the kinetic factors. A slower cooling rate may allow the enthalpy to follow the equilibrium curve path to a lower temperature. Due to this, the glass transformation region could shift to the lower temperature until the formation of completely frozen liquid or glass at a lower temperature. Such process obtained glass could have lower enthalpy than that obtained using a faster cooling rate. Therefore, the glass transformation occurs over a range of temperature and cannot be characterized by any single temperature. This kind range of the temperature is termed either, the glass transformation (T_g). The glass transition temperature is rather vaguely defined by changes in either thermal analysis curves or thermal expansion curves. Traditionally glass transition temperature has been defined as the temperature at which viscosity becomes $\sim 10^{12}$ Poise. Commonly the glass transition temperature is defined in term of heat capacity (C_p^{onset}), that is correspond to the temperature at which molecular liquids have viscosity $\sim 10^{10}$ Poise. Another widely used definition is the "$C_p^{midpoint}$" determined during heating where the viscosity is 10^9 Poise. All these kinetic parameters depend on the way in which the system is prepared.

Thus, the considering significant importance of glass transition kinetic parameters to define their different physiochemical properties for prospective applications this work predominately devoted to glass-formation and structural modifications of the chalcogenide glasses. A write-up on the topic glass-formation and infringements in chalcogenide systems has been provided in the beginning followed a separated section on the criteria of glass-formation is given in detail by discussing kind of existing approximations or theories on this topic. Moreover, the periodic law and glass-formation of the chalcogenide systems are also discussed. The structural characterizations of the different kinds of materials are also valuable to know the various technical parameters, therefore, a complete segment devoted on the structural characterizations of glassy solids including glass formation and phase diagrams, qualitative and quantitative criterions, energetic and kinetic aspects, liquidus temperature effect and stable and metastable phase equilibriums of the chalcogenide. In chalcogenide glassy materials knowledge of atomic ordering is also a significant thing to explore the structure of various systems, the short and medium range ordering in these materials are also addressed. A view is also emphasized on chalcogenide glassy materials rings and isolated molecule properties. To explore the structural property of any kind of the solid selection of the methods of the characterization are also an important thing. Specifically, methods or approximations selection for the structural characterizations of disordered materials are the challenging issue, therefore, we should devote a completed segment on this topic in this study. In this work we have accommodated two most acceptable methods or approximations; eutectoidal model for the stable electronic configurations, glassy state with the experimental verification and physicochemical analysis of vitreous semiconductor chalcogenide systems.

GLASS FORMATION AND INFRINGEMENTS IN CHALCOGENIDE SYSTEMS

Well-established concepts, theories, criteria, semi-empirical rules and models of glass formation can be divided into three important groups: (1) structural–chemical, (2) kinetic and (3) thermodynamic. In this order Uhlman (1977) [1] pointed out the differences between these groups. Very often these concepts can overlap from one group to another. As an example, in 1967 [2] Rawson did not differentiate the thermodynamic group as a separate from the others: despite the contrast in the chemical bond energy and the energy of the system at the crystallization (melting) temperature. Rawson had introduced the thermodynamic (energetic) aspect in his structural–chemical criterion of glass formation. Along with the statement, "an acceptable theory of glass formation cannot be created exclusively on the basis of one of the aspects". Furthermore, Tammann (1935) [3] was the first scientists who tried to characterize the glass formation process. By combining thermodynamic and kinetic descriptions of the process together with the structural relationship, Tammann was able to investigate the glass structure and chemical bonding between constituent atoms. In addition to this, an important harmonic combination concept was introduced for all three group theories to present in the form of a three-in-one concept [1]. Here the glass formation from the well-accepted physical–chemical view for the field creation in chalcogenide glass-forming materials are described.

Goryunova and Kolomiets in 1958, 1960 [4, 5] presented a pioneer research on the glassy semiconductors to demonstrate the regularity of Glass Forming Ability (GFA). They found the size of the glass formation region in two and three components of chalcogenide alloys can be decreased by replacing one of the components of 4th (Ge, Sn), 5th (As, Sb, Bi) or 6th (S, Se, Te) main subgroups of the periodic table, the elements having a greater atomic number. They also demonstrated that such a decrease in GAF due to increase in the metallization degree of the covalent bonds with the increasing elements atomic number. In the 1966, Hilton et al. [6] also came to a similar conclusion to explain the GFA of the glassy semiconductors. The only difference was that they compared regions of the glass formations for various ternary systems. Hilton et al. also lined up VI, V and IV groups elements in a decreasing order of the glass formation tendency: S . Se . Te, As . P . Sb, Si . Ge . Sn. In 1972, Borisova [7] studied several compositions including B, Ga, In with the different alloying percentages and concluded that GFA in group III of the periodic table also decreases with the increase in atomic numbers of the elements. Borisova's study excluded the element thallium from group III. This result had established that the element containing compositions of arsenic selenides and sulfides can significantly contribute in wider glass formation region.

However, an anomalous behavior was also noticed in phosphorus and thallium, etc. glassy semiconductors. This led to the conclusion that investigators should pay attention on the decreasing order GFA for the higher atomic numbers among components of main subgroups of the periodic table. This is one of the main regularities of glass formation in chalcogenide glasses or amorphous semiconductors. Thus, to qualitatively predict the relative GFA of glasses for a system under test of an unknown glass formation region. The GFA should be expressed as the size of the glass formation region with other similar systems in which one of the elements of the investigated system is sequentially replaced by the same subgroup elements from larger or lesser atomic numbers. With this in mind, the glass formation regions can be evaluated for various systems.

Unfortunately, the concepts described above are not a universal rule because several violations have been noted even in simplest binary semiconducting glassy systems, therefore, it turns out to be more complicated in practice. The violations in the projected regular decrease of the GFA with higher atomic numbers even in some binary compositions led to additional research to determine the more common root for the variances. Considering these difficulties in glassy semiconducting materials, Minaev carried out extensive research between 1977 to 1991 [8–10].

Thus, research investigators managed to reveal the inversion nature of glass formation in glassy chalcogenide systems for several individual elements from the even groups of elements of the periodic table in a connected manner with the secondary periodicity of elemental properties. These redefined regularities consistently violate earlier discovered regularities for those connected with the increased atomic number. The newly discovered regularities are limited by showing a qualitative agreement of the glass formation of the chalcogen group elements. The quantitative determination problem of GFA was resolved by the Minaev in 1980 [11]. To solve this problem, it is useful to apply the experience of investigators who developed various theories and concepts of glass formation and analyzed various factors of the glass formation.

CRITERIA OF GLASS FORMATION

The structural and chemical concepts analysis of glass formation of the disordered network was chronologically investigated by various researchers [12–15]. While, the kinetic theory of glass formation was developed by Stavely, Turnbull, Cohen in 1952, 1961 and extensively reviewed by the Rawson in 1967 [2], who suggested that there is no possibility of a concrete quantitative prediction method for the GFA of substances.

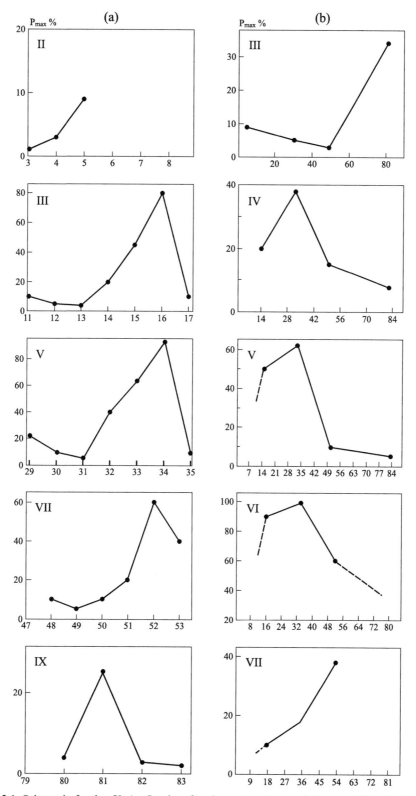

Figure 2.1. Schematic for the X–As–Se glass-forming system Pmax of elements (X) with the atomic number Z in the periodic table [41]: **(a)**: (II, III, V, VII, IX); **(b)**: (III, IV, V, VI, VII) groups.

In order to define the criteria of GFA, Goldschmidt proposed an empirical criterion for the ability of glass formation ability in 1926 [16]. According to this concept, the ratio of the radii of cations and anions in glass-forming systems lies between 0.2–0.4 region, typically anion locations in vertexes of tetrahedrons. The disordered locations of atoms concept indicate that such configurations must remain unchanged after cooling of the melt and formation of the glass, but it is incompatible with exact data of lengths and angles. In order to explain this, Smekal introduced an idea of the presence of 'composed' chemical bonds for the glass formation [17]. Stanworth demonstrated a correlation between the tendency of glass formation and the degree of ionicity or covalencity of the bond [18]. Pauling used the electronegativities quantitative expression of Stanworth [17]. It was concluded in the study that with this expression, the differences in electronegativities of elements and the degree of ionicity (covalencity) of adjustment bonds can be evaluated.

Furthermore, Myuller and his coworkers (Myuller, Baydakov and Borisova) presented detailed investigations of chemical bonds in glasses and glass-forming liquids [19, 20]. They demonstrated that the type of the main structural unit and nature of the chemical bond could have enormous significance in the formation of the glassy state. Myuller connected these substances with the glass formation directional bonds and the reduced radius of action, whereas, the first turn were predominantly powerful covalent bonds. These dominating covalent bonds can play an important role in alternation of valence of the elements that establish trigonal and tetrahedral configurations of chemical bonding. Thus, covalent bonds in the atomic network at moderate temperatures can be attributed to the reduction of the vibrational amplitude of atoms compared to the vibrational amplitude of ions in the ionic lattice. In Myuller's view, the cause of high viscosity and the increased activation energy of the atomic regrouping of substances can be prepared for glass formation.

Moreover, Kolomiets, Goryunova (1955), Myuller (1965), Kokorina (1971) and Borisova (1972) scientists from Leningrad connected the glass formation phenomenon with the theories for the chalcogenide systems for the main subgroups of III–V groups elements of the periodic table. It was found that those having a predominance of directional localized bonds from shared electron pairs covalent bonds, whereas, the portion of ionicity determined from electro negativities of elements is in the range 3–10%. These investigations have shown significant portion of ionic chemical bonds containing systems those can form well glasses. As an example Chuntonov, Kuznetsov, Fedorov, Yatsenko and Fedorov, Chuntonov, Kuznetsov, Bolshakova and Yatsenko presented the Cs–Te, Cs–Se system's equilibrium phase diagrams [21, 22]. The ternary system $Cs_2S–Sb_2S_3$ glass formation region was presented by the Salov et al. [23].

The outcome of various studies indicated that glass formation can be characterized not only by 'pure' covalent (S–S, Se–Se) or predominant covalent (As–S, P–Se) bonds, as well as through the covalent–ion bonds with the ionicity degree equal to $\leq 55\%$ for Cs–S, and equal to $\leq 40\%$ for Cs–Te. The bond's dependence through the ionicity degree on the difference of electro negativities of elements formed the chemical bond was established by Pauling [24]. To compare the oxide glass-forming systems, ion portions of chemical bonds can be investigated between ≤ 45 to $\leq 51\%$ [25]. Subsequently, halide glassy systems can have greater ionicity upto 80%, such as BeF_2.

After extensive examinations in various studies it was concluded that the key role of the covalent bond in glass formation should be revised. The glass-forming chalcogenide systems contain covalent-ionic chemical bonds which follow the common rule under the predominant covalent component role. However, there are also some exceptions like Cs–S glassy system. Only those glass-forming chalcogens (sulfur and selenium) can have chemical bonds 100% covalent, while, S–Se chalcogen glassy system can have some ionic components (note that the electronegativity of sulfur is 2.5, selenium 2.4). This is also an important point for the chalcogens chains in chalcogen and chalcogenide glasses through interconnection by van der Vaals bonds. Considering all this evidence on various glass forming systems, a generalized point of view was presented by Smekal; it was pointed out that the necessity of the presence of composite bonds for glass formation is also applicable to chalcogenide glasses as well as other kinds of glasses.

Polymerization of structural fragments is the most important feature in glass formation. The polymeric structure of the concept of glass was introduced in the second half of the 19th century by Mendeleev [26]. He stated that 'the glass structure is polymeric'. This concept provided novel practical and theoretical confirmation works by Sosman, Zachariasen, Kobeko, Tarasov and Myuller [15, 27, 28–31]. Poray-Koshits studied the most important part of the polymeric–crystalline concept of the glass structure [32]. Kokorina defined the necessary and sufficient condition for glass formation [33], which is as follows:

(i) The localized paired electrons bonds are present in the structure.

(ii) An endless polymeric complex exists in the main polymeric network construction.

(iii) The structural complex connection through only a single bridge bond. Occurrence of such bonds in the structure is called swivel bonds.

In this order Winter connected the GFA with the number of p-electrons in the external atom shell per one atom [34]. In his concept for glass formation, the most favorable number of p-electrons per atom to four. The minimum number of p-electrons for glass formation is two.

Later Sun [35] demonstrated the criterion of the bond strength based on the idea that the stronger the bonds between atoms for easy glass formation. According to his concept, the intensity of the process of atomic regrouping during crystallization of material should be accompanied by rupture of individual bonds and formation of new inter-atomic bonds dependent on the strength of bonds. Therefore, the glass formation ability of the system can be connected to increase the strength of chemical bonds. Hence the strength of the chemical bonds 'metal-oxide' can be determined by dividing the oxide dissociation energy by the number of oxygen atoms surrounding the atom in the glassy crystal. This can also be connected through the coordination number.

In 1956 Rawson modified the glass forming criterion by taking account of the liquidus temperature effect [36]. He connected the glass formation process not only with the bond strength, but also with the thermal energy present in the system. This allows it to be measured as the energy of melting temperature (for an elementary substance or a compound) or the liquidus temperature (for a multi-component system) in Kelvin degrees. This glass formation criterion is the ratio of the bond strength corresponding to the melting temperature. This criterion also permits a sharper frontier between glass forming and non-glass-forming oxides. However, this model did not apply to the glass forming criterion for the multicomponent glasses. This model successfully explains the liquidus temperature effect and existence of conditional glass formers, in which systems can form glasses, but they are not the usual glass formers. A condition reduction of the liquidus temperature due to the presence of the second oxide, therefore leads to the reduction of the thermal energy providing glass formation due to insufficient rupture of the existing bonds. Therefore, formation of other bonds occurs in the process of the atom regrouping to lead to the crystallization. Further, Rawson also indicated that phase diagrams can be helpful for the understanding of glass formation processes in binary and ternary-component systems.

Although earlier to Rawson in 1947, Kumanin and Mukhin had also come to almost the same conclusions with the different systems crystallization tendency. They demonstrated that the glass-forming systems, in the region of crystallization of a certain chemical compound (usually for the compounds with congruent melting) have a progressive reduction in the crystallization tendency when their compositions are moved away from the compound composition (i.e., with the liquidus temperature reduction). The glass forming systems crystallization tendency reaches up to a minimum in regions of the cooperative crystallization of the specific compound as well as other examined chemical compositions compounds [37].

Thus, general physical-chemical accepted glass formation criteria were developed between 1903 to 1956 by Tammann, Kumanin and Mukhin, Rawson (1956) and others considered the low-temperature eutectic points in the phase diagrams. Alternation in liquidus temperature also affects eutectic points, this concept was proposed by Rawson. Later Dembovsky connected the glass formation process with the phase diagram by providing the examples of various chalcogenide systems [38].

The evaluation criterion of disposition to glass formation with the reduced thermodynamic crystallization temperature was suggested by Turnbull and Cohen with the following relationship [39]:

$$\theta_c = \frac{kT_c}{h} \qquad (2.1)$$

Here k is the Boltzmann's constant, T_c is the equilibrium crystallization temperature, h is the evaporation thermal energy per molecule (it reflects the bond strength in a substance) and θ_c is the disposition unit for glass formation, its smaller value leads a greater disposition.

The evaluation of the glass forming ability with the various thermal analysis methods had been proposed by Hruby, an empirical relationship was developed as follows [40]:

$$\text{GFA} = \frac{T_c - T_g}{T_m - T_c} \qquad (2.2)$$

here T_g, T_c, T_m are the glass crystallization, glass-transition and glass melting temperatures.

Since the Turnbull–Cohen and Hruby criteria evaluations are based on the experimental data availability of the synthesized glass forming systems; therefore, making these methods limited. Considering this difficulty, Funtikov introduced the electronic configuration model: according to this the disposition to glass formation and the properties of chalcogenide glasses depend on features of electronic configurations of initial atoms [41]. He analyzed the maximum content X of various elements in glass-forming alloys with the As–Se–X systems and concluded that the GFA depends periodically on their atomic number. By using this approach, they examined the periodic table III, IV and V rows elements characters, and showed the minimum and maximum dependency for the III and VI rows, this model obtained maxima is represented in Fig. 2.1. It demonstrates the creation of GFA is greatly influenced by stable electronic configurations d^0, d^{10}, f^0, and f^{14}.

Accordingly Dembovsky and Ilizarov introduced the number of Valence Electrons (*VE*) of an element formula for the GFA [42, 43]. They developed an empirical theory of glass formation for chalcogenide glasses.

$$\text{GFA} = \gamma \frac{(A+E)(VE-K)}{2} \qquad (2.3)$$

here $\gamma = \Sigma_i T_i X_i / T_{\text{liq}}$, A is the number of atoms of different types, E the number of structural nodes, K the coordination number, T_i the melting point of the i component, X_i the mole fraction of the i component and T_{liq} is the liquidus temperature of the alloy. Later Ovshinsky also defined the important parameter for stability of non-crystalline materials by demonstrating the total constraint in them, this important parameter is the covalent connectivity of their atomic network. The glasses connectivity can be determined from the number of neighboring atoms by the 'average' atom of the covalent bonds (or the average Covalent Coordination Number (CCN)) under 8-*N* rule.

In the early 1980s, Phillips and Thorpe established a correlation between the degree of network reticulation and physical properties. Considering average coordination $\langle r \rangle$ or mean coordination as the single structural parameter, they described the following relationship [44, 45]:

$$\langle r \rangle = \Sigma r_i a_i \qquad (2.4)$$

here a_i is the molar fraction and r_i is the covalent coordination of atom i. The average coordination number or mean coordination provides a direct estimate of the number of topological constraints (or rigidity of percolation) in the glassy network. Considering the rigid bonds between two atoms each generate one bonding constraint and their fixed bond angles between three atoms; the generated single angular constraint has been described in Fig. 2.2(a, b c). This most extensively studied AsSe$_3$ trigonal pyramid, the average number of constraints per atom n can be defined by the following relationship:

$$n = \frac{\langle r \rangle}{2} + (2 \langle r \rangle <) \tag{2.5}$$

Since each atom of the three-dimensional glass has three degrees of freedom ($d = 3$) and their rigidity of the network can be defined by comparing n and d. The floppy or underconstrained, overconstrained or stressed rigid and isostatic can be correlated with the $n < d$, $n > d$, $n = d$.

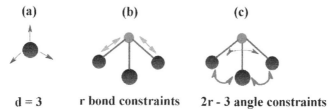

(a)   **(b)**   **(c)**

$d = 3$ **r bond constraints**   **2r - 3 angle constraints**

Figure 2.2 **(a)** In a three-dimensional glassy network an atom degree of freedom, **(b)** Typical bond stretching constraints of AsSe$_3$ trigonal pyramid, **(c)** Typical bond bending constraints of the AsSe$_3$ pyramid under three independent angular constraints.

In an ideal condition three-dimensional ($n = 3$) glassy networks have an average coordinate $\langle r \rangle = 2.4$. At this threshold, topological glasses have particular physical properties such as a high glass-forming ability with the onset of rigidity that leads to an increase in elastic modulus. This also demonstrates that glassy networks having low coordination can undergo deformations at no cost of energy, therefore, the bond angles and bond lengths remain unchanged. The number of such deformations per atom called zero frequency modes is equal to the difference between d and n. The fraction of zero frequency mode f for likely deformation can be related to underconstrained glasses when $f > 0$, as shown in Fig. 2.3(a).

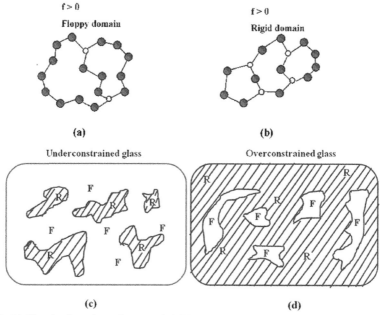

(c)   **(d)**

Figure 2.3 **(a, b)** Sketch of topology floppy and rigid domains containing zero frequency modes deformation and degree of freedom, **(c, d)** Comparative schematics of underconstrained glass and overconstrained glass.

However, with increasing bonds and constraints numbers, the f decreases and rigid domains are form the structure as shown in Fig. 2.3(b) [45, 46]. Further, with increasing $\langle r \rangle$, the three-dimensional ring structure of the network decreases; therefore, rigid elements can be eventually interconnected. Furthermore, at $\langle r \rangle = 2.4$ the rigidity is said to percolate through the structure, as per shown in Fig. 2.3(c, d). During this stage, the underconstrained network may be composed of isolated rigid domains surrounded by a continuous floppy matrix, while the overconstrained network in a continuous rigid domain containing floppy inclusions. Such a transition between two configurations is often associated with a change in their physical properties [46].

Angell's Fragility

Key parameters of glass formation and their large relevance to the relaxation processes, such as fragility have a strong correlation with the mean coordination $\langle r \rangle$ [47–50]. This usually occurs when overconstrained and underconstrained systems tend to be fragile glass formers; however, ideally constrained systems tend to be strong glass formers. To describe the strong/fragile glass formers Angell's has introduced their classification [51–53].

Kinetic Fragility

Kinetic fragility of glass-forming liquids can be related to their Arrhenius behavior of the viscose temperature. This is well recognized by comparing the liquids with different glass transition temperatures (T_g). The fragile systems usually showed higher departure with the Arrhenius behavior due to a rapid collapse of the amorphous network under the small increment in temperature above the T_g. Strong systems tend to retain their network character even in the liquid state and remain viscous over a wider temperature range with the Arrhenius behavior. Generally, molecular liquids structural integrity reflects the weak van der Waals like interactions and tends to collapse easily with temperature; therefore, they exhibit typical fragile behavior. However common covalent networks can remain connected at higher temperatures and exhibit a strong behavior.

Thermodynamic Fragility

The fragile systems rapid structural degradation can be characterized from the thermodynamic signature by measuring entropy of the systems [53, 54]. Fragile systems can gain in structural degrees of freedom at a higher rate and show steeper changes in entropy with temperature [54]. Earlier innovations also demonstrated greater structural mobility at higher temperatures. This can result in faster diffusion (lower viscosity) to generate new degrees of freedom that contribute in heat capacity and raise the entropy of the system [54]. Therefore, the equivalent crystalline phase of the same stoichiometry has been well established. However, in several cases chalcogenide systems form glass over large continuous compositional domains without corresponding crystalline phases. Such useful compositional comparisons within the limitation of several glassy systems entropy–temperature profiles can be postulated if equivalent crystalline phases occur. These kinds of systems can have a very similar vibrational entropy. Therefore, entropy-temperature profiles of various systems could basically normalize. Moreover, with the help of system liquid/glass thermodynamic data the effect of mean coordination on fragility can also be explored.

Fragility Index

The viscosity–temperature profile can provide a significant estimate of the fragility. Therefore, it is useful to establish a parameter that provides a quantitative measure of fragility. The commonly used fragility index 'm fragility' or 'steepness index' can be defined by the following relationship [52, 55]:

$$m = \frac{1}{T_g} \frac{d \log(\eta)}{d\left(\frac{1}{T}\right)} T = T_g = \frac{E_a}{\ln(10)RT_g} \tag{2.6}$$

where η is the viscosity, R is the gas constant and E_a is the activation energy for the viscose flow, i.e. the alternative for enthalpy relaxation. Usually the fragility index m reflects the onset of viscosity–temperature and provides a reliable quantified fragility with the single parameter. Therefore, the kinetic of fragility can be defined from the viscosity data. In the case of viscous flow the activation energy (E_a) can be replaced from the E_η, since activation energy of the enthalpy relaxation (E_H) and activation energy to viscous flow E_η can have a strong correlation. Considering this occurs in most cases of E_H, it can be replaced by E_η without introducing a large error in determining m. The main advantage of examining this fragility index in a suitable way, that E_H can be easily determined by Differential Scanning Calorimetry (DSC) [55–57]. Thus, E_H is a useful parameter to investigate the correlation between fragility and mean coordination.

Other Fragility Determination

Glass-forming liquids fragility is widely controlled from the features of the heat capacity jump throughout the glass transition. Therefore, the shape of the glass transition curve can be analyzed to estimate the fragility. Usually fragile systems have a tendency to undergo large gains in degrees of freedom during the glass transition, this interprets into a large jump in heat capacity within a short temperature range. In contrast to this, strong systems can retain high viscosity over wider temperature ranges, consequently, they exhibit shallower heat capacity gains spread out in temperature. Usually width and height of the glass transition, ΔT_g and ΔC_p, are used to measure the fragility of the glassy systems [58, 59]. With the help of the DSC measurement, differences between the heat capacity (ΔC_p) of the solid and liquid can be achieved. The ΔC_p is usually normalized from the melting entropy (ΔS_m) to associate variations in the entropy of the corresponding crystalline phase [58]. However, ΔS_m is not a frequently available quantity for several chalcogenide glasses, but its value should be more or less constant within the chalcogenide glass system. In a similar way, the width of the glass transition ΔT_g can be obtained from the DSC data by measuring the temperature of onset and completion of the glass transition [57, 59]. To get better fragility results, the ΔT_g width should be normalized in terms of onset T_g, for the types of glass covering a wide range of transition temperatures [60]. Thus, overall three independent measures of fragility can be defined from simple DSC analysis of glass-forming materials. Generally, these methods are used to describe the effect of average coordination on the physical properties of chalcogenide amorphous networks systems [61, 62].

Periodic Law and Glass Formation in Chalcogenide Systems

Glass formation analysis in chalcogenide systems can be divided according to the principle of participation of elements from 1–7 groups of the periodic table. Therefore, some general principles concerning the problem of glass formation and its relation with the geography of individual elements of the periodic table will be discussed here.

According to Minaev's study [63], all groups and elements on the periodic table can be a part in the glass formation through ternary chalcogenide systems, except the eighth group (inert gases) elements and the first and seventh elements. This kind of frequent occurrence of elements gives the possibility of consideration of glass formation in chalcogenide systems under the periodic system. It also leads glass formation of the alloys in which their constitutes depend on the elements periodic property. This is an important approach for the prediction about new glass-forming systems and seeking glasses with previously unknown combinations of physical and chemical properties. A number of elements has been used to constitute binary chalcogen systems, those distributed in groups of the periodic table, according to Minaev [64, 65] is listed in Table. 2.1.

Table 2.1 Chalcogene systems distribution in groups of the periodic table for the binary systems (in brackets—the expected number of elements with the systems) [64, 65]

Groups	I	II	III	IV	V	VI	VII	Total
The number of elements	2(+6)	–(+1)	5	4	3	3	2(+3)	19(+10)
The number of systems	3(+17)	–(+3)	8	8	7	2(+1)	3(+12)	32(+33)

The known phase diagrams of glass formation regions and the properties of binary glasses plots have been demonstrated by various investigators to describe the direct genetic relationship between them. On the basis of this concept a number of new binary glass-forming chalcogenide systems genetic relationship have been made, therefore, hundreds of new ternary and scores of tetrad glass-forming systems have been created. In addition from Table 2.1, it is clear that the number of glass-forming systems is the least for II group. When one move towards groups I to VII, it was found that III to V groups are almost equal and VI to VII groups show a decreasing order. This interpretation led investigators to take into account the different systems and predict various systems with VII group elements; as a consequence a number of systems can be increase significantly. A typical periodic distribution of the number of elements and numbers of systems is listed in Table 2.2. It can be seen that the glass forming systems number increases from the second period to the maximum fourth, later decreasing with further movement for the fifth and sixth periodic number tables. This behavior may remain unchanged in addition to the other systems, like, H–S, H–Se and H–Te [66].

Table 2.2 Periodic number distributions and number of elements with the number of systems

Periodic number	1	2	3	4	5	6
The number of elements	–(+1)	1(+2)	5 (+1)	5 (+2)	5 (+2)	3(+2)
The number of systems	–(+3)	2 (+6)	8 (+6)	10 (+7)	7 (+5)	5(+)

The analysis of the real system GFA and the invented system GFA can be best described with the help of the SRM criterion, by considering the framework of the individual time of the periodic table. The glass formation in binary chalcogenide systems tend to decrease with the movement from the first group to second group elements systems, then it grows with the increasing atomic number of the non-chalcogen element for the third, fourth and fifth groups, respectively.

Minaev extensively worked on glass formation between the 1980's to 1991 [63, 64, 66, 67]. According to him, most existing chalcogenide systems whose main features had been observed earlier in individual systems, for example, existence of direct relation in-between the structures of phase diagrams and the glass formation ability of alloys. His description also indicated that, the minimum tendency of glass formation is apparent usually for the alloys corresponding to the chemical compound composition (excluding peritectical alloys). This means that the glass formation ability can be expressed by the rule "glass forming ability increases with the decrease in the liquidus temperature". This is typical for the chalcogen-enriched alloys and often maximum for the chalcogen-enriched eutectic alloys.

For distinct chalcogenide systems, such as $A^{IIIA}–B^{VI}$, $A^{IVA}–B^{VI}$ and $A^{VA}–B^{VI}$, the general tendency was described by Goryunova and Kolomiets. They stated that, the decrease in the glass formation ability of alloys with an increase in atomic numbers of elements in groups of the periodic table become apparent and it can correlate with the increase and decrease in metallization and chemical bond energy. Accordingly several investigators have described various binary chalcogenide systems, including systems with Al, Ga, In, Sn, Pb, Cu, Ag, Au, Hg, Bi [66, 68, 69]. Specifically, it was observed that in these systems the inversion phenomenon of the regular decrease in the glass formation ability with the increase in atomic numbers in chalcogen alloys, particularly, compared to tellurium alloys with selenium and sulfur (Te–Se inversion) when using the Sun–Rawson–Minaev criterion. It was concluded that the existence of inversion behavior in a specific glassy system is responsible for their non-glassy phase diagrams.

The regular inversion was also noticed in telluride systems when they formed alloys with elements of IA and IB subgroups of the periodic table. It demonstrated a regular decrease in the glass formation ability with increasing atomic numbers of elements [66, 70]. Thus, their glass formation ability grows with the increase in atomic numbers of elements of the first group. It was also predicted that the existence of inversion can be in several alloys of the alkaline metals with sulfur and selenium. In a similar way the binary systems, A^{VIIA}- tellurium (where A^{VIIA} can be Cl, Br, I) have also reported inversion of GFA decrease with increasing atomic numbers. Their glass formation regions can be bigger for the row of systems.

Accordingly Minaev reports [64, 65] that in the binary systems other types of the inversion exist. Three types of them are present with elements of the fourth period which offers larger glass formation regions compared to their analogs in groups of the third period. As an example, these types of the inversion can exist in systems such as, Ge–Si–S, Se–As–P, S–Te, Se–S–P.

Prior to inversions in the regular decrease of the glass formation ability, one of the forms of the secondary periodicity was manifested. Mendeleev (in 1864, 1947), demonstrated the periodicity in binary systems based on the properties of elements for the individual substances and compounds. Such complications of periodicity of glass formation can also be demonstrated in binary chalcogenide glass-forming systems. In 1915 Biron [71] demonstrated the logic line of Mendeleev's consideration about properties' alterations of elements of the same group in the periodic table. This attracted much attention due to the absence in some cases of a monotonic character of alterations of one or another property, when moving in a group from one element to another with the increase of their atomic numbers. Such a non-monotonic character is known as secondary periodicity. Shchukarev and Vasilkova introduced the non-monotonic character of alteration of ionization potentials in their sums (in electron-volts) for the IIIA and IVA groups elements with the increase in atomic numbers (eV): (1) B–69.97; Al–53.74; Ga–57.02; In–52.37; Tl–56.27; (2) C–147.17; Si–102.62; Ge–103.24; Sn–93.27; Pb–96.71 [72].

Several investigators determined that in both rows a common tendency of decreasing the sum of ionization potentials can be achieved. However, the sum of gallium, thallium, germanium and lead can be larger than those preceding aluminum, indium, silicon and tin, respectively. A similar situation was later also described on introduced glass formation ability of approaches, specifically for the systems with thallium, indium, lead and tin under the inversion manifestation.

In 1954 Schukarev [73] studied the secondary periodicity responsible for properties of electronic shells of atoms; under the prominent role of s-electrons and less important role of the p-electrons. Therefore, the plunging of elliptical s-orbits of the 10 d-electrons shells may be significant. They have also related the appearance of secondary periodicity with d- and f-strengthening considering the diving electron bond and their electron shells compression. The most common example is increasing ionization potential of thallium and lead.

These explanations are more acceptable in the case of glass formation because it can relate to the strengthening of chemical bonds. It also leads to an increase in the glass formation ability from the framework of the structural–energetic concept introduced by Minaev [66, 67]. In 1953 Shchukarev and Vasilkova reached the conclusion that periodicity can be related to the structure of the system of elements itself in which periods beginning from the second reiteration by pairs. The first pairs (2nd and 3rd periods) without containing d-electrons, while, the second (4th and 5th periods) and third (6th and 7th periods) pairs electrons contains both d- and f- electrons. This inference can directly correlate to the inversion of 4th–3rd period and 6th–5th period under alteration of properties. Thus, glass formation is specifically due to turning from one pair of time of the periodic table to another pair.

It was also noticed that in the systems A^{VA}–B^{VI}, there are four pairs of binary systems revealing the inversion in the main regularity of the glass formation [74, 75]. This has been also supported by the decrease in glass formation ability with increase in the atomic number [76, 77]. The pairs nclude P–Se and P–S, As–S and P–Se, As–Se and As–S, As–Te and P–Te. However, in the group of systems A^{IVA}–B^{VI} there are six pairs of inverted systems (Si–Se and Si–S, Ge–S and Si–S, Ge–Se and Si–S, Sn–Te and Sn–Se Pb–Te and Pb–Se, Pb–Te and Sn–Te). Additionally, in the group seven pairs of inverted systems are B–Se and B–S, Al–Te and Al–Se, Ga–Te and Ga–Se, In–Te and In–Se, Tl–S and In–S, Tl–Se and In–Se, Tl–Se and Tl–S.

At the same time several investigators presented the experimental data of binary and ternary chalcogenide systems with alkaline metals, based on the SRM criterion [70, 78] and predicted that A^{IA}–B^{VI} group of systems collectively represent the tendency of inversion with the regular decreasing and increasing order GFA and atomic numbers. Furthermore, in the row of binary chalcogenide systems elements of VA, IVA, IIIA, and IA subgroups have a decreasing order in the degree of appearance of the tendency to the regular decrease in GFA with increase in the atomic number of elements.

Hence both the 4–3 and 6–5 class inversions revealed that binary systems of those can also be manifested in ternary/ or multi-component chalcogenide systems according to their genetic relations between phase equilibrium diagram structures, the glass formation and glass properties. The most common examples of 4–3 inversion of the ternary systems are Si–P–Te and Si–As–Te, Ge–P–Te, and Ge–As–Te [79, 80]. Earlier glass formation regions with arsenic were considered were larger than the corresponding regions with phosphorus. Similarly, 6–5 inversion were also manifested for the several systems, such as, Ge–Bi–Se and Ge–Sb–Se ('Bi–Sb') [81]. While Goryunova and Kolomiets had earlier shown earlier, that in Sn(Pb)–As–S and Sn(Pb)–As–Se compositions glass forming system lead can enter in a larger amount than tin. Later several innovations were carried out on inversion in different forms such as Ge–Sn–S and Ge–Pb–S, As–Sn(Pb)–Te, Si–Sn(Pb)–Te, Ge–Sn(Pb)–Te, Pb–Sn and Sn(Pb)–Ge–As–Se [82–85].

Therefore there is no doubt that the different types of inversions mentioned above, act in the same way including binary and multi-component systems. However, it was concluded that the inversion property exists in several ternary and multicomponent chalcogenide glassy systems such as lithium, sodium, potassium, rubidium, cesium [86, 87]. Similarly, the glass formation region of the system Cs_2S_3–Sb_2S_3 can be significantly larger than those in the system Rb_2S–Sb_2S_3.

It is interesting that later the inversion manifests itself also in some binary, ternary and other oxide systems [88]. Thus, glass formation in several ternary sulfide systems with rare earth elements and Ga falls in the exhibited inversion Several researchers systematically studied the chalcogenide systems, such as Ln_2S_3–Ga_2S_3, where Ln–La, Ce, Pr, Nd, Sm, Eu, Gd, Tb, Dy, Ho, Er and the system Y_2S_3–Ga_2S_3. It is found that these glass-forming systems become smaller with the increase in atomic numbers of lanthanides [89]. Nevertheless, there are a few exceptions, like europium, that does not form glass with this particular system.

STRUCTURAL CHARACTERISTICS OF GLASSY SOLIDS

The glass formation phenomenon in different systems is not an exception to the rule. Most of the theories and practices do not suggest any specific rule that could successfully explain glass formation regions in unexplored binary, ternary and multicomponent systems. Though several investigators such as Goldshmidt, Zachariasen, Lebedev, Smekal, Steanworth, Winter, Phillips and others have presented concepts and theories to explain glass formation and formulate it in generalized form. However, as yet none of the suitable concepts and theories provide satisfying guidelines for the prediction of glass formation regions in concrete chalcogenide systems or other systems.

Considering this Dembovsky and Ilizarov presented an empiric glass formation theory of GFA [48]. Based on the qualitative interpretation of structural nodes in multi-component systems, the Sun–Rawson criterion was introduced to the GFA criterion [2, 36]. These criteria revealed the cause of formation is due to two key factors 'the structural–chemical factor and energetic factor' with an additional important kinetical' factor [36]. The additional third factor only contributes when two key factors originate within the substance. This is an important parameter for suitable the glass formation phenomenon initiation. Therefore, the glass formation kinetic factor can be correlated to the practical glass formation conditions. The concept of the kinetical factor usually allows in describing used melt quenched glasses, and glass formation regions with a different size depending on kinetics of the cooling rate of melt. Considering this, Goryunov et al. demonstrated relaxation of GFA with the increase in atomic numbers of elements based on the structural–chemical and energetic factors of glass formation [74, 75].

Glass Formation and Phase Diagrams of Chalcogenide Glasses

According to Fig. 2.4 the binary and ternary chalcogenide systems are regarded as mixtures of melts of the nearest congruent compounds. Using this approach structure and properties of the chalcogenide system can be predicted [90, 91].

Several phase diagrams of binary, ternary and multicomponent chalcogenide systems data on glass formation in these systems were demonstrated according to their likelihood to obtain glasses [63, 64, 66, 67, 68, 70, 76, 92]. The simplest binary chalcogenide systems eutectics phase diagram was extensively studied at low temperature. As an example, binary systems Al–Te, Ge–Se, Si–Te, As–S, P–Se, Cs–Te, as well as other systems glass forming region with the adjoining chalcogen are represented in Fig. 2.4(a). The segregated regions of adjoining chalcogens low temperature eutectic phase region for the Cs–S, K–Se, Tl–S, Tl–Se, Sb–S, and others are given in Fig. 2.4(b). In order to explore the binary glasses, the glass formation of eutectic phase diagram of the two chalcogen systems (like S–Se and S–Te etc.) have also been extensively studied [91–93]. The non-glass forming binary chalcogenide systems phase diagrams are given in Fig. 2.4(c, d). The first diagram shows a sharp rise in the liquidus temperature close to the adjoining point of the chalcogen, while the second diagram exhibits the phase segregation in such systems.

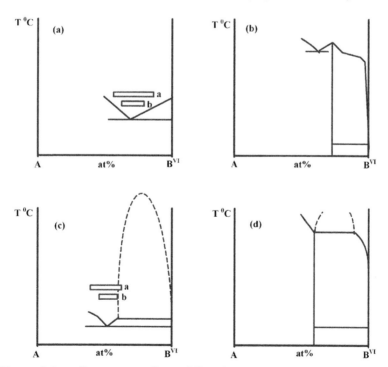

Figure 2.4 Types of phase diagrams according to Minaev's proposed model for the binary chalcogenide: **(a)** chalcogens-enriched eutectic glass forming system; **(b)** the eutectic phase liquation in the chalcogens-enriched region formation; **(c)** the non-glass-forming type under the sharp liquidus rise in chalcogens enriched region; **(d)** phase liquation in such systems.

Innovations on the phase diagrams of the glass formation of chalcogenide systems are not limited to the systems discussed above, a large number of data are available in the form of binary, ternary and multicomponent compositions. Here the simplest form of chalcogen glass forming, and non-forming phase diagrams that were demonstrated by the various investigators are described.

Glass Formation Qualitative Criterion

Several investigators have extensively studied the temperature effect on glass formation by introducing corresponding chemical compounds with the help of binary, ternary and multi-component systems phase diagrams. The phase diagrams of such glassy systems can be obtained by considering where the process of glass formation is more or less probable, for binary, ternary and multicomponent systems. A typical ternary glassy system phase diagram can be constructed to define the probable glass-forming region.

A schematic diagram is shown in Fig. 2.5(a), here A, B and C at the corner of the triangle represent the pure elements, which examine the composition and the corresponding age to define the atomic percentage of the alloying concentration of a specific element. Further, it can be deduced that a particular ternary system has experimental glass formation data point as shown in Fig. 2.5(a). Then the composition can be described by considering the cut points of the parallel extended from the specific data point. The total sum of the system of each individual element concentration should be 100. In a similar way if we have several experimental data points for a chalcogenide system, then a glass forming region can be drawn for that system by interpreting the data points. Since such systems usually show the eutectic phase with temperature, therefore, they can have eutectic points in different separated regions. The eutectic points are shown in Fig. 2.5(b), though in this simple presentation temperature is not mentioned. Thus with the help of the experientially obtained eutectic data points, a chalcogenide system can define the glass formation region.

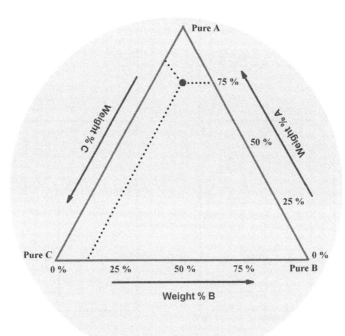

Figure 2.5(a) Schematic of the ternary system phase diagram.

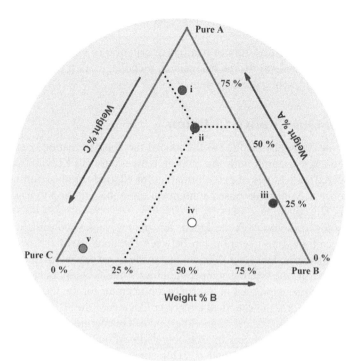

Figure 2.5(b)   Schematic of the ternary system under the phase diagrams – II.

Considering the eutectic points discussed above, the concept of the glass formation region under the liquidus temperature effect, the most probable locations for the various binary, ternary, multi-component systems have been described. Minaev extensively discussed the monovariant curves connecting binary and ternary eutectics [67]. Although several common factors were earlier introduced by Kumanin and Mukhin, this study was further enlarged by Mukhin and Gutkina [37, 94]. They demonstrated that "the crystallization ability reaches a minimum in regions of the cooperative crystallization of the given compound with compounds of different chemical composition". To continue this approximation of the glass formation region, Minaev et al. [8, 95] proposed the concept of the line of dilution of binary, ternary and multi-component eutectics for the third, fourth, etc. components. In a ternary system, the binary eutectic dilution line is the line of data point connection of the vertex of triangle corresponding to 100% of the third component content [as described above in Fig. 2.5(a, b)]. As an example, there are some exact analogies of the line of the binary to ternary eutectic line of dilution that have been extensively studied telluride systems, like system Ge–As–Te, Te–As_2Te_3–GeTe, etc. [96]. In such a system, calculations can be carried out according to the rule of the dilution line of binary eutectic and their glass formation regions indicate the propriety of replacement of monovariant curves by dilution lines of binary eutectics. Therefore, systems glass formation regions can be formulated on the basis of the qualitative criterion.

Application of the qualitative criterion of glass formation can also be verified for the prediction of several hundreds of ternary chalcogenide systems and in the several achieved ternary telluride systems based on elements of I^A, I^B, II^B, III^A, IV^A, V^A, VII^A subgroups of the periodic table. The experimental evidence has successfully demonstrated the synthesis of materials at 1000°C in rotary evacuated quartz ampoules during 12 hours. And their subsequent quenching in the cold water (cooling rate around 10–20°C s^{-1}) can provide such formation possibility in a few systems; like Cu–Si–Te, Cu–Ge–Te, Ga–Si–Te, Ga–Ge–Te, Ga– Pb–Te, Ga–As–Te, In–Si–Te, In–Ge–Te, In–As–Te, Tl–Si–Te, Tl–Ge–Te, Si–Ge–Te, Si–Sn–Te, Si–Pb–Te, Si–Sb–Te, Ge–Pb–Te, Ge–Sb–

Te, Sn–As–Te, Pb–As–Te, Ge–Pb–Te, Al–Si–Te, Al–Ge–Te, Al–Pb–Te, Al–As–Te, Se–Zn–In, Se–Te–Ge, etc. [67, 84].

Hence, the successful interpretation of qualitative criterion of glass formation has provided a path for their extension in general for the halogenide, metallic, oxides, etc. systems. This can also be a candidate for the multi-component systems.

Glass Formation Quantitative Criterion

In the years 1956 and 1967 Rawson [2, 36] introduced the glass formation criterion based on the ratio of the bonds energy strength with the melting temperature (in Kelvin degrees). Using this concept, he generalized the features of glass formation for all kinds of glass-forming compositions. To continue this concept, Minaev proposed a modification in this criterion to make it more general for multicomponent chalcogenide glasses [8, 67, 95].

Initially Rawson's criterion was designed for oxide glasses considering their chemical compositions. This can provide a relationship between the quotient of the oxide's bond energy and their melting temperature. However, Minaev had made a correction, instead of oxide's melting temperature, by considering the liquidus temperature for the multi-component systems. This allows the single bond energy to be expressed in a form of a numerator for the individual system in Sun–Rawson's criterion. Therefore, the chemical energy or Covalence–Ion Binding (CIB) of substance per atom average can be taken. The corresponding sum of the energies of specific system chemical bonds can be expressed as E_i. Additionally, the bounded portion of the atoms and their half-value of valence (CN) can be defined as M_i and K_i. This approximation is defined for the case when each individual atom chemically binds with other atoms; one is considered for two atoms, energy value per atom and then it should be divided by the factor of two. Thus an empirical relation can be expressed as:

$$E_{CIB} = \frac{\Sigma_i E_i M_i \left(\frac{K_i}{2} \right)}{\Sigma_i M_i} \qquad (2.7)$$

The presence of chemical bonds and their quantitative ratio can be determined in terms of an atomic connection substance structure. Therefore, in general the glass structure can be characterized from the chemically ordered continuous random network of atoms, according to their chemical bonds with the valence CNs in chemical order (K) dictation network. Hence according to the modified criterion of glass formation, the ratio of covalence–ion binding of atoms in multi-component alloy value can be correlated to their melting temperature or liquidus temperature T_{liq} in Kelvin, it can be taken as: $\frac{E_{CIB}}{T_{liq}}$. The glass formation ability can be achieved from the following relationship:

$$GFA = \frac{E_{CIB}}{T_{liq}} \qquad (2.8)$$

The chemical bond energy (E_{CIB}) can be obtained according to the Pauling formula of interatomic bond energies, the updated data can also obtain from the collection of the Batsanov's monograph [97]. The empirical Pauling formula is as follows:

$$E_{A-B} = \frac{1}{2}(E_{A-A} + E_{B-B}) + 100(X_A - X_B)^2 + 6.5(X_A - X_B)^4 \qquad (2.9)$$

here E_{A-B}, E_{A-A}, and E_{B-B} are the bonds energies between the atoms A and B, A and A, B and B, respectively; X_A and X_B are the electronegativities of A and B atoms according to Pauling's explanation. Using this expression, various investigators calculated the bond energy (kJ mol^{-1}) by using different approaches, some of the key calculations are listed in Table 2.3.

Table 2.3 Different investigators reported values E, kJ mol^{-1} for the same bonds

System	Myuller	Ioh and Kokorina	Pauling
As–S	255.4	202.6	224.4
As–Se	217.7	159.1	174.8
Ge–S	–	230.3	259.0
Ge–Se	234.5	180.0	205.7

Therefore, both Sun–Rawson criterion and its modification by Minaev are constructed by considering both approaches to resolve the glass formation problem. The first one is structural–chemical (the CN, the chemical bond), and the second is bonding energetic (the thermal energy of substance at crystallization).

To characterize the glass formation concretion in a certain system, the area of the glass formation region depends on the alloy's cooling conditions. This was neither given in the Sun–Rawson criterion nor in its modification—the Sun–Rawson–Minaev criterion (the SRM criterion). Therefore, the third kinetic approach reporting the alloys cooling rate on which the size of the concrete glass formation range can be described. In both the Sun–Rawson criterion and SRM criterion, the value for each composition remains constant. These criteria at cooling of such alloys at a rate higher than the critical one (V_{cr}), formed glass. While, at a cooling rate lower than the critical one, such systems can be characterized as crystallized melt under thes $\frac{E_{\mathrm{CIB}}}{T_{\mathrm{liq}}}$ value. This is the major difference in previously described criterions.

According to Minaev, the Sun–Rawson and SRM criteria can measure the glass formation ability of the physical–chemical essence of a substance which is independent of the conditions of the concrete glass formation or crystallization [66, 77, 95]. They had not considered the glass formation ability dependence on the cooling rate or on the intensity of other external factors (pressure, electromagnetic radiation, etc.). This is the inherent property of the substance and is determined from the physical–chemical nature. SRM criterion can also simultaneously reflect both structural–chemical and energetic approaches to the glass formation problem and demonstrate that in some cases insufficiency, like the effect of the liquidus temperature with the eutectic law [2, 98].

In such a system, usually the glass formation region is located near the eutectic and expands in the direction of the chemical composition connected with the eutectic by a gentle sloping liquidus curve. The GFA can increase (and the crystallization ability can decrease) when moving from the eutectic to this chemical composition. However, if the covalent–ion binding of alloys increases in this direction to a greater extent than the thermal energy of the system, then the rate of increase can be determined from the steepness of the liquidus curve.

The SRM glass forming ability criterion can also be used for creating concrete glass formation by taking into account an additional factor that reflects specific conditions of glass formation. Usually the factor used is the cooling rate, which reflects the kinetic approach to the glass formation problem. As an example, comparing glass formation in telluride systems, it appears to be convenient to use the cooling rate of $\approx 180°$ Cs^{-1}. According to several studies it is possible that glass formation in many systems at this temperate and substances quantity is sufficient for measurements and practical applications.

Glass Formation Energetic and Kinetic Aspects

Sun–Rawson and Sun–Rawson–Minaev established the glass forming criteria and provided the foundation for the structural–chemical and energetic aspects of glass formation [9, 65], to consider the chemically ordered structural network of the atoms of the substance and the energy of this structure (E_{str}) [2, 66, 67, 95]. This glass formation criterion is defined as $E_{\mathrm{CIB}} = E_{\mathrm{str}}$, in which T_m and T_{liq} can be considered as the criteria of thermal energy E_{therm} expressions. This expression is required for the rupture of chemical bonds and reconstructing the considered structure, to transform it from the liquid to solid-state structure [2].

Beside exploitations, both criteria have suffered from the lack of a kinetic component. Therefore, without this component they could not be used to evaluate dimensions of glass formation regions and even to predict glass formation itself in given conditions. However, both criteria can be used to determine the glass forming ability of smaller as well as larger compositions concentration.

The structural–energetic concept of glass formation was introduced by Minaev [67, 77, 95], who considered the fixed cooling rate of glass forming alloys equal or greater than a certain value. Later the critical cooling rate V_{cr} become the known characteristic of the glass. At that cooling rate of the given alloy, composition is still able to form glass. In addition, at a lower cooling rate a crystal can be formed. Therefore, for the considered system, each alloy composition can have its own V_{cr}. Moreover, the glass formation ability of all alloys of a specific system may exceed the V_{cr} to form glass.

Usually higher value of V_{cr} expands the glass formation region, while the lower value of V_{cr} narrows it. Nevertheless, V_{cr} is not the basic parameter to determine the glass forming ability. On the other hand, the glass forming ability of each individual alloy expresses its own genuine physical–chemical behavior. This is desired for definite physical–chemical conditions to form glasses at a certain V_{cr} under the precise external pressure (1 atm) which acts on the undercooling material. For example, the evaluated value of telluride alloys is 0.270 ± 0.010 kJ mol^{-1} K^{-1} at a cooling rate ≈ 180 K s^{-1}, 0.250 ± 0.01 kJ mol^{-1} K^{-1}, at a cooling rate 10^6 K s^{-1}, and 0.230 ± 0.01 kJ mol^{-1} K^{-1} at the cooling rate 10^{10} K s^{-1} [66, 77]. The interpretation of these results can also be verified from the Sun–Rawson–Minaev criterion for the prognostic evaluation of the possibility of glass formation at a certain cooling rate. Furthermore, to verify this criterion sulfide and selenide systems have also been used. However more research is desired to explore the possibility of glass formation at different cooling rates, particularly statistical data concerning critical cooling rates and the comparison with the calculated values of the glass formation ability.

Hence, in the area of glass-forming ability different concepts have been presented by investigators, accordingly Rawson introduced the E_{therm} with the factor T_m and constant $R = 8$: 3143 kJ mol^{-1} K^{-1}: Subsequently. Minaev replaced the T_{liq} with $RT_{\text{liq}} = E_{\text{therm}}$ and defined the following relationship:

$$\text{GFA} = \frac{E_{\text{CIB}}}{RT_{\text{liq}}} = \frac{E_{\text{str}}}{E_{\text{therm}}} \tag{2.10}$$

In this relationship, the ratio of two energies E_{str} (energy of structure binding) and E_{therm} has been established to obtain the glass-forming ability of the substances. Using this relationship, the binary composition As–Te parameters have been calculated. Note that the obtained glass-forming ability values have now related their compositions with the cooling rates (i.e. $\approx 10^2$ Ks^{-1}). The notable feature of this interpretation is that the calculated values of glass-forming ability can be achieved when the covalent-ion binding energy is several times larger then the thermal energy RT_{liq} during crystallization of the systems.

Rawson's et al. anaylsis [99, 100] is an interesting result as the kinetic theory of glass formation has shown that liquid without foreign crystallization centers could not crystallize if the activation energy is higher than $30\,RT_m$. Based on the Rawson's criterion, it was calculated that the activation energies expatiation can determine the origin of nucleus and crystal growth in glass-forming liquids (such as SiO_2, GeO_2, B_2O_3). This would apparently be the same order of value as the free activation energy of viscous flow (25–$30\,RT_m$), which causes both crystallization and viscous flow break M–O bonds. Similar explanations were provided for chalcogenide glasses; the activation energy of viscous flow is apparently a part of the energy of the covalent–ion binding that conforms to the fact that glass formation in the As–Te system is at $E_{\text{CIB}} > 33{:}2\,RT_{\text{liq}}$.

Therefore, in view of Rawson's application of the kinetic theory of glass formation to more complex systems would not successfully describe the difficulties of such complications [2]. However, he concluded that this criterion is more complicated and accurate in term of the kinetic theory of glass formation. He also stated the "the glass formation depends on relative values of strength of bonds (which must be ruptured at the crystallization), and the required thermal energy for the rupture". Their melt crystallization rate guided from the kinetic theory to glass formation is proportional to $\exp(-B_{M-O}/RT_m)$, here B_{M-O} is the strength of the chemical bond in the B_xO_y oxide, and R is the gas constant. Thus, according to Rawson when the ratio of B_{M-O}/RT_m is large, the crystallization rate can be small and stable glass could form.

In further developments [63, 66, 67], exploitation was provided for the glass formation ability of complex substances. This relationship depends on the ratio of energies of chemical (for most glasses—covalent-ion) binding of the structure of the given substance and the thermal energy of the system, according to the relationship GFA $= E_{CIB}/RT_{liq}$, which is required to destroy this binding. Considering this relationship, several systems of glass-forming ability exist with different cooling rates of the melt. The outcome of this study revealed the exponential dependence of the critical cooling rate on the glass formation ability of alloys that directly corresponds to the Rawson's criterion (B_{M-O}/RT_m).

Thus Rawson and Minaev developed the structural–energetic concept through the concrete quantitative evaluations that allowed exploring new glass formation regions with certain cooling rates. This concept also provided simultaneous consideration of the glass formation process from the structural–chemical, energetic (thermodynamic) and kinetic positions in a rather simplified form.

Liquidus Temperature Effect

Increase in the glass formation ability regions with the decrease in liquidus temperature was demonstrated by Tammann, Lebedev, and Kumanin and Mukhin [37, 101, 102, 103]. Proceeding with this basic concept of liquidus temperature of glass-forming ability regions, Rawson [2] described the temperature effect. Considering this point of view; it could be most likely for eutectic compositions. Considering the eutectic law Cornet [98] demonstrated that the glass-forming ability of binary telluride systems with elements of IIIA, IVA and VA subgroups of the periodic table can be maximal for the compositions located near eutectic ones. These systems glass forming ability emphasized the importance of the use of phase diagrams for the evaluation. But the role of liquidus temperature effect had been proved insufficient to describe other criterions.

To provide more evidence, Cornet presented the glass formation region in the system Ga–Te as 15–25 at% Ga [98]. His study established the eutectic in the system can be located at the point $Ga_{14}Te_{86}$, due to which this system could not form glass. Moving further downward of the liquidus line from pure tellurium in the direction of the chemical compound GaTe$_3$ (Fig. 2.6), the glass formation could not be formed even at the lowest eutectic temperature. This means no liquidus temperature effect was observed. The glass formation can be observed at the point $Ga_{15}Te_{85}$, with the increasing liquidus temperature. It reveals that the liquidus temperature effect as well as eutectic law of Cornet is uneffective. To overcome this difficulty Cornet made a statement, that the GFA can be maximal for compositions 'near eutectic ones.' However, he could not provide a concrete explanation about why there is no glass in this eutectic. Later, the anti-eutectic phenomenon was proved by the study of the liquidus temperature anti-effect. When such systems were analyzed from the standpoint of the quantitative determination of the glass formation ability of compositions containing more than one component from the Sun–Rawson–Minaev criterion, the following expression could be considered:

$$GFA = \frac{E_{CIB}}{T_{liq}} \tag{2.11}$$

Using this expression one can obtain the glass forming ability superposition values of the phase diagram of Ga–Te system. It was noticed that under the influence of the liquidus temperature effect the glass forming ability gets larger with increasing gallium content with pure tellurium. This increase is up to a certain eutectic point ($Ga_{14}Te_{86}$), whereas the GFA reaches value 0.265 kJ mol^{-1} K^{-1}, and further increase in the liquidus temperature it becomes against the liquidus temperature effect.. When the glass formation ability increases to the value 0.267 kJ mol^{-1} K^{-1}, for the $Ga_{15}Te_{85}$ composition, the glass formation region begins. According to Cornet's glass-forming ability, this value is maximum for the composition $Ga_{20}Te_{80}$. This calculation had shown that the glass-forming ability increases from the eutectic to this composition. It further increases and then decreases simply in the region of 24–25 at% Ga, this may be due to the liquidus temperature effect. Here noticeable things point just begins to liquidus temperature and increase sharply, while, it was earlier demonstrated that the increase is like a flat slope.

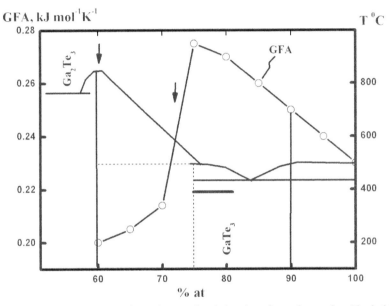

Figure 2.6   Typical phase diagram schematic associated the glass formation region (the bold line), and glass formation ability for the Ga–Te system.

Thus, the increase in glass-forming ability in the eutectic range from 0 to 14% Ga can be explained when one side decreases in the liquidus temperature, and the other side increases in the covalent–ion binding of the alloy. The covalent–ion binding can be determined with the help of the chemical bond's energies of the alloy. Using Pauling's expression, the calculated bond energy for the system Ga–Te was 177.3 kJ mol^{-1} K^{-1}, and the bond energy Te–Te is 168 kJ mol^{-1} K^{-1}. Therefore, it is clear that with the increase in the Ga content the E_{CIB} also gets larger. They also reported that when the Ga content reaches at 14% the E_{CIB} continues to grow actively, and the liquidus temperature also increases simultaneously. This is to demonstrate that the effect of increasing liquidus temperature is insignificant due to the flat slope of the liquidus temperature. As a consequence, the structural–chemical factor predominates over the thermal factor, this expands the glass-forming ability despite the increase in the liquidus temperature.

Vengrenovich et al. also contributed in defining the glass formation regions of different systems [103]. The binary systems such as Al–Te and Ga–Te illustrate the easiest glass formation located outside the eutectic alloy, whereas the chemical compounds were shifted to form the eutectic with tellurium. This means that in the system Al–Te, the initial flat and consequently sharp slope of both actions of the liquidus temperature effect and the eutectic effect could not be observed, only the flat slope liquidus rule was observed. This result favors the components of the structural–energetic concept of the glass formation in chalcogenide systems to explain the action role in the systems such as, As–Te, Si–Te, In–Te, Au–Te, etc.

Stable and Metastable Phase Equilibriums of the Chalcogenide Systems

The physicochemical approach of the structural transformations in semiconductors can be described from the melts transitions. Since the semiconductor–metal transition verification from the melts is of particular interest, in view of modeling the formation of medium-range order in vitreous alloys with covalent and metallic bonds. Due to the fact that both covalently bound bonds and metallic glasses can be produced from the same melt. Usually glasses (chalcogenide) are formed at a cooling rate of 180 K s^{-1}, while metallic glasses can be obtained from completely metalized melts at extremely high rates (of an order of 10^6 K s^{-1}) of cooling from the above temperatures of upper boundary temperature range of the semiconductor-metal transition.

There are several approaches presented by investigators, in which predominantly the semiconductor-metal transition has been considered based on their physical standpoint. Keeping in mind, the shortcoming of previous research, an important physicochemical approach was introduced by the investigators. This approach is based on the consideration of two types of the semiconductor-metal transitions under isobaric conditions. The phase transition under these conditions can be directly related to a number of properties, such as electric conductivity, thermal emf, relative density, magnetic susceptibility, viscosity and heat capacity [104–109]. According to this approach, the first type of the semiconductor–metal transitions (such as, melting of germanium and silicon) can occur at specific temperatures [104]. While in the second type of semiconductor–metal transitions can be smeared over a sufficiently wide range of temperatures, as an example, chalcogenide glass-forming melts [105–109].

Later, investigators considered whether the semiconductor–metal transition can be referred to as a smeared first-order phase (polymorphic) transition [107–109]. The conclusion confirmed the example of As_2Te_3 glass formation through the melt. They also measured the thermal effect of the semiconductor–metal transition. This outcome demonstrated the semiconductor–metal transition involvement in elementary transformations within microregions, consisting of several dozens of atoms across only a few nanometers away. According to this approach, the estimated average number of atoms in a microregion of the elementary structural transformation can be 33–60 [109]. Moreover, the differential scanning calorimetry analysis demonstrated that the released heat is over the entire range of the semiconductor–metal transition occurrence in the As_2Te_3 melt. This system's maximum thermal effect was observed at a temperature close to the midpoint (780 K) in this range [109]. The total thermal effect was estimated equal to 25 kJ kg^{-1} (0.6 kcal mol^{-1}). These experimental facts of semiconductor–metal transition in arsenic telluride melt had established the first-order phase transition through the thermal effect. In such systems, second-order phase transitions did not release or absorb the heat [110].

The lower boundary of the colloidal state belongs to the particle dimensions in the range of 1 nm. Their micro-inhomogeneity region can correspond to the particle size of 0.1 nm at high temperatures. This can be associated with the early stage of microemulsion formation in a melt due to the phase separation processes occurrence. This may fulfill the requirement of the formation of emulsion or partial immiscibility of the liquid phases. Such systems can make a big difference in types of the chemical bonding. In later studies, investigators introduced the existence of the metallic and covalently bound phases in these systems. According to this approach the emulsions can be coarsely dispersed in the systems due to small drops rapidly disappearing under the isothermal distillation. There are two possibilities; in the first case the microemulsions can appear spontaneously with rather stable system formations [111, 112]. Particularly in the binary systems (free of an emulsifying agent), it can be noticed at temperatures somewhat below the critical temperature. Under these conditions the interfacial tension was very small ($< 0.1 \times 10^{-3}$ J m^{-2}) since they were completely compensated by the entropy factor. To access the particle sizes and the absence of the third component it can be assumed that critical microemulsions with particle sizes vary from several nanometers to several dozens of nanometers in the temperature range of the semiconductor-metal transition in the melt. This is in agreement with the average particle sizes estimated by Tver'yanovich et al. [109]. Such critical microemulsions reflect in a variety of the lyophilic colloidal systems [113]. Since the individual particles sizes are generally too small, therefore, the interfaces between them can be smeared, thus, the critical emulsions can be designated as quasi-heterogeneous systems. In the specific case, the microinhomogeneity on the medium-range level can be retained with an alternation of the medium range orders of the covalently bound and metallic phases.

In view of the several drawback of the pseudobinary model, investigators described three fundamental principles; continuity, correspondence and compatibility [114]. The principle of continuity can correlate the composition and continuous changes until a new phase arises in their alloy's properties. The correspondence principle can be described as a distinct geometrical image or a combination of such images in the phase diagram resembling to an individual chemical. The third compatibility principle received less attention with limited use; although this approach may have promising applications. This principle describes an individual set of components which are irrespective of their number and properties. This can directly correlate to a constituted physicochemical system.

Since the variations in the substance masses are independent, all possible variations in the composition of a system can be determined [115–117]. In such systems under chemical transformations the number of components can be equal to the difference between the number of particle types in a system and the number of independent reactions. Usually the number of components depends on the conditions in which a system occurs. In the case of no occurrence of the reversible chemical reactions in a system, the number of components is equal to the number of substances. As an example, substances with different compositions and molecular structures (such as molecular oxygen and ozone) can have independent components [114]. Each component can be kinetically independent of other particles under certain conditions with the particular phase formation. In the essence of metastable components (quasi-components), there is no difference from the stable components, excluding those participating in metastable phase equilibriums.

However, classical physicochemical analysis based on equilibrium reversible systems, by assuming the basic components should be universally unaltered. A close examination of the practical compounds revealed that it can be possessed only for a few systems. Usually most kinds of practical systems can be separated into two groups: first, those systems having a total number of components remain constant under variations of the temperature and pressure, and their components can undergo a chemical modification. Second, for systems whose number of components and their chemical nature changes during the transition from one state of aggregation to another state. It can also be same for the aggregation within the state.

Through the chemical modifications of the bonding and structure of the particles, the components of the system can be defined. Usually this kind of problem does not appear in thermodynamics. The basic division of science covers only the macroscopic parameters of systems and is not useful to examine the structure of substances. Therefore, this approach is good only for ideal systems under certain conditions. The involved interactions reflect only components of those exhibiting absolute minima of the free energy. However, in actual practice, occurrence of several structural–chemical modifications for the particular component can arise in metastable phase equilibrium, therefore, the corresponding phase diagram may differ substantially from the stable phase equilibrium.

The useful classification of physicochemical systems due to chemical modifications in the initial components for different states of aggregation in a system is listed in Table 2.4. Here K^0 is the number of initial components in a system, N_i is the number of the structural-chemical modifications of the ith component in the crystalline, liquid and gaseous states. This reveals that in these systems, initial components can exist in various forms due to the association of dissociation, association, etc. processes. Thus, the chemical equilibrium can be achieved between new and initial forms of components. According to this concept as long as such an equilibrium occurs in a melt, it can easily re-establish the variations in temperature and pressure. However, under the rapid freezing the equilibrium can be form new components, therefore, $K > 1$. This can be acceptable because even a single chemically independent molecule (unlike a chemical individual whose identification requires the phase formation) may make a component. In the glassy system usually substances in the molten state undergo an abrupt semiconductor-metal first-order phase transition, it is shown as type I systems in Table 2.4. However, type II second materials can have smeared phase transitions, due to the microparticles with covalent and metallic chemical bonds coexistence of the transition range. This phenomenon can correlate to the nuclei of two-phase modifications corresponding to two different components.

Table 2.4 Physicochemical systems classifications [96]

		State of aggregation (N_i)		
Type I		**Crystalline**	**Liquid**	**Gaseous**
1	$K = \text{const} = K^0$	$N_i = 1$	$N_i = 1$	$N_i = 1$
2	$K = \text{const} = K^0$	$N_i = 1$	$N_i = 1$	$1 < N_i < \infty$
3	$K = \text{const} = K^0$	$N_i = 1$	$1 < N_i < \infty$	$1 < N_i < \infty$
4	$K = \text{const} = K^0$	$1 < N_i < \infty$	$1 < N_i < \infty$	$1 < N_i < \infty$
Type II				
5	$K \neq \text{const} < K^0$	$1 < N_i < \infty$	$1 < N_i < \infty$	$1 < N_i < \infty$
6	$K \neq \text{const} > K^0$	$1 < N_i < \infty$	$1 < N_i < \infty$	$1 < N_i < \infty$

Usually the analysis of semiconductor-metal transition has been described from the eutectoidal and other subsequent models like vitreous state generalized components. According to the generalized component model, the simple substances (or chemical compounds) correspond to one component in melts until the melts remain homogeneous and all the products of their dissociation in a chemical equilibrium. In contrast to this, if the microregions coexist due to a microinhomogeneous structure and different types of medium-range ordering in a melt, then it is not possible for a one-component melt ($K = 1$). Further, in the case when the transition occurs between the phase characterized by covalent medium-range order and the phase with metallic medium range order in the microregions of a melt at temperatures corresponding to the semiconductor-metal transition, then the melt must be considered as a two-component melt ($K = 2$). These two statements demonstrate that formally a one-component melt can possess covalent and metallic components. The semiconducting (covalent) or metallic modifications of the substance is stable only at certain temperatures, this means it is possible to choose such conditions at which the temperature range of their steady-state can be extended. If we remove these conditions, then a corresponding modification either exists in a metastable state or immediately transforms into a thermodynamically more stable phase.

Considering the concepts described above on meta-stability of the systems, several experiments were conducted. One of the key methods were based on the interaction between components with covalent and metallic bonds that were from a pseudobinary phase diagram under the phase separation. Similarly, the phase diagrams for systems whose initial components can be expressed from the metallic simple substances and semiconducting chemical compounds. The typical examples of such systems are, $Cu-Cu_2S$, $Cu-Cu_2Se$, $Cu-Cu_2Te$, $Ag-Ag_2S$, $Ag-Ag_2Se$, $Ag-Ag_2Te$, $Tl-Tl_2S$, $Tl-Tl_2Se$, etc. [118]. In general, in these systems a cupola-shaped phase diagram corresponds to the liquid-liquid phase separation can be achieved.

The temperature dependence magnetic susceptibility (χ) of the vitrifying melts with the temperature range of the semiconductor-metal transition is an important parameter. The usual variation schematic is given in Fig. 2.7(a). This type of dependency has been noticed for the high-conductivity semiconductors such as As_2Te_3 and $TlAsTe_2$ [108]. The semiconductor-metal transition temperature can occur at the maximum value of $d\chi/dt$, here around 50% of bonded electrons can be delocalized, it can be designated as T_{S-M}. The initial and final temperatures of the semiconductor–metal transition are in their usual notation T_b and T_f, respectively.

The hypothetical phase diagram for a semiconductor-metal pseudo-binary system is given in Fig. 2.7(b). Here the line ABC corresponds to the states of a melt in the range of the semiconductor–metal transition. The S-shaped form of this line demonstrates d their inverse dependence of the fraction of delocalized bonded electrons as a function of the melt temperature in the range of the semiconductor-metal transition [108].

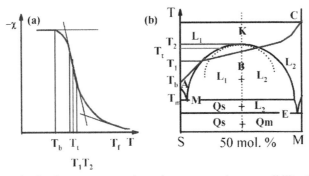

Figure 2.7 **(a)** Schematic for the temperature dependence magnetic susceptibility (χ) to vitrifying melts under the semiconductor-metal transition, **(b)** the well described hypothetical phase diagram of semiconductor-metal pseudobinary system. Here the line ABC belongs to the actual states of a melt for semiconductor-metal transition, while T_b and T_f are the onset and completion temperatures of the system, T_m is the temperature of the semiconductor-metal transition, T_1, T_2 are the boundary temperatures, M is the line of a monotectic equilibrium, E is the line of an eutectic equilibrium, L_1, L_2 are the covalently bound liquid and metallic liquid phases, Q_s, Q_m are the covalently bound crystalline and metallic crystalline phases, the dashed line represents the spinodal.

Figure 2.7(b) represents a phase diagram in which line M shows monotectic equilibrium in a case when two liquid phases and one solid phase are under equilibrium ($L_1 \Leftrightarrow\Leftrightarrow L_2 + Q_s$) [116]. The eutectic equilibrium line for the two solid phases and one liquid phase are in equilibrium ($L_2 \Leftrightarrow\Leftrightarrow Q_s + Q_m$), this line is also denoted as E [116]. In this diagram notations L_1, L_2, Q_s, Q_m represent covalently bound liquid, metallic liquid, covalently bound crystalline and metallic crystalline phases of the system. According to this phase diagram $T_m < T_b$, (here T_m is the melting temperature of a simple substance or the liquidus temperature of an intermediate alloy). The useful parameters like $T_m(M) < T_m(S)$ and $T_m(S) < T_b$, can be taken at the critical point K. At which the homogeneous melt structure is usually stable, the corresponding reference point lies between the binodal curve (a cupola of phase separation) and the monotectic line M with the phase separation of the liquid into two liquid phases L_1 and L_2. In this phase diagram, points A, B, and C represent the equilibrium states of an alloy at specified constant pressure as well as other external conditions. The occurrence of point B below the critical point K leads to the corresponding condition for spontaneous formation of critical microemulsions. This can also favor the metastable states of homogeneous melts, which cannot occur due to location of point of B inside the spinodal (dashed line). Therefore, only the pseudo-binary transition is possible in which micro-inhomogeneous structure of melts occur in a narrow temperature range from T_1 to T_2, whereas the ABC line intersects the binodal. This approach demonstrates that the particle sizes may become maximum at the semiconductor–metal transition temperature T_{S-M}.

This concept was extensively explored with different conditions, like an increase in pressure gives a decrease in the binodal size in the phase diagram for a pseudo-binary system, therefore, a gradual transition to the state when the critical point K lies below point 50. At that point semiconductor–metal transitions can occur with continuous change in the melt structure rather than through a phase transformation. This approach can also be useful to solve the problem of preparation of the metallic phase in a metastable crystalline state. In a similar way if one attempts to produce such a phase (like $TlAsTe_2$) from ultrafast cooling of the melt to above temperature T_f, then crystallization of resulting metallic glass most likely cannot end at the stage of an intermediate metallic phase due to very low energy barrier and considerable heat release of the crystallization. Further, in the case of the metallic modification, this is an essential requirement to decrease the size of crystallized particles therefore, stabilizers inhibition and their rapid growth can reduce heat release from crystallization. Under these circumstances, crystallization of melts in the temperature range corresponding to the semiconductor–metal transition can meet the desired requirements. A microheterogeneous material can also produce a cooling rate intermediate between 10^2 and 10^6 Ks^{-1}. Such a semiconducting configuration which incorporates metallic crystalline inclusion of the metastable particle sizes is quite suitable for X-ray diffraction analysis. In the case of higher degrees of the systems (ternary systems) in which two of the three components are immiscible. However, their individual components may be completely miscible in pairs with the third component. This indicates that the concept discussed above can also be useful for ternary systems [113]. Therefore, it can be stated that a system possesses only a larger number of degrees of freedom; then the critical state approach can be from a binary system owing to the change in both temperature and composition of a system.

SHORT-RANGE ORDERING

The absence of a long-range order of atomic arrangement in crystalline or fluids is called local ordering. Usually in a crystalline substance the order of atomic arrangement is pre-defined by the translational symmetry. On the other hand to understand the local ordering in disordered solids requires some specifications, like dimension of fields of local order in atomic arrangement as well as parameters which are required and can provide sufficient descriptions.

The non-crystalline materials elemental properties could be explored in terms of their short-range atomic ordering to determine the chemical nature of atoms (such as, bond length, bond angle, valency) within the substance. Therefore, the short-range order describes the involvement of the atoms that are nearest to the atom chosen as a central one and that form the first coordinate sphere, as presented in Fig. 2.8(a, b). In the first representation [Fig. 2.8(a)] atoms 1 and 3 in respect to atom 2, while in the second [Fig. 2.8(b)] atoms 1, 3, 4, 5 in respect to atom 2 [119]. The number of the nearest neighbor atoms (first coordination number) is the key parameter for the short-range ordering.

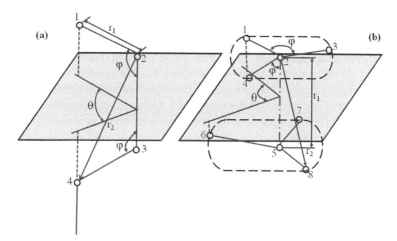

Figure 2.8 **(a)** Schematic for the linear relative disposition atoms characteristics, **(b)** tetrahedral structures, here r_1, r_2 are the first and second coordination sphere radii, φ is the bond angle, θ is the dihedral angle.

Along with their type, the distance between them and the central atom (radius of the first coordination sphere, r_1) and their angle position in respect to the central atom of the bond angles (valency angles φ) are equally important to define the short-range order of a system. These parameters could help to define only the short-range order to the first coordination sphere. If the second coordination sphere radius (r_2) is considered, then the radius of first coordination sphere as well as valency angles can be provided. An empirical relationship between the first and second coordination sphere radius and their corresponding valence angle can be described as:

$$r_2 = 2r_1 \sin\left(\frac{\varphi}{2}\right) \tag{2.12}$$

To resolve the short-range ordering problems Popov and Vasil'eva [120], proposed an amendment stating that one could pass the geometric parameters of short-range order to the power parameters of interaction among the atoms. Later, in the field of short-range order the atoms strong interactions and their respective position concept was also included. This permitted semiconductors having predominance of covalent type of chemical bonds under the strong interactions to be defined from the parameters of covalent relations (bond length-energy of interaction (v_s) and bond angle-energy interaction (v_b)). Thus, the modern concept of the short-range order field can be incorporated involvement of first coordination sphere and second coordination sphere of the atoms with their position with respect to the chosen central atom as well as the covalent interaction [121].

MEDIUM-RANGE ORDERING

Introduction of the short-range ordering of the non-crystalline materials could not successfully describe the local order in the atomic arrangement. This concept could not address several questions, such as how the short-range ordered fields are connected to each other. Also, it could not provide any concrete information about the considerable length of the ordered fields in non-crystalline materials. Experimental evidence showed long ordered fields, therefore, investigators triggered the introduction of the notion of medium-range order in atomic arrangement in non-crystalline materials. As in various microcrystalline and cluster models, the structure of these substances provided the presence of medium-range order.

It is believed that the first implication of a medium-range order in non-crystalline solids was studied by Vaipolin and Porai-Koshits [122]. This study successfully demonstrated the existence of some medium-range structural ordering at scales of the 0.5–3 nm. Though in this study, some presented orders were controversial, due to lack of the experimental evidences of the atomic structure for those having such scales in disordered materials. Later, this became one of the biggest subjects covering structures and properties in non-crystalline solids [121].

According to the modern concept of medium-range order, it can be linked with the rule of the distribution of dihedral angles. In several innovations, this concept was not comprehensive therefore, a concrete definition of this term is required. Considering this problem Lucovsky [123] introduced the concept of the medium order in terms of their regular distribution of dihedral angles up-to around 10 atoms. He also suggested that in the case of linear polymers, like chalcogens, it can apply only to the atoms belonging to one molecule (chain or ring) and other molecules atoms even with their position being closer to the atom chosen as the central one should be excluded from the elements of medium-range order. In a subsequent study Elliot [124] divided the medium-range order field into three levels; i) the field of local medium-range order (mutual disposition of neighboring structural units), ii) common medium field of medium-range order (mutual disposition of clusters), iii) long field of medium-range order linked to spatial order of different fields of a structural network. Later various studies [125] demonstrated the paracrystalline atomistic model of amorphous silicon was justified to be the medium-range order. Generally inconsistent in the medium-range order was noticed for materials whose chemical bonds differ, like linear polymers or chalcogenides. This is due to insufficient characteristics to consider even with the second and third neighbours, for those atoms belonging to other molecules.

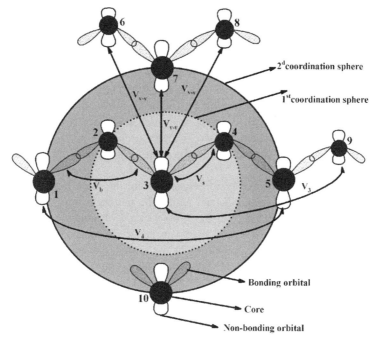

Figure 2.9 Schematic for the interatomic interaction in case of the linear polymer.

A more rational approach for the definition of medium as well as short-range order of atomic arrangement is transitional from geometrical characteristics to energy characteristics of mutual atom interaction. As discussed earlier, the short-range order can be from the strongest interaction between atoms, a schematic of such atomic interactions v_s and v_b is given in Fig. 2.9.

Figure 2.9 demonstrates that linear polymer (selenium, sulfur) short-range order includes first the nearest neighbors (atoms 1 and 3, if atom 2 is considered as the central one) and atoms associated to the second coordination sphere for those are of the same molecule as the central atom (atom 4 position with respect to atom 2 can be defined through the interaction v_s and v_b). The second order interaction between the atoms can also be related to the long-pair electrons of atoms, they may be in the same or different molecules. The Van der Waals's interaction between atoms of the neighboring molecules v_3, v_4, v_{v-v}, can also be formed. The existence of such second order interactions reveals the medium-range order atomic arrangement. Hence, the medium-range order is formed by atoms due to their partial positions in the second coordination sphere as well as higher orders coordination spheres of atoms, as an example the atoms of neighbor molecules in a linear polymer.

Thus, no long-range order atomic arrangement in non-crystalline materials has been established. It has been demonstrated only short-range or medium-range ordering can exist in these materials. Additionally, in the amorphous semiconducting materials, predominance of the covalent kinds of the chemical interaction can also exist. The short-range order can be determined by the interaction of covalent bonded atoms which includes the first and partially second coordination sphere. While, medium-range order can be explored by the interaction of long-pair electrons, Van der Waals's interaction as well as partial position formation of atoms in the second coordination sphere or higher orders coordination spheres.

Rings in Chalcogenide Glasses

Amorphous semiconductors materials networks have a large ring structure that can be described accurately from the concept of the rigidity percolation model. In such materials the number of topological constraints can be obtained with the help of Eq. (2.5), when the ring size is in the order of 10–12 atoms. While with the decreasing ring size below six atoms, some constraints become linearly dependent, therefore, a correction is required [45]. As an example, a triangle can be defined by three independent variables (two lengths and one angle), Eq. (2.5) can yield $n = 2$ constraints per atom for the total of six constraints. Therefore, the value obtained from Eq. (2.5) overestimates the number of constraints by the factor of three. In the case of the four-member ring system can generate two edge-sharing tetrahedral, as shown in Fig. 2.10. Here the number of constraints is overestimated by two. The occurrence of such constraints in a system can be demonstrated in terms of the tetrahedra edge-sharing from the comparison corner-sharing tetrahedral at which no rings exist.

It can be noted that the number of bond constraints are the same between tetrahedral edge-sharing and tetrahedral corner-sharing clusters, while, the number of angular constraints can be smaller for tetrahedral edge-sharing. This is because, a single variable can sufficiently define the bridging angle, here only four outer selenium remain and each can contribute one half angular constraint in comparison to six outer selenium for the corner-sharing tetrahedral. This concept leads to a total of 12 angular constraints for tetrahedral edge-sharing in contrast to 14 tetrahedral corner-sharing constraints. This contradictory result implies that the presence of tetrahedral edge-sharing can reduce the network rigidity. This is a significant outcome for the existence of the edge-sharing tetrahedral constraints which appeared in several chalcogenide glasses. $GeSe_2$ can contain large fractions of edge-sharing tetrahedral constraints and remain in the glass even for very low germanium content [126]. The existence of edge-sharing tetrahedral constraints has also been demonstrated for the ternary $Ge_xAs_ySe_{1-x-y}$ system [127]. Thus, the rigidity percolation threshold can be defined well from the average number of constraints n (all zero frequency modes vanish when $n = d$), with the accurate estimate of the threshold.

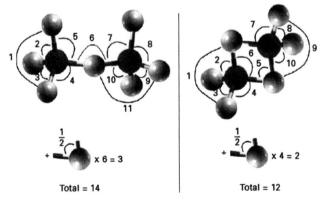

Figure 2.10   Mechanism of the corner and edge sharing tetrahedral pair angular constraints evaluation, under each outer bridging Se having half angular constraint (as defined by P. Lucas [46]).

Isolated Molecules in Chalcogenide Glasses

In chalcogenide glasses isolated molecular fragments are not covalently linked to the backbone of the glassy network as it does not contribute in the rigidity percolation; only the central forces can contribute to this concept. According to this model, the existing molecular species can only interact through weak van der Waals forces that are negligible to constraint counting. Though rigidity can remain, the backbone network may still be described from the percolation model due to its effective average coordination, which may significantly deviate from that derived from

Eq. (2.5). It is well established that sulfur-rich glasses can contain significant amounts of S_8 rings; this can provide a significant underestimation of the rigidity about the backbone of the glassy network. In contrast to this, arsenic-rich glasses can produce molecular species such as As_4Se_4, As_3Se_4 or As_4 (isomorphism of P4 tetrahedra) to keep arsenic out of the main network. This gives significant overestimation of its effective average coordination [128]. It was also demonstrated this may have a significant impact on obtained physical properties like, elastic modulus [129]. Therefore, this may predominantly describe the dynamic behavior of isolated molecular species that largely decouple from that of the backbone network. A direct evidence has been provided from the interpretation of As_6Se_4 and As_2Se_3 Raman spectroscopic study. In which it was demonstrated that arsenic-rich glass can contain large amounts of molecular species, while the stoichiometric glasses can have pyramidal modes within the network. Therefore, in such systems two different structures with the entirely distinct structural dynamics can exist in the heat capacity spectroscopy investigation. Similarly, the As_6Se_4 molecular network can have two distinct dynamic contributions as opposed to a single contribution for the As_2Se_3 glassy network. In this order, Bustin and Descamps [130] also concluded that the heat capacity spectroscopy analysis can permit the response of the structure to a temperature oscillation to be monitored. This outcome revealed the two distinct responses in the As_6Se_4 glassy network, one corresponding to the molecular species and the other corresponding to the glassy backbone. However, it was noticed that As_2Se_3 glassy network can have a single narrow response consisting of a uniquely composed uniform structure from pyramidal units. The existence of such molecular species in chalcogenide glasses is common since experimental evidence has been provided from the interpretation of GeAsS and PSe glasses network [131, 132]. Subsequently, this outcome also demonstrated the decoupled dynamics with polyamorphism [133, 134].

METHODS OF DISORDERED STRUCTURE INVESTIGATION

To investigate the disordered structure of chalcogenide materials some important models were proposed. In order to do this, a key demonstration based on the possibility of glass formation in elemental substances and their alloys can be related to the specific features with the electronic structure of the atoms, such as the stable electron configurations p^0, p^3, d^0, d^5, d^{10}, f^0, f^7 and f^{14}. The glass formation in the alloys can be influenced by the existing structural–configurational equilibriums in the melt at temperature between clusters during the synthesis. Owing to electronic configurations of atoms in the chemically bound states, this can be close in terms of energy that varies widely in the degree of polymerization. In such materials, the glass formation parameters can be quantitatively defined. Using the capacity of atoms of the chemical elements, they can be characterized to what make up the melt to form a vitreous network. These materials have shown the dependence on parameters based on nuclear charge of the elements primarily and secondary on the periodicity specifically in the case of sulfide, selenide, telluride and oxide systems. It is predicted that electron configuration model concept, can also be applicable for the halide, diamond and metallic systems.

Eutectoidal Model for the Stable Electronic Configurations, Glassy State and Experimental Verification

Stable Electronic Configurations

Goryunova and Kolomiets, Winter-Klein were the first to draw attention on the eutectoidal model of the glassy state and their experimental evidence [4, 34]. They established the relation between the capacity of substances to form glasses and the number of valence p-electrons per effective atom. According to Winter-Klein's criterion, alloys which have two to four valence p-electrons in atoms have shown the strongest tendency to form glass. However, the drawback of this criterion is that it could not be generalized, limiting it to some specific cases. The Winter-Klein's concept could not attract attention of investigators for future progress. In further development, the valence capabilities of various atoms were analyzed considering not only partly filled valence orbitals, but also vacant and completely filled orbitals close to their valence counterparts in terms of energy.

This is pivotal as a principal condition for the glass formation. According to theirs concept, the substances structural–configurational equilibrium can occur between the low and high-molecular-weight forms of the atomic groups in the synthesized melts (solutions). Moreover, such an equilibrium can be related to the electron configuration equilibriums in the atoms that are made up from all these groups. The most significant salient features of the approach are to provide the examples with selenide and telluride systems [41, 135–137]. Later investigators considered the different systems glass forming abilities by providing several experimental verifications.

One important development is to establish the relationship between the electronic structure of atoms and their ability to form a vitreous matrix that gives a reasonable beginning from the elemental substances. This also indicated that it was possible to take out a fragment of the periodic system that consists of the most easily vitrified four elements. Due to this different modifications could be denoted as Low-Molecular (LM) and High-Molecular (HM) [41]. Elements such as sulfur, selenium, phosphorus, arsenic are the best examples that can form such elements. More accurately, it can define such simple molten substances and produce both types of molecular groups possessing the same free energy and equilibrium to each other. The experimental evidencs for elements such as sulfur and selenium has been provided by Addison and Feltz, demonstrating the existence of cyclic X_8 and chain X_n molecules in the molten and vitreous states [138, 139]. Cyclic and chain molecules equilibrium in the chalcogens is the specific case of n LM $\Leftrightarrow$ HM. They may also contain their ionic and radical decompositions. The approach of the equilibrium particles in the involvement of the different degree of polymerization under the medium-range order character to form kinetic barrier between them allows in defining a mandatory condition for the increase in the relaxation for the corresponding equilibriums, while a possible decrease can allow in substance crystallization. According to well-established solution theories of glass-forming melts, it can have high-molecular substances in low-molecular solvents. This is a significant distinction between glass-forming melts and normal solutions of high-molecular compounds with the possibility of intraconversion between low-molecular and high-molecular particles. Additionally, the structural groups can have a wider range of distribution in glass-forming melts [140].

The structural–configurational equilibrium may exist in various substances, especially in the case when corresponding atoms rearrange their electron configurations during the synthesis. This is the case of chalcogens in which the structural–configurational equilibrium n LM $\Leftrightarrow$ HM occur during glass formation and their equilibriums electron configuration can be types of $s^1 p^x + (5-x)$ $e^- \Leftrightarrow s^2 p^4$, $x = 1, 2, 3$. In chalcogen group from oxygen to tellurium this kind equilibrium shifts to the right for the identical conditions. This equilibrium shift may take place in a non-monotone manner due to the orbitals of free sublevels, d^0 and f^0 [41, 135–137]. In isolated atoms the outer s and p orbitals may have different energy [141]. However, in case of sulfur and selenium the difference decreases to 10.0 and 10.1 eV [142]. Moreover, the multiparticle systems sublevel splitting leads to the decreasing energy gap between them. Thus, the usual energies of the various electron configurational states come closer to one each other, this fact has been recognized by researchers [143]. However, the single-bond energy has a larger tendency with the increasing number of atoms per molecule. But it is restricted up to a few kJ per mol, this could be due to a gradual change in the type of hybrid state. In chalcogens, the change of cyclic and chain of the molecules can also affect the valence orbitals, as a consequence, changing the electronic configurations from the sp^3 to the p-electronic states. As the chain of the molecules becomes stable in selenium and tellurium, they do not have a tendency of sp^x hybridization. The closeness of the energy between different molecular species in a glass-forming melt can lead to their distributions by size as well as other parameters [144].

According to this approach, the glass forming capacity of the multicomponent alloys can be described by considering two key factors, first is the limiting concentration P_{lim} of the element in the glass former (amount of the atomic percentage at the preserve vitreous state). Second is the fraction S that can be expressed as a percentage of the glass formation region in the entire concentration space. To validate this concept, Funtikov presented a study on glass forming capacity of several alloys, like $A^{III-V}-E$ systems ($E = S$, Se, Te). His study demonstrated that a correlation for atoms of this period can acquire a free acceptor inner 4 f sublevel (4 f^0), which is responsible for metallization of the chemical bonds as well as proportionate decrease in the tendency of alloys to form glass. Thus, an increase in the number of components in an alloy can be accompanied by an increase in the glass-forming capacity. This is due to fragments of the vitreous network of atoms formation in several alternative types. Therefore, the specific features of the electronic structure of atoms can affect the glass-forming capacity predominantly in ternary alloys.

Specifically in amorphous semiconductors the d-elements can be divided into two groups: (i) the d-elements whose concentration in the glass ranges from a few hundreds to 1%, (ii) the d-elements whose concentration in the glass may be as high as 30%. [135]. The first group elements have an electron configuration of atoms in the isolated state can be correlated as d^n ($1 \leq n \leq 9$). The second group elements stable configuration belongs to d^{10}. Additionally, when atoms of d-elements are combined together with the d^{10} and f^{14} configurations then they can have a positive effect on the concentration in chalcogenide glasses. In the case of the f-elements, they may act similar to d-elements.

Glassy State

Eutectoidal model of the glassy structure is based on Smits's concept of the pseudobinary systems. According to this model all glasses can be described as a variety of eutectics formation under the interaction of the pseudophases. The pseudophase may be equivalent to the nucleus of the ordered particle that appears in the supercooling melt.

The scientific community is not aware of proponents on the homogeneous and the microinhomogeneous structure of vitreous alloys. However, the experimental evidence learns more toward microinhomogeneous structure in glasses. Although the eutectoidal model prevents the ideal glass non-uniformity in which their ensemble can have sticking microspheres. According to the eutectoidal approach, equilibrium between low and high molecular clusters is the necessary condition for stable glass formation. The existing clusters may be neutral containing charge particles [145]. This allows the availability of a large number of internal reactions along with an appropriate chemical equilibrium to promote the formation of glasses. It can be characteristic for one-element substances as well as for their compounds. The interaction between clusters of the different degrees of the poly regularity may be predominant for elementary substances and steady compounds. In the case of quasi-independent components, the co-existence of the low and high-molecular clusters can be considered. Such existing components can be capable for stable and metastable alloys. This concept can also be applicable for glass formation from the multicomponent alloys melt. Under these circumstances, formed pseudophase can be equivalent to nucleus of the ordered particle that appears in the supercooling melt. As an example, sulfur and selenium are the simple pseudobinary systems. The sulfur pseudobinary system consists of the molecules S_8 and S_n (n = const). The eutectic approach also consists of S_8-S_n and similar systems with the limited solubility in the solid state. According to this approach, special conditions (P, T) must exist for the formation of the eutectic melt. Such melt components lose a degree of the freedom during solidification in the slow relaxation to reach a corresponding equilibrium. Hence, glasses synthesized from the melts can be microdispersed into metastable eutectic even for elementary substances. A pseudobinary system based on the elementary substance is exhibited in Fig. 2.11. The sets of equilibrium between the low- and high-molecular clusters (n LM $\Leftrightarrow$ HM) can exist in the melt at the synthesis temperature (T_s). The high viscosity of these kind of melts can be determined from the power generating barrier between the low and high molecular clusters transition. During the cooling process the temperature (T_s) of the melt, the relaxation of the chemical equilibriums becomes very weak, therefore, original solution (melt) can be turned to lyophilic colloid L_c, as shown in Fig. 2.12. For optimum velocity of cooling of melts can be correlated like $V_c^{opt} = V_c^3$. At this condition the maximum dispersity of the glass particles can be formed due to equilibrium between L_T, L_c and solid solutions $S\alpha$ and $S\beta$.

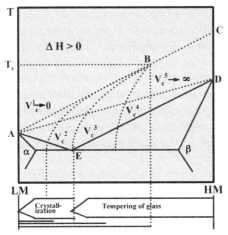

Figure 2.11 Schematic of the pseudobinary system states hypothetic diagram in term of the elementary substance. The equilibrium between low- and high-molecular clusters can be obtained (nLM ⇔ HM) to melt temperature T_S, the change of enthalpy can be determined from direct reaction $\Delta H > 0$, while change free energy can be expressed as $\Delta G = 0$. In case of low and high molecular clusters at $T = 298$ K ($G_{LM} < G_{HM}$) and corresponding velocity of the cooling of melts $V_c^1 < V_c^2 < V_c^3 < V_c^4 < V_c^5$, however, the optimum velocity of cooling meltse $V_c^{opt} = V_c^3$. corresponding to maximum dispersity.

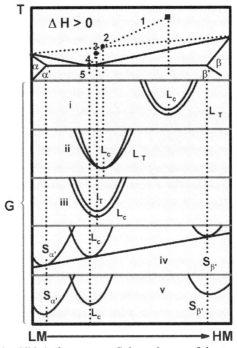

Figure 2.12 Schematic of the Gibbs's free energy G dependences of the true solution (melt) L_T, lyophilic colloid L_c, solid solutions S_α, S_β to composition of alloys for the low and high molecular components pseudobinary system of the elementary substance. Here diagrams i-v represents the hypothetic states of the pseudobinary system based on the elementary substance eutectic approach.

Therefore, the topological approach based on the extremely small volumes of space to form continuous ordered structure permit all possible elements of the symmetry with first infinite orders. However, the conventional approach of the glasses structure analysis is usually described from the laws of crystallography. According to these laws only 1, 2, 3, 4 and 6 orders can permitted from the theoretical opportunity. This can fulfill the desire for the formation of the ideal homogeneous glass. The topological model also demonstrates that only the large power-generating barrier of the transition from microscopic particles, having infinite set elements of the symmetry for the crystalline state with the limited symmetry may permit a substance to pass the glassy state from the melt, therefore, they can form the amorphous film from a vapor. A schematic of the transitions of glasses and amorphous films in the crystalline state through a stage of the formation of quasicrystals is exhibited in Fig. 2.13. According to this interpretation, quasicrystals should have an axis of symmetry of the fifth order and the icosahedral structure for the metal alloys and the dodecahedral structure for the semiconductors and dielectrics. In the case of glass structures, the structural elements can have the most geometrical variant that all elements of the symmetry of icosahedron and dodecahedron $(6L_510L_315L_215PC)$ may dominate.

Experimental Verification

The presence of nanostructures in glasses and their eutectic interaction has been experimentally demonstrated in various studies. The accurate glassy state of gases from the eutectoidal model could be described by considering the microscopic aspect of the glass formation process. This indicates that it is next to impossible to create a uniform idea about the glass, while not considering the nature of a chemical bond and other features of vitrescent substances. This also provides a method of physicochemical model operation of the process of a glass transition, allowing to construct glasses chemically. Therefore, various differential thermal analysis methods can be used for the investigations of metastable chalcogen systems.

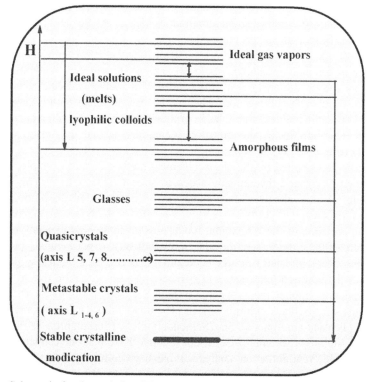

Figure 2.13 Schematic for the enthalpy (H) corresponding to different states of substances, as well as formation of glasses and amorphous films.

Accordingly the modern concept of glass formation usually describes it from the kinetic theory of the glass transition related to relaxation processes in vitrescent melts. The kinetic theory reveals that a glass is a supercooled fluid having very high viscosity to acquire the properties of a solid body. This kinetic theory description could not describe the process occurring at an atomic and molecular level, as it is based on the macroscopic concept. Therefore, the mechanisms of glass structure formation could not effectively describe it, as well as there were no generally accepted definitions given for a glassy state substance. The subsequent modern concept of glass formation has been breaking ground for the development of representations about glass and principles of a series of technological processes in the production of glasses by covering the new classes of glasses such as chalcogenide and metallic glasses. Modern principles of a glassy substance take into account both the macroscopic and microscopic aspects, but could not provide a concerted and a concrete method for the reception of glasses. Considering the drawbacks of the previous concepts of glassy structures, an advanced eutectoidal model of a glass transition of substances was introduced by researchers [146]. The concept of this model demonstrates that glasses can have a type of the ultradispersed multicomponent eutectoidal structures. This is not structurally homogeneous material in the limits of the medial order. The medial order in glasses can exist due to topological order at a level of second, third and other spheres of atoms interaction. According to this model, the upper sphere of such an interaction of atoms can lead to the size of the nanostructures in composition in a glassy network. Such nanostructures formations may exist in more than two types of glasses. However, for simple substances and stable chemical combinations, a topological order within a frame of one nanostructure should be homeomorphous with a geometrical order in crystals of some modification of a substance [147].

Therefore, the existence of the nanostructure in chalcogenide glasses can be described from a system of chemical and electrochemical equivalent measurement [148]. It has been demonstrated that the composition of nanostructures in glasses can correspond to the composition of simple substances with stable and metastable compounds. It was also well established that nanostructures can be formed by selenium, such as stable compounds $GeSe_2$, As_2Se_3 and metastable compound from $GeTe_2$ as well as with other compounds As–Se, As–Te, Ge–Se, Ge–Te, Tl–Ge–Te, Tl–As–Te, etc. [149–151]. Further, an additional chemical and electrochemical equivalent concept was also introduced for their measurements [152]. This relatively new method is based on the law of equivalents and on the analysis of a dependence of a chemical equivalent of the glasses and the ceramics from their composition. This model can correlate to the glassy and ceramic insulators, the glassy semiconductors and the amorphous semiconductor films microheterogeneous alloys structures. The significant outcome of this approach is that the selective dissolution of the separate fragments of alloys in solutions of acids, alkalis, oxidizers, reductants and organic compounds, can be investigated by the electrochemical method. As an example, the dependence of the experimental CE^{exp} and theoretically accessed sizes CE_I^{theor}, CE_{II}^{theor}, CE_{III}^{theor} of the molar weight of a chemical equivalent for the system Ge–Se composition dissolution in 1 M KOH solution was also demonstrated [149].

Based on the principle of the eutectoidal model several individual elements simulating glass formation structure of the glasses were introduced. Specifically, experimentally proven process of glass transition has been validated from the eutectoidal concept. This model validated chalcogens elements, sulfur, selenium and tellurium, and is receiving great attention due to their successful interpretations. Using this model, the metastable systems Se_8 (Semonocline)- Se_n (Sehexagonal), Se_8 (Sered amorphous)–Se_n (Seblack amorphous), Se_8 (Semonocline)–Te_n (Tehexagonal), Se_8 (Sered amorphous)–Te_n (Teblack amorphous), Se_8 (Semonocline)–S_8 (Serhombic), Se_8 (Semonocline)–I_2 as well as stable system S_8 (Srhombic)–Se_n (Sehexagonal) have been investigated and interpreted with the help of the differential thermal analysis. It has been commented that in the case of glassy selenium, the choice of the physicochemical model operation for the study of the process of glass transition of substances, this could be a perspective method.

The eutectoidal model of a glassy state of substance prevents the formation of glasses that can be connected with a variable number of components of the vitrescent melts and their special role for metastable phase equilibriums [146]. A simple metastable physicochemical system using the experimental and theoretical modeling analysis is preferred. This enables one to find the role of chemical and phase equilibriums in the formation of the medium order and physical and chemical properties of the glasses. One of the best examples of this condition is selenium (Se_8–Se_n) systems. Four elements (P, As, S, Se) can exist in the form of low and high molecular modifications. A chemical equilibrium between the monomeric and polymeric molecules can be established for their melts. The molecules of the Se_8 and Se_n in the structural chains of (Se_n) form a basis of the stable phase of the gray hexagonal selenium. The cyclic Se_8 molecules due to the three modifications of the metastable red selenium for that monocline modification can be considered the most stable modification.

According to this concept, the physicochemical interaction between crystals of the gray and red selenium can be analyzed from differential thermal analysis. Using this technique, the structure and physical–chemical properties of glassy selenium can be determined. This outcome can be interpreted in terms of the special role of metastable phase equilibriums between gray and red modifications of selenium. Accompanied with the necessary condition, the chemical equilibrium between the molecules Se_8 and Se_n should be infringed at the time of experiment, to create sharp cooling of the vitrifying melt.

The diagram of fusibility of a system component of the chemical composition of selenium can be investigated. As per co-constancy with the experimental evidence to the eutectoidal model of a glassy state of a substance, it can be stated as that the performances of the curves of the differential thermal analysis of alloys Se_8 (Semonocline)–Se_n (Sehexagonal) change non-linearly when the transformation occurs from low molecular weight to a high-molecular component. As shown in Fig. 2.14, with the increasing mass of the hexagonal gray selenium (Se_n), the minimum melting temperature is around 10–20°C initially.

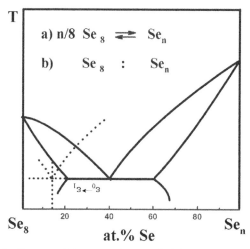

Figure 2.14 Schematic diagram of the states for pseudo-binary system $Se_8(\beta - Se)-Se_n(\gamma -Se)$.

Moreover, to magnify the dispersity of the starting stable and metastable (quasi) components, the red and black amorphous modifications of selenium can play an important role. Due to this modification, the structure of red and black amorphous selenium can differ significantly. Therefore, red amorphous selenium can consist of cyclic molecules Se_8, while the black amorphous selenium may form Se_n chain molecules. The system Se_8 (Se-red amorphous)–Se_n (Se-black amorphous) specifically with the surplus red modification, can reveal curves of the differential thermal analysis when melting of two modified amorphous selenium coincide. Furthermore, the curve of the differential thermal analysis can be obtained during heating of glassy selenium.

Physicochemical Analysis of Vitreous Semiconductor Chalcogenide Systems

Theories of glass have established that it is an alloy close to metastable eutectics composition exhibiting a high glass-forming ability [153]. Therefore, it is important to know at which compositions of eutectic alloys metastable phase diagrams differ from the equilibrium eutectics. Thus, a tendency of glass formation implies that a correlation exists between compositions and their glass-forming ability. The eutectoidal model can be used for this purpose to describe the structural properties. According to this model, a hypothesis "the melts of all substances prone to glass transition should feature a quasi-eutectic structure" can be adopted [146, 154, 155]. Considering this, it was demonstrated that glasses of any composition (even elemental) can be considered as a modification of the multicomponent melts, as a consequence highly disperse eutectics may be achieved. This concept is based on Smith's idea of pseudobinary systems and the notion of pseudophases introduced by Porai-Koshits [156].

It is also desirable to define the term 'components'. According to its definition, the components of a system are called substances if they have independent mass variations to determined all possible variations in the composition of a system [116, 117, 157]. Systems under chemical transformations can have a number of components equal to the difference between the number of particle types in the system and the number of independent reactions. Usually the number of components depends on the conditions under which a system occurs. If in such a system the reversible chemical reactions do not occur then the number of components may be equal to the number of substances. As an example, some simple substances with different compositions and molecular structures, like molecular oxygen and ozone having independent components can be considered [114].

There is 1 to ∞ components of the system that should be verified. The individual component should consist of particles that are kinetically independent to other particles. In principle those capable of forming a certain phase. This leads to ions of the solutions or melts that may not serve as components. Although, molecules having different degrees of polymerization and the same type atoms can fulfill the role of components in the melts, additionally, their interconversion would be impossible; such behavior can be observed in sulfur. In equilibrium, the element sulfur can be treated as a single component. Therefore, the element sulfur equilibrium corresponds to only one modification. However, when temperature is higher (around 159°C) then the molten melt can consider the solution of polymer in monomer. In this case, the ratio between two forms of the melt can be temperature-dependent [138]. In the case of rapid cooling, the temperature can drop in the range 159 to 119°C. Therefore, the total number of components can sharply increase in the melt due to the formation of molecules that kinetically become virtually independent. Moreover, further rapid cooling of the sulfur melt can allow frequent component interactions even in multi-components system. This situation can be correlated in a single system being in a multi-component system form. The number of the quasi-components of vitrified melt can be equal to the number of different molecules for those holding their identity during the cooling. This can lead to the highest homogeneity order in the vitrifying melt at freezing; it is a desirable condition for those particles associated to different quasi-components to grow simultaneously. Therefore, according to physicochemical analysis, this can be used to describe the eutectic mechanism of the growth ordering of the nuclei. Hence, the eutectic of a system from the number of components can be addressed as a solution in which the corresponding equilibrium to solid phases and their number is equal to the number of components in a system. Moreover, the vitrifying melts can have considerable numbers of quasi-components, but it cannot form actual phases in principle due to the deficit of the material. Therefore, ordering within the substance may form the pseudophase stage. According to the eutectoidal model, the optimum glass transition can be achieved after cooling and total N components may be created from the N-component of the metastable highly disperse eutectic phase. The eutectic particles may be very small as they are likely to be liquid instead of solid. This exploitation of the eutectic particles is consistent with the kinetic theory of glass transition. Thus, the number of quasi-components (N) of the good glass-forming melt must be comparable to the number of structural elements. This is also related to the question why the actual phases could not form, and glass transition occurs. This also explains the fact that chemical ordering of all nuclei should grow virtually simultaneously to get maximum homogeneity in a glass. To obtain such a condition the eutectoidal mechanism of the solidification of a melt could be the optimum approach. Therefore, ideally glass is a multicomponent eutectic containing a total number of components comparable to the order of magnitude. In addition, the feasible total number of structural units of the elements may also be formed under short-range order.

Table 2.5 Structural–configurational equilibriums under the eutectoid interactions between components and glass transition of melts for the binary A–B systems. Here ΔH_{I}^{from}, ΔH_{II}^{from} and ΔH_{III}^{from} are representing the heat formation for the $(A_n)_m kA_1$, $(B_n)_m kA_1$, $(B_n)_m kB_1$ and solvates [96].

A		B
$2A_1 \Leftrightarrow A_2$ $A_2 + A_1 \Leftrightarrow A_3$ $A_3 + A_1 \Leftrightarrow A_4$ $A_{n-1} + A_1 \Leftrightarrow A_n$		$2B_1 \Leftrightarrow B_2$ $B_2 + B_1 \Leftrightarrow B_3$ $B_3 + B_1 \Leftrightarrow B_4$ $B_{n-1} + B_1 \Leftrightarrow B_n$
I. $(A-A)$ $(A_n)_m kA_1$	II. $(A-B)$ $(B_n)_m kA_1$	III. $(B-B)$ $(B_n)_m kB_1$
$\Delta H_{I}^{from} < 0$	$\Delta H_{II}^{from} < 0$ $\left| \Delta H_{II}^{from} \right| > \frac{1}{2} \left\| \Delta H_{I}^{from} + \Delta H_{III}^{from} \right\|$	$\Delta H_{III}^{from} < 0$

The eutectic melts can be also considered as lyophilic colloidal solutions containing the disperse particles for those completely dissolved through a dispersion medium [158]. Beside the other aspects of the glass's modification, the frozen lyophilic colloidal solutions can also be considered. This indicates that the vitrified melts can be treated as lyophilic colloidal solutions in which spherical micelles may be formed in the initial stage. However, with the increasing concentration of the molten solution the micelles may play an important role to transform into anisometric (liquid-crystalline) micelles [159]. Various reports of the feasible structural–configurational equilibriums of the melts for binary A–B systems are given in Table 2.5.

As shown in Table 2.5, homosolvates and heterosolvates can be formed. From the accepted concept, heterosolvates can be more stable due to the fact that the interconversion between the dispersed phase and dispersion medium is almost impossible. Both kinds of the neutral molecules as well as their molecular products dissociation can act as quasi-components, as summarized in Table 2.6. It clearly demonstrates that all the charged products of this dissociation cannot form the quasi-components, therefore, only a complicated configuration can result in the glass structure.

Hence, heterogeneous interactions could predominate among a wide variety of interactions. Under the equilibrium, a melt crystal can be different owing to the components of the composition. Tables 2.5 and 2.6 summarize the diagrams enthalpy for the transitions between equilibrium crystals to the melts that may occur along the non-equilibrium pathway due to retention and formation of the metastable crystalline phases. However, at high temperatures such phases correspond to rather stable components under normal conditions. Therefore, it is appropriate to assume that eutectoidal hypothesis can be used for the composition–property diagrams to vitreous systems for both quasi-homogeneous systems as well as context of the quasi-heterogeneous systems [160].

Table 2.6 Structural–configurational equilibriums of the melts to their vitrifying elemental substances and compounds. Here LM and HM are representing low- and high-molecular neutral (or charged) particles [96]

Elemental substances and dystectic compounds	Peritectic compounds
$n\text{LM}_i^0 \Leftrightarrow \text{HM}_i^0$	$\text{HM}_j^0 \Leftrightarrow \text{LM}_k^0 + \text{LM}_p^0$
$2\text{LM}_i^0 \Leftrightarrow \text{LM}_i^+ + \text{LM}_i^-$	$2\text{LM}_k^0 \Leftrightarrow \text{LM}_k^+ + \text{LM}_k^-$
$2\text{HM}_i^0 \Leftrightarrow \text{HM}_i^+ + \text{HM}_i^-$	$2\text{HM}_p^0 \Leftrightarrow \text{HM}_p^+ + \text{HM}_p^-$
$n\left(\text{HM}_i^{0(+-)}\right) + m\left(\text{LM}_i^{0(+-)}\right)$ $\Leftrightarrow \left(\text{HM}_i^{0(+-)}\right).\,m\left(\text{LM}_i^{0(+-)}\right)$	$n\left(\text{HM}_p^{0(+-)}\right) + m\left(\text{LM}_k^{0(+-)}\right)$ $\Leftrightarrow \left(\text{HM}_p^{0(+-)}\right).\,m\left(\text{LM}_k^{0(+-)}\right)$

The composition–property diagrams (isotherms of properties) under the thermodynamic equilibrium multicomponent homogeneous and heterogeneous systems theoretically can be analyzed for true equilibrium alloys. Various rigid solubility problems may arise in the thermodynamic non-equilibrium systems. These kinds of systems predominate to real systems as well as correspond to crystalline, liquid and vitreous states. Specifically, inorganic vitreous alloys do not support this direction due to the existence of the thermodynamic equilibrium state. The inorganic vitreous alloys composition–property diagrams have been interpreted in different way by various investigators. Thus, the outcome depends on the model used for glass transition and glass structure. However, in agreement with one widely accepted model, the same synthesis procedure can be used for all glasses of a particular system. The pseudomolar, volume-additive and mass-additive properties can be verified from this model. This allows the desired appropriateness and correct physicochemical analysis of the pseudomolar properties. In a simpler way, these properties subsequently can also be designated as molar properties. Hence, this indicates that it is almost impossible to calculate truly molar properties for thermodynamic non-equilibrium systems. In the case of glassy systems, such properties can be calculated per mole of atoms with the accurate stoichiometric compositions that are not typical of vitreous alloys. Particularly the molar properties can play a vital role when the system components do not chemically interact, therefore, these properties linearly depend on the composition in terms of mole fractions. Further, the value of mass-additive property can be converted into a mole-additive modification in a simpler form. To accomplish this, a value of mass-additive property is multiplied by the mass of one mole of glass atoms. The volume-additive property can be obtained from their value multiplied by a molar volume, typical in the millimeter range and can be calculated from a molar mass and experimentally determining the density of the substance. The last case described above reveals the parameter that is the product of values of two properties which are capable of reflecting the chemical processes that advance in a system as a function of mole fraction. On the basis this description they can have three types of the composition – property diagrams (i) entire features of the diagrams (extremums and inflections) can be more pronounced, (ii) entire features can be less pronounced, (iii) entire features of the diagrams may completely be mutually neutralized. The third condition is the most critical as it provides a space for the incorrect conclusion of the system. Considering this key issue investigators suggested that in specific cases, more objective information concerning the interactions in a system is desirable to analyze the volume-additive property (such as, density ρ, refractive index n, optical density D, integrated intensity of the absorption bands B, dielectric constant ε, etc.) in the mole fraction diagrams. In terms of applications, the correct conditions for physicochemical properties diagrams can be described by the following relationship.

If one considers the simplest binary A–B system with the following designations: N^V is the volume fraction, N^m is the mole fraction, V^m is the molar volume, P^V is the volume additive property, and P^m is the molar property. If we considered the $V_A^m = V_B^m$. k then:

$$P^V = P_A^V N_A^V + P_B^V N_B^V = P_A^V + \left(P_B^V - P_A^V\right)N_B^V \tag{2.13}$$

In the case of weak interaction between the A and B

$$N_B^V = \left(\frac{V_B}{V_A + V_B} \right) = N_B^m + \left(\frac{1}{N_B^m(1-k)+k} \right) + N_B^m \gamma \qquad (2.14)$$

here

$$\gamma = \frac{1}{N_B^m(1-k)+k}$$

Thus

$$P^V = P_A^V + \left(P_B^V - P_A^V \right) N_B^m \gamma \qquad (2.15)$$

The function is linear for $\gamma = 1$ and $k = 1$. The composition $N_B^m = 0.5$ with the k value ($k = 1{:}1$) can deviate from additivity of about 0.3%. For the case of $k = 1.4$, the deviation can be about 3%. From the ensuing approach, it was demonstrated that the extensively studied As_2Se_3–As_2Te_3 system diagram can be illustrated and interpreted.

The obtained equilibrium phase diagram of this system can have regions of solid solutions as well as the region of their coexistence that recognizes the crystalline state as shown in Fig. 2.15 [93]. By using this data, the density and dielectric constant can be obtained [161]. The interpreted plots in straight-line portions are the concentration dependences of these parameters just corresponding to the above three regions in the equilibrium phase diagram. This makes it evidently clear that the interactions between equilibrium components hardly differ from those between predominant metastable components in the system. The points of inflections in the corresponding curves can be correlated to the change in the type of glass microinhomogeneity.

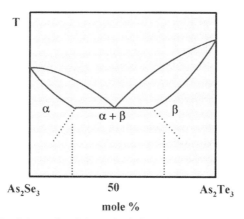

Figure 2.15   Schematic of the typical phase diagram with concentration
for the vitreous As_2Se_3–As_2Te_3 alloys.

CONCLUSIONS

In the conclusive words this work demonstrates the key structural parameters modification that can extensively influenced the technical performances of the devices based on chalcogenide glassy materials. Considering this key issue of such glassy materials, the basic concepts of glass formation based on structural modifications of the alloys composition have been described. The structural transformations of these materials also depend on their infringements, therefore, both structural transformation and infringements formation have been examined with the help of available theoretical descriptions. In a separate segment different kinds of glass formation criterion were studied to define the basic concepts of glass formation in chalcogenide systems. For technological applicability it is also important that these materials should be characterized to define different physical parameters, therefore, it is significant to pay an attention to their illustration by using various existing criterions. With this in mind a segment based on the evaluation of physical parameters of such glassy materials were added. Moreover, determination of a solid as the glassy material is an important property for this kind of system, therefore, short and medium ranges ordering should be defined of such systems for their better technological uses. Hence, a brief description on the short and medium range ordering in chalcogenide glasses have also been incorporated in the separate segments. This includes descriptions on key concepts of rings and isolated molecules in chalcogenide glasses. These kinds of disordered solids experimental physical parameters evaluations are equally important with theoretical interpretations. Therefore, one of the well accepted eutectoidal models for the amorphous semiconductors interpretation has also been discussed. An extensive description of the stable electronic configurations of the glassy state with their experimental verification and physicochemical analysis of vitreous semiconductor chalcogenide systems have also been interpreted in separately .

Hence, a detailed description on chalcogenide glassy materials to define their structural transformations and associated physical properties variations have been presented. The interpretation of structural transformations and their physiochemical parameter changes are based on different theoretical interpretations of glassy materials.

◼ References

1. Uhlman, D.R. (1977) Glass-formation. J. Non-Cryst. Solids, 25, 43–85.
2. Rawson, H. (1967) Inorganic Glass-Forming Systems. Academic Press, London.
3. Tammann, G. (1935) Vitreous State. ONTI, Moscow.
4. Goryunova, N.A., Kolomiets, B.T. (1958) Glassy semiconductors. IV. On the problem of regularities of glass-formation. J. Tech. Phys., 28, 1922–1932.
5. Goryunova, N.A., Kolomiets, B.T. (1960) Glassy semiconductors. IX. Glass-formation in compound chalcogenides based on arsenic sulfide and selenide. Solid State Phys., 2, 280–283.
6. Hilton, A.R., Jones, C.E., Brau, M. (1966) Non-oxide IVA–VA–VIA chalcogenide glasses. Phys. Chem. Glasses, 7, 105–126.
7. Borisova, Z.U. (1972) Khimiya Stekloobraznykh Poluprovodnikov (Chemistry of Vitreous Semiconductors). Leningr. Gos. Univ., Leningrad.
8. Minaev, V.S. (1977) On the problem of prediction of glass-formation regions in ternary chalcogenide systems and some "violations" of glass-formation regularities. Deposit Fund No. DE-2213, Publication of Central Science and Information Institute Electronics, Moscow.
9. Minaev, V.S. (1980) Prediction and seeking for new chalcogenide glassy semiconductors. Electronics Industry, 8–9, CSII Electronics, Moscow, 67–71.
10. Popov, A. (2018) Disordered Semiconductors Second Edition: Physics and Applications. Pan Stantfort Publishing Pte. Ltd.
11. Minaev, V.S., Timoshenkov, S.P., Oblozhko, S.A., Rodionov, P.V. (2004) Glass formation ability: Is the rawson's "liquidus temperature effect" always effective? J. Opto. and Adv. Mater. 3, 791–798.
12. Frankenheim, M.L. (1835) Die Lehre von der Cohäsion umfassend die Elastizitat der Gase, die Elastizitet und Cohärenz der Flüssigen und festen Körper, und die Kristalkunde. A. Schulz und Komp., Breslau.
13. Lebedev, A.A. (1921) On polymorphism and annealing of glass. Proc. State Optic. Inst., Sanct-Peterburg, 2(10), 1–21.
14. Lebedev, A.A. (1924) On annealing of optical glass. Proc. State Optic. Inst., Leningrad, 3(4), 1–24.
15. Zachariasen, W.H. (1932) The atomic arrangement of glass. J. Am. Chem. Soc., 54, 3841–3851.

16. Goldschmidt, V.M. (1926) Geochemische Verteilungsetze der Elemente. VIII. Untersuchungen über Bau und Eigenschaften von Kristallen. I. Math. Naturvidenskapelig Klassen. Skrifter Norske Videnskaps Akad., Oslo.

17. Smekal, A.G. (1951) The structure of glass. J. Soc. Glass Technol., 35, 411–420.

18. Stanworth, J.E. (1952) Tellurite glasses. J. Soc. Glass Technol., 36, 217T–241T.

19. Myuller, R.L. (1940) Structure of solid glasses on electro-conductivity data. Proc. AS USSR. Ser. Phys., 4, 607–615.

20. Myuller, R.L., Baydakov, L.A. ; Borisova, Z.U. (1962) Electro-conductivity of As–Se–Ge system in vitreous state. Bull. Leningrad State Univ., 17(10, Part 2), 94–102.

21. Chuntonov, K.A., Kuznetsov, A.N., Fedorov, V.M., Yatsenko, S.P. (1982) Synthesis of alloys and phase diagram of Cs–Te system. Proc. AS USSR. Non-Org. Mater., 18, 1108–1112.

22. Sangster, J., Pelton, A.D. (1997) The Cs–Se(Cesium-Selenium) System. Journal of Phase Equilibria, 18, 1054–9714. —JPE 18, Article number: 173.

23. Berul, S.I., Lazarev, V.B., Salov, A.V. (1971) Thermo-graphic investigations of Me2X–Sb2X3 phase diagrams. J. Non-Org. Chem. (in Russian), 16, 3363–3366.

24. Pauling, L. (1970) General Chemistry. W.H. Freeman and Company, San-Francisco.

25. Pye, L., Steven, H.J., LaCourse, W.C. (eds) (1972) Introduction to Glass Science: Proceedings of a Tutorial Symposium at the State University of New York. Plenum Press, New York.

26. Mendeleev, D.I. (1864) Glass Production. Sankt, Petersburg.

27. Sosman, R.D. (1927) Properties of Silica. American Chemical Society Monograph Series No. 37, 37, Chemical Catalog Co. Inc., New York.

28. Kobeko, P.P. (1952) Amorphous Substances. AS USSR Publishers, Moscow.

29. Tarasov, V.V., Savitskaya, Ya.S. (1953) Proc. AS USSR, 88, 1019–1022.

30. Myuller, R.L. (1960) Chemical peculiarities of polymeric glassforming substances and nature of glassformation. Vitreous State, AS USSR Publishers, Moscow, pp. 61–71.

31. Myuller, R.L. (1965) Chemistry of solid state and vitreous state. Chemistry of Solid State, Leningrad State University Publishers, Leningrad, pp. 9–18.

32. Poray-Koshits, E.A., Besborodov, M.A., Bobkova, N.M., Brekhovskikh, S.M., Ermolenko, N.I., Mazo, E.E. (1959) Glass Structure. Diagrams of Vitreous Systems. Belarusian State University Publishers, Minsk, pp. 13–35, Chapter I.

33. Kokorina, V.F. (1971) Influence of chemical bond on glass-formation and glass properties. Vitreous State, Nauka Publishers, Leningrad, pp. 87–92.

34. Winter, A. (1955) The glassformers and the periodic system of elements. Verres Refract., 9, 147–156.

35. Sun, K.H. (1947) Fundamental condition of glass-formation. J. Am. Ceram. Soc., 30, 277–281.

36. Rawson, H. (1956) The relationship between liquidus temperature, bond strength and glass-formation. Proceedings of IV International Congress on Glass, Impremenie Chaix, Paris, pp. 62–69.

37. Kumanin, K.P., Mukhin, E.Ya. (1947) On control of crystallization ability of industrial glasses. Optic.–Mech. Ind., 1, 21–25 (Leningrad, USSR (in Russian)).

38. Dembovsky, S.A. (1978) Some aspects of glass-formation in chalcogenide systems. Phys. Chem. Glass, 4, 522–528.

39. Turnbull, D., Cohen, M. (1959) Molecular transport in liquids and glasses. J. Chem. Phys., 31, 1164–1169.

40. Hruby, A. (1972) Evaluation of glass-forming tendency by means of DTA. Czech. J. Phys., B22, 1187–1193.

41. Funtikov, V.A. (1987) Stable electronic configurations and properties of glassy semiconductors. *In*: Matveev, G.M. (ed.), New Ideas and Physics of Glass. Moscow Mendeleev's Chemical–Technological Institute, Moscow, pp. 141–148

42. Dembovsky, S.A. (1977) Empiric theory of glass-formation in chalcogenide systems. J. Non-Org. Chem., 22, 3187–3199.

43. Dembovsky, S.A., Ilizarov, L.M. (1978) Application of empiric theory of glass-formation to calculations of region borders in chalcogenide systems and coordination of melts. Proc. AS USSR. Non-Org. Mater., 14, 1997–2006.

44. Phillips, J.C. (1979) Topology of covalent non-crystalline solids I: Short range order in chalcogenide alloys. J. Non-Cryst. Solids, 34, 153–181.

45. Thorpe, M.F. (1983) Continuous deformations in random networks. J. Non-Cryst. Solids, 57, 355–370.

46. Lucas, P. (2014) Mean coordination and topological constraints in chalcogenide network glasses. In: Adam, J.-L., Zhang, X. (eds), Chalcogenide Glasse. Woodhead Publishing, Cambridge, UK. pp. 58–81.

47. Tatsumisago, M., Halfpap, B.L., Green, J.L., Lindsay, S.M., Angell, C.A. (1990) Fragility of germanium-arsenic-selenium glass-forming liquids in relation to rigidity percolation, and the Kauzmann paradox. Phys. Rev. Lett., 64, 1549–1552.

48. Senapati, U., Varshneya, A.K. (1996) Viscosity of chalcogenide glassforming liquids: An anomaly in the 'strong' and 'fragile' classification. J. Non-Cryst. Solids, 197, 210–218.

49. Lucas, P., Doraiswamy, A., King, E.A. (2003) Photoinduced structural relaxation in chalcogenide glasses. J. Non-Cryst. Solids, 332, 35–42.

50. Stolen, S., Grande, T., Johnsen, H.B. (2002) Fragility transition in $GeSe_2$–Se liquids. Phys. Chem. Chem. Phys., 4, 3396–3399.

51. Angell, C.A. (1991) Relaxation in liquids, polymers and plastic crystals–strong/fragile patterns and problems. J. Non-Cryst. Solids, 131-133, 13–31.

52. Angell C.A. (1995) Formation of glasses from liquids and biopolymers. Science, 267, 1924–1935.

53. Martinez, L.M., Angell, C.A. (2001) A thermodynamic connection to the fragility of glass-forming liquids. Nature, 410, 663–667.

54. Wang, L.M., Velikov, V., Angell, C.A. (2002) Direct determination of kinetic fragility indices of glass forming liquids by differential scanning calorimetry: Kinetic versus thermodynamic fragilities. J. Chem. Phys., 117, 10184–10192.

55. Moynihan, C.T., Easteal, A.J., Wilder, J., Tucker, J. (1974) Dependence of the glass transition temperature on heating and cooling rate. J. Phys. Chem., 78, 2673–2677.

56. Moynihan, C.T., Easteal, A.J., DeBolt, M.A., Tucker, J. (1976) Dependence of the fictive temperature of glass on cooling rate. J. Am. Ceram. Soc., 59, 12–16.

57. Moynihan, C.T., Lee, S.K., Tatsumisago, M., Minami, T. (1996) Estimation of activation energies for structural relaxation and viscous flow from DTA and DSC experiments. Thermochim. Acta, 280–281, 153–162.

58. Wang, L.M., Angell, C.A., Richert, R. (2006) Fragility and thermodynamics in nonpolymeric glass-forming liquids. J. Chem. Phys., 125, 074505/1–074505/8.

59. Moynihan, C.T. (1993) Correlation between the width of the glass transition region and the temperature dependence of the viscosity of high-T_g glasses. J. Am. Ceram. Soc., 76, 1081–1087.

60. Angell, C.A., Green, J.L., Ito, K., Lucas, P., Richards, B.E. (1999) Glassformer fragilities and landscape excitation profiles by simple calorimetric and theoretical methods, J. Therm. Anal. Calorim., 57, 717–736.

61. King, E.A. (2011) Structure and Relaxation in Germanium Selenide and Arsenic Selenide Glasses, Ph.D. thesis. The University of Arizona.

62. Wang, R. (ed.) (2013) Amorphous Chalcogenides: Advances and Applications. CRC Press, Taylor & Francis, Boca Raton, FL.

63. Minaev, V.S. (1982) Classification of glass-formation diagrams of ternary chalcogenide systems. Electronics Engineering Series: Materials, CSII Electronics, Moscow, 7, 47–53.

64. Minaev, V.S. (1987) Periodic law and glass-formation in chalcogenide systems. Collected Reports of All Union Seminar on New Ideas in Physics of Glass, 2, 125–132.

65. Minaev, V.S. (1987) Structural–energy concept of glass-formation in chalcogenide systems. Abstracts of Seminar on Structure and Nature of Metal and Non-Metal Glasses, Udmurtian State University, Izhevsk,USSR, 68.

66. Minaev, V.S. (1991) Glass-Forming Semiconductor Alloys. Metallurgy Publishers, Moscow.

67. Minaev, V.S. (1980) Structural–energy concept of glass-formation in chalcogenide systems. Electronics Engineering Series: Materials, CSII Electronics, Moscow, 9, 39–48.

68. Minaev, V.S. (1981) Glass-formation in the main subgroup of the sixth group of the Periodic Table. Electronics Engineering Series: Materials, CSII Electronics, Moscow, 11.

69. Fedorov, V.M., Chuntonov, K.A., Kuznetsov, A.N., Bolshakova, G.A., Yatsenko, S.P. (1985) Phase diagram of the cesium–selenium system. Proc. AS USSR. Non-Org. Mater., 21, 1960–1961.

70. Minaev, V.S. (1985) Classification of varieties of solid state of substances based on differences of shortrange orders. Abstracts of All-Union Conference "Glassy Semiconductors", AS USSR, Leningrad, pp. 137–138.

71. Biron, E.V. (1915) Phenomena of secondary periodicity. J. Russ. Phys.-Chem. Soc., 47, 964–988.

72. Shchukarev, S.A., Vasilkova, I.V. (1953) The phenomenon of secondary periodicity on the example of magnesium compounds. Bull. Leningrad State Univ., 2, 115–120.

73. Shchukarev, S.A. (1954) The periodic law of D.I. Mendeleev as the main principle of modern chemistry. J. Gen. Chem., (in Russian), 24, 581–592.

74. Minaev JSC, V.S., Timoshenkov, S.P. (2004) Glass-formation in chalcogenide systems and periodic system. Semiconductors and Semimetals, 78, 1–50.

75. Blinov, L.N. (2016) Glass chemistry: Problems, perspectives, and application. Glass Phys. Chem., 42, 429–439

76. Minaev, V.S. (1979) Regions of existence of glassy semiconductor materials in systems A^{VA}–B^{VIA} and their relation with phase diagrams. Electronics Engineering Series: Materials, CSII Electronics, Moscow, 1, 52–60.

77. Minaev, V.S. (1980) Peculiarities of glass-formation in systems $A^{III}A$–$B^{VI}A$ and their relation with phase diagrams structure. Electronics Engineering Series: Materials, CSII Electronics, Moscow, 9, 29–38.

78. Minaev, V.S. (1985) Glass-formation in chalcogenide systems A^{IA}–B^{VIA}. Experimental data and prognosis. Electronics Engineering Series: Materials, CSII Electronics, Moscow, 9, 44–50.

79. Morigaki, K., Kugler, S., Shimakawa, K. (2017) Amorphous Semiconductors: Structural, Optical, and Electronic Properties. John Wiley & Sons, UK.

80. Borisova, Z.U. (1972) Chemistry of Glassy Semiconductors. LGU Publishers, Leningrad, 248.

81. Savage, J.A., Nielsen, S. (1964) Preparation of glasses transmitting in the infrared between 8 and 15 m. Phys. Chem. Glasses, 5, 82–86.

82. Feltz, A., Achlenzig, E., Arnold, D. (1974) Glasbildung im System SnS–GeS–GeS_2 und die Mossbauer Spektren der Glaser. Ztschr. Anorg. und Allg. Chem., 403, 243–250.

83. Feltz, A., Foigt, B. (1974) Bleithiogermanat (II, IV)-Glaser und ihre Eigenschaften. Ztschr. Anorg. Und Allg. Chem., 403, 61–67.

84. Minaev, V.S. (1983) New glasses and some peculiarities of glass-formation in ternary telluride systems. Phys. Chem. Glass, 9, 432–436.

85. Pazin, A.V., Morozov, V.A., Borisova, Z.U. (1979) Glass-formation region and some properties of lead selenide–germanium selenide–arsenic selenide (PbSe–GeSe–AsSe) glasses. Bull. Leningrad State Univ. (22, Part 4), 84–89.

86. Chechetkina, E.A. (2012) Crystallization in glass forming substances: The chemical bond approach. In: Andreeta, M. (ed.), Crystallization. InTech Publishing, London, pp. 1–28.

87. Borisova, Z.U. (1971) Glass-formation in chalcogenide systems and Periodic Table of elements. Proc. AS USSR. Non-Org. Mater., 7, 1720–1724.

88. Minaev, V.S., Timoshenkov, S.P. (2002) On the general regularity of glass-formation in binary chalcogenide (AI–BVI) and oxide (A^I2–B_xO_y) systems where AI–Li, Na, K, Rb, Cs. Pros. of 6th ESG Conference Glass Odyssey, Montpellier, France, CD-ROM.

89. Cervelle, B.D., Jaulmes, S., Laurelle, P., Loireau-Lorach, A.M. (1980) Variation with composition of refraction indexes of lanthanum sulfide and gallium sulfide glasses and indexes of some related glasses. Mater. Res. Bull., 15, 159–164.

90. Huang, S.J., Wang, W.C., Zhang, W.J., Zhang, Q.Y., Jiang, Z.H. (2018) Calculation of the structure and physical properties of ternary glasses via the phase diagram approach. J. Non-Cry. Solids, 486, 36–46.

91. Bletskan, D.I. (2006) Glass formation in binary and ternary chalcogenide systems. Chalcogenide Lett., 3, 81–119.

92. Goncalves, C., Kang, M., Sohn, B.U., Yin, G., Hu, J., Tan, D.T.H., Richardson, K. (2018) New candidate multicomponent chalcogenide glasses for supercontinuum generation. Appl. Sci. 8, 2082, 1–20.

93. Vinogradova, G.Z. (1984) Glassformation and phase equilibriums in chalcogenide systems. Binary and Ternary Systems. Nauka Publishers, Moscow.

94. Mukhin; E.Ya., Gutkina, N.G. (1960) Crystallization of Glasses and Methods of Its Prevention. Gosoboronizdat, Moscow.

95. Minaev, V.S. (1978) Criterion of glass-formation in chalcogenide systems. Proceedings of International Conference on Amorphous Semiconductors—78, AS ChSSR Publishers, Pardubice, Prague, 71–74.

96. Fairman, R., Ushkov, B. (2004) Semiconducting Chalcogenide Glass I. Elsevier Ltd.

97. Batsanov, S.S. (2000) Structural Chemistry. Book of Facts (in Russian). Dialog-MGU Publishers, Moscow, pp. 292.

98. Cornet, J. (1976) The eutectic law for binary Te-based systems: A correlation between glass-formation and eutectic composition. In: Kolomiets, B.T. (ed.), Structure and Properties of Non-crystalline Semiconductors. Proceedings of the Sixth International Conference on Amorphous and Liquid Semiconductors, Leningrad, USSR, Nauka Publishers, Leningrad-1975, 18–24, 72–77.

99. Staveley, L.A.K. (1955) Homogeneous nucleation in supercooled liquids. The Vitreous State, Glass Delegacy of the University of Sheffield, 85–90.

100. Turnbull, D., Cohen, M. (1958) Concerning reconstructive transformation and formation of glass. J. Chem. Phys., 29, 1049.

101. Tammann, G. (1903) Kristalliziren und Schmelzen. Leipzig.

102. Lebedev, A.A. (1910) Proc. St. Peterburg Polytechnic Institute, 13, 613.

103. Vengrenovich, R.D., Lopatnyuk, I.A., Mikhalchenko, V.P., Kasian, I.M. (1986) Production and crystallization of amorphous Te-based alloys. Proc. II All-Union Conf. Material Science and Technology of Chalcogenide and Oxygen-Containing Semiconductors, Chernovtsy, USSR, 1, 150.

104. Glazov, V.M., Chizhevskaya, S.N., Glagoleva, N.N. (1967) Zhidkie Poluprovodniki (Liquid Semiconductors). Nauka, Moscow.

105. Mustyantsa, O.N., Velikanova, A.A., Melnik, N.I. (1971) Electric conductivity of melts in the As–Te system. Zh. Fiz. Khim., 45, 1738–1739.

106. Shmuratov, E.A., Andreev, A.A., Prokhorenko, V.Ya., Sokolovskii, B.L., Balmakov, M.D. (1977) Thermoelectromotive force and electric conductivity of arsenic chalcogenide melts upon high-temperature transition to metallic conductivity. Fiz. Tverd. Tela (Leningrad), 19, 927–928.

107. Tver'yanovich, Yu.S., Borisova, Z.U., Funtikov, V.A. (1986) Metallization of chalcogenide melts and its relation to the glass formation, Izv. Akad. Nauk SSSR, Neorg. Mater., 22, 1546–1551.

108. Tver'yanovich, Yu.S., Gutenev, M.S. (1997) Magnetokhimiya stekloobraznykh poluprovodnikov (Magnetochemistry of Vitreous Semiconductors). St.-Peterb. Gos. Univ., St. Petersburg.

109. Tver'yanovich, Yu.S., Tver'yanovich, A.S. and Ushakov, V.M. (1997) Microregions of cooperative structural transformations at the semiconductor–metal transition in As_2Te_3 melt. Fiz. Khim. Stekla, 23, 55–60 ([Glass Phys. Chem. (Transl.), 23, 36–39]).

110. Karapet'yants, M.Kh. (1975) Khimicheskaya Termodinamika (Chemical Thermodynamics). Khimiya, Moscow.

111. Frolov, Yu.G. (1982) Kurs Kolloidnoi Khimii (A Course of Colloidal Chemistry). Khimiya, Moscow.

112. Fridrikhsberg, D.A. (1984) Kurs Kolloidnoi Khimii (A Course of Colloidal Chemistry). Khimiya, Leningrad.

113. Shchukin, E.D., Pertsov, A.V., Amelina, E.A. (1982) Kolloidnaya khimiya (Chemistry of Colloids). Mosk. Gos. Univ., Moscow.

114. Goroshchenko, Ya.G. (1978) Fiziko-Khimicheskii Analiz Gomogennykh I Geterogennykh Sistem (Physicochemical Analysis of Homogeneous and Heterogeneous Systems). Naukova Dumka, Kiev.

115. Gibbs, G.W. (1928) Longmans Green. New York; Translated under The Title Termodinamicheskie Raboty (1950). Gos. Izd. Tekhniko-Teor. Lit., Moscow.

116. Anosov, V.Ya., Ozerova, M.I. and Fialkov, Yu.Ya. (1976) Osnovy Fiziko Khimicheskogo Analiza (The Foundations of Physicochemical Analysis). Nauka, Moscow.

117. Khimicheskaya entsiklopediya (1992) Chemical Encyclopedia, Bol'shaya Ross. Entsiklopediya, Moscow, Vol. 3.

118. Abrikosov, N.Kh., Bankina, V.F., Poretskaya, L.V., Skudnova, E.V., Chizhevskaya, S.N. (1975) Poluprovodnikovye Khal'kogenidy i Splavy na Ikh Osnove (Semiconducting Chalcogenides and Alloys on Their Base). Nauka, Moscow.

119. Aivasov, A.A., Budogyn, B.G., Vikhrov, S.P., Popov, A.I., (1995) Disorded Semiconductors (Neuporyadochennee poluprovodniki). Vesshaya shkola, Moscow (in Russian).

120. Popov, A.I. and Vasil'eva, N.D. (1990) Ordering criteria for the atomic structure of non-crystalline semiconductors. Fisika Tverdogo Tela, 32, 2616–2622 (in Russian).

121. Tanaka, K., Shimakawa, K. (2011) Amorphous Chalcogenide Semiconductors and Related Materials. Springer Science-Business Media, New York.

122. Vaipolin, A.A., Porai-Koshits, E.A. (1963) Structural models of glasses and the structures of crystalline chalcogenides. Sov. Phys.-Solid State 5, 497–500.

123. Lucovsky, G. (1987) Specification of medium-range order in amorphous materials. J. Non-Cryst. Solids, 97-98, 155–158.

124. Elliot, S.R. (1987) Medium-range order in amorphous materials: Documented cases. J. Non-Cryst. Solids, 97–98, 159–162.

125. Voyles, P.M., Zotov, N., Nakhmanson, S.M., Drabold, D.A., Gibson, J.M., Treacy, M.M.J., Keblinsky, P. (2001) Structure and physical properties of paracrystalline atomic models of amorphous silicon. J. Appl. Phys., 90, 4437–4451.

126. Lucas P., King E.A., Gulbiten O., Yarger J.L., Soignard E., Bureau, B. (2009) Bimodal phase percolation model for the structure of Ge-Se glasses and the existence of the intermediate phase. Phys. Rev. B: Condens. Matter Mater. Phys., 80, 214114.

127. Wang R.P., Smith A., Prasad A., Choi D.Y., Luther-Davies, B. (2009) Raman spectra of $Ge_xAs_ySe_{1-x-y}$ glasses. J. Appl. Phys., 106, 043520.

128. Yang, G., Bureau, B., Rouxel, T., Gueguen, Y., Gulbiten, O., Roiland, C., et al. (2010) Correlation between structure and physical properties of chalcogenide glasses in the As_xSe_{1-x} system. Phys. Rev. B: Condens. Matter Mater. Phys., 82, 195206.

129. Halfpap, B.L., Lindsay, S.M. (1986), Rigidity percolation in the germanium-arsenicselenium alloy system. Phys. Rev. Lett., 57, 847–849

130. Bustin, O., Descamps, M. (1999), Slow structural relaxations of glass-forming Maltitol by modulated Dsc calorimetry. J. Chem. Phys., 110, 10982–10992.

131. Soyer-Uzun, S., Sen, S., Aitken, B.G. (2009) Network vs. molecular structural characteristics of Ge-doped arsenic sulfde glasses: A combined neutron/x-ray diffraction, extended x-ray absorption fne structure, and Raman spectroscopic study. J. Phys. Chem. C, 113, 6231–6242

132. Georgiev, D.G., Mitkova, M., Boolchand, P., Brunklaus, G., Eckert, H., Micoulaut, M. (2001) Molecular structure, glass transition temperature variation, agglomeration theory, and network connectivity of binary P–Se glasses. Phys. Rev. B: Condens. Matter Mater. Phys., 64, 134204.

133. Sen, S., Gaudio, S., Aitken, B.G., Lesher, C.E. (2006) Observation of a pressure-induced frst-order polyamorphic transition in a chalcogenide glass at ambient temperature. Phys. Rev. Lett., 97, 025504.

134. Gjersing, E.L., Sen, S., Yu, P., Aitken, B.G. (2007) Anomalously large decoupling of rotational and shear relaxation in a molecular glass. Phys. Rev. B: Condens. Matter Mater. Phys., 76, 214202.

135. Funtikov, V.A. (1977) Impact of stable electron configurations on the glass-forming capacity of chalcogenide semiconductor alloys, Amorfhyei Stekloobraznye Poluprovodniki (Amorphous and Vitreous Semiconductors). Kaliningradskaya Pravda, Kaliningrad, Russia, 42–50.

136. Funtikov, V.A. (1984) Effect of stable electron configurations on the formation of chemical bonds and structure in chalcogenide vitreous semiconductors, Mezhdunarodnaya Konferentsiya Amorfhye Poluprovodniki-84 (Int. Conf. on Amorphous Semiconductors). Gabrovo, Bulgaria, 165–167.

137. Funtikov, V.A. (1989) An electron configuration model of glass formation, Mezhdunarodnaya Konferentsiya Nekristallicheskie Poluprovodniki-89 (Int. Conf. on Non-Cryst. Semiconductors-89). Uzhgorod, 58–60.

138. Addison, W.E. (1964) The Allotropy of the Elements. Oldbourne, London.

139. Feltz, A. (1983) Amorphe Glasartige Anorganische Fest-Korper. Akademie-Verlag, Berlin.

140. Balmakov, M.D. (1989) On the variety of structures. Fiz. Khim. Stekla, 15, 293–295.

141. Samsonov, G.V. (1965) Influence of stable electron configurations on the properties of chemical elements and compounds. Ukr. Khim. Zh., 31, 1233–1247.

142. Akhmetov, N.S. (1975) Neorganicheskaya Khimiya (Inorganic Chemistry). Vysshaya Shkola, Moscow.

143. Shukarev, S.A. (1974) Neorganicheskaya Khimiya (Inorganic Chemistry), Vol. 2. Vysshaya Shkola, Moscow.

144. Balmakov, M.D. (1988) Defects and fluctuations of structure in disordered systems. Fiz. Khim. Stekla, 14, 801–809.

145. Funtikov, V.A. (1993) On the structure of chalcogenide glasses. Glass Phys. Chem. (Transl.), 19, 111–115.

146. Funtikov, V.A. (1995) Eutectoidal Model of Glassy State of the Substance. Proc. of XVII Int. Congress on Glass, Beijing, China, 2, 256–261.

147. Funtikov, V.A. (1996) Topological model of formation of structure of glasses. Proc. of Int. Symposium on Glass Problems, Istanbul, Turkey, 2, 367–371.

148. Funtikov, V.A. (1996) Electrochemical dissolution of glasses. Proc. of Int. Symposium on Glass Problems, Istanbul, Turkey, 2, 105–108.

149. Funtikov, V.A. (1996) Use of cemical equivalent for estimation of chemical stability of glasses. Proc. of Int. Symposium on Glass Problems, Istanbul, Turkey, 2, 380–382.

150. Funtikov, V.A. (1996) Chemical equivalent and its application as a parameter for estimating the structural features of chalcogenide glasses. Glass Phys. Chem. (Transl.), 22, 215–218.

151. Funtikov, V.A. (1998) A pseudobinary model of the semiconductor–metal transition in vitrifying melts. Glass Phys. Chem. (Transl.), 24, 427–431.

152. Funtikov, V.A. (1998). Equivalent measurement of glasses. In: Proc. of the XVIII Int. Congress on Glass. Section D8 of CD-ROM Proceedings, San Francisco, CA, USA.

153. Comet, J.B. (1976) The Eutectic Law for Binary TE-Based Systems: A Correlation between Glass Formation and the Eutectic Composition, Strukturai svoistva nekristallicheskikh poluprovodnikov (Structure and Properties of Noncrystalline Semiconductors). Nauka, Leningrad, 72–77.

154. Funtikov, V.A. (1990) Eutectoidal Model for the Vitreous State of a Substance, Tezisy konferentsii Stroenie, svoistva i primenenie fosfatnykh, ftoridnykh i khal'kogenidnykh stekol (Proc. Conf. on Structure, Properties and Application of Phosphate, Fluoride, and Chalcogenide Glasses). Latv. Akad-Nauk, Riga, 235–236.

155. Funtikov, V.A. (1991) Eutectoidal Model for the Structure of Noncrystalline Solids, Tezisy II Vsesoyuznoi konferentsii po fizike stekloobraznykh tverdykh tel (Proc. II All-Union Conf. on Physics of Vitreous Solids). Latv. Akad-Nauk, Riga, 39.

156. Porai-Koshits, E.A. (1985) Structure of glasses: The struggle of ideas and prospects. J. Non-Cryst. Solids, 73, 79–89.

157. The Collected Works of G. Willard Gibbs: In Two Volumes (1928). Longmans Green, New York; Translated under the title Termodinamicheskie raboty. (1950). Gos. Izd. Tekhniko-Teor. Lit., Moscow.

158. Zalkin, V.M. (1987) Priroda evtekticheskikh splavov i effekt kontaktnogo plavleniya (The Nature of Eutectic Alloys and Effect of Contact Melting). Metallurgiya, Moscow.

159. Shchukin, E.D., Pertsov, A.V. and Amelina, E.A. (1982) Kolloidnaya khimiya (Chemistry of Colloids). Mosk. Gos. Univ., Moscow.

160. Gutenev, M.S. (1993) On the use of physicochemical analysis with reference to vitreous studies. Fiz. Khim. Stekla, 19, 375–383, ([Glass Phys. Chem. (Transl.), 19, 186–189]).

161. Kasparova, E.S., Gutenev, M.S., Baidakov, L.A. (1984) Dielectric investigation of glasses in the $AsSe_{1.5}$– $AsTe_{1.5}$ system. Fiz. Khim. Stekla, 10, 541–548.

Chapter 3

Introduction of Carbon Nanotubes

INTRODUCTION

Carbon ability to form bond with itself and with other atoms in endlessly varied combinations of chains and rings on basis of the sprawling scientific discipline is modern organic chemistry. Yet known two types of all carbon crystalline structures have been wieldy explored; one is the naturally occurring allotropes diamond and graphite. Despite of the serious efforts of various world leading synthetic scientists to prepare the novel forms such as molecular or polymeric carbon are not much impassive. The significant all carbon structures proposed by Roald Hoffman, Orville Chapman and others. Ultimately, the discovery which breakthrough revolutionized carbon science that came not from synthetic organic chemistry but from the experiments on clusters that formed by the laser vaporization of graphite. Indeed, Harry Kroto and Richard Smalley university of Sussex and university of Rice Houston had different purpose to synthesis of the carbon clusters. Kroto had interested in early 1960s to explore the processing occurring on the surfaces of stars, therefore, he believed the experiments of the vaporization of graphite might be provide significant insides into these processes. On other hand, Smalley had been working since several years on semiconductors like silicon and gallium arsenide. He had also interest to what will happen if one vaporizes carbon. Around 1985 a group of the scientist came together at Rice university with a group of colleagues and students and performed the famous series of experiments on the vaporization of graphite. They were struck with a surprising result. They had analyzed the distribution of gas – phase carbon clusters using mass spectrometry, there result showed that 'C_{60}' was far the dominant species. The dominance of the species became more marked under the conditions such as maximized the amount of time when the clusters annealed in the helium. Even they were successfully demonstrated the experimental findings but not able to provide immediately explanation about the open structures containing 60 atoms. The delightful moment came when they realized a closed cluster containing 60 carbon atoms could have a structure with the unique stability and symmetry. However, further progress in this was slow, the key reason beyond that the amount of C_{60} produced from Kroto–Smalley experiments was minute. It is desired if C_{60} curiosity more than laboratory, some way must be produced in bulk. Ultimately it was achieved using a technique simpler than that of Kroto–Smalley approach. Wolfgang of the Max Planck institute at Heidelberg, Donald Huffman of the university of Arizona and his co-workers used a simple carbon arc to vaporize graphite in an atmosphere of helium and collected the soot which settled on the walls of the vessel. Further dispersion of the soot in benzene produced the red solution, this could be derived the production of beautiful plate like crystals of fullerite.

Since carbon nanotubes are key subject of this book chapter, therefore, some most fruitful scientific research outcomes are outlined here. That discovered by the electron microscopist Sumio Iijima, in 1991, of the NEC laboratories, Japan. According to him molecular carbon fibers can consist of tiny cylinders of graphite that closed at each end with the caps which precisely contains six pentagonal rings. Their structure can be illustrate considering two archetypal carbon nanotubes. That formed due to cutting of a C_{60} molecule in half and a graphene cylinder placed between the two halves. Usually C_{60} divided parallel to one of the three-fold axis to make zig-zag nanotubes, while, C_{60} bisect along one of the five-fold axis to construct the armchair nanotubes. The terms zig-zag and armchair correlates to arrangements of the hexagons around the circumference. Another third class of structure that contains hexagons helical arrangement around the tube axis. But in practice experimentally fabricated nanotubes are usually much less perfect to the structure of the carbon nanotubes due to formation of either multilayered of single layered. Nonetheless, carbon nanotubes have found much attraction of the imagination of physicists, chemists and materials scientists. Physicists have paid much attention due to their extraordinary electronic properties, chemists due to their potential to make nanotest tubes. While, materials scientists due to interest of their amazing properties such as stiffness, strength and resilience. But the key issue with the nanotube research is method that described by Iijima gave relatively poor yield, to make further better research into their structure and properties is a challenging task. A significant advancement came in this area in 1992 when Thomas Ebbesen and Pulickel Ajayan had described a for making gram quantities of nanotubes. This was considered as the serendipitous discovery in the area of nanotube research. They were also tried to make fullerene derivatives and found with the increasing pressure of helium in the arc -evaporation chamber dramatically improved the yield of nanotubes formed in the cathodic soot. This finding gave the ability of nanotubes bulk formation that enormously boost the pace of nanotube research worldwide. Ajayan and Iijima had also paid attention on the molecular containers using the carbon nanotubes and nanoparticles. In this field landmark was demonstrated by them, the nanotubes could be field by the molten materials, hence the moulds can be used to fabricate nanowires. In more precise way the opening and filling the nano tubes have been developed that enable to make a wide range of materials including biological ones which to be placed inside. This kind of opened or filled carbon nanotubes may have fascinating properties for the various applications such as catalysis or as biological sensors. Similarly, the filled nanoparticles can also have a numerous important scientific and technical utility in diverse areas such as magnetic recording and nuclear medicine. A large volume of research of carbon nanotubes have been devoted on their electronic properties. After the discovery of the carbon nanotubes MIT group and Noriaki Hamada and colleagues from Iijima laboratory in Tsukuba carried out band structure calculations of the narrow tubes by using the tight-binding model and demonstrated the electronic properties in terms of both tube structure and diameter. This theoretical demonstration gave a great boost the research in carbon nanotubes. But experimental determination of electronic properties of carbon nanotubes were still suffer from the great difficulties. However, in 1996 the experimental measurements have been carried out for the individual carbon nanotubes; this experimental outcome has confirmed to the theoretical predictions. These experimental and theoretical finding have promoted the speculation that carbon nanotubes could be potential candidate for the future nanoelectronics devices. In sequence explored different properties of the carbon nanotubes mechanical properties determination was also a challenging task, but experimentalist have proved the difficulties. This measurement was carried out using the transmission electron microscopy and atomic force microscopy to demonstrate the mechanical characteristics of carbon nanotubes. The findings were demonstrated with the words mechanical characteristics of carbon nanotubes may be just as exceptional as their electronic properties. Therefore, a great interest has grown to using nanotubes with the different compositions. At present a variety of other possible utility of carbon nanotubes makes them exciting interest in this material. As an intense, several groups have been explored using the idea nanotubes as tips for the scanning probe microscopy with their elongated shapes pointed caps. The high stiffness of the carbon nanotubes would be ideally suited to this purpose. In the experimental verifications in this area have demonstrated with the impassive results. Moreover, carbon nanotube could also be useful for the field emission properties. Hence, the volume of the carbon nanotube research is rising with the astonishing rate and their commercial applications will not be far behind.

Considering impassive and exciting features of the carbon nanotube this study intended to describe the basics of the carbon and its hybridizations with a short note on graphite. The different kinds of carbon nanostructures and fullerene including multi and single walled carbon nanotubes are discussed. A complete section of electronic properties of carbon nanotubes has also been provided. In which the structural parameters, electronic structure, curvature effects of the carbon nanotubes are addressed. The defects in carbon nanotubes may also play in their technical performances, therefore, a separate segment is also devoted on this topic, by describing the distinct types of defects formations in carbon nanotube. That can be create desired defects in carbon nanotube for their specific purpose applications. Moreover, distinct physical properties of the carbon nanotubes are also significant parameters to assess the material technical utility for the general or specific purposes. Information about the crucial mechanical, thermal, optical and electrical physical parameters of the carbon nanotubes are important for their suitable and target applications. Thereby, herein the mechanical properties (including elastic properties), thermal properties (including specific heat and thermal conductivity) optical properties (including selection rules and density of states, antenna effect, absorption transitions, exciton, binding energy and band gap, excitons transitions, metallic nanotubes excitons, exciton size as well as oscillator strength) and electronic properties (including electrical transport in single walled carbon nanotubes, transport properties in multi-walled carbon nanotubes, magneto resistance and superconductivity) have to be taken under discussion. The descriptions of these crucial physical properties of the carbon nanotubes would help to make a concrete view on their possible fascinating utitlites as well as to fabricate distinct possible organic-organic or organic- inorganic, organic-biological etc compositions as per desired scientific purposes to achieve the technical applications goal.

CARBON AND ITS HYBRIDIZATIONS

Carbon is a unique chemical element. It can form a variety of architectures in all possible dimensions from macroscopic to nanoscopic scales. During the last two decades, various new forms of carbon have been revealed. Thus, carbon is one of the most versatile elements in the periodic table that can create a number of compounds by forming different types of bonds. The formed bonds may be single, double and triple bonds by adjoining the number of different atoms in bonding. If we examine carbon ground state (i.e. lowest energy state) electronic configuration $1s^2 2s^2 2p^2$, this reveals that carbon can possess two core electrons (1s), they are not available for chemical bonding and their 2s and 2p four valence electrons can contribute in bond formation [Fig. 3.1(a, b, c)]. Since two unpaired 2p electrons are present, carbon should normally form only two bonds from its ground state.

A possible hybridization can consist by the mixing four atomic orbitals in which one electron of the 2s orbital and three electrons of 2p orbitals may contribute to form four electron possessed sp^3 hybrid orbitals. Each orbital in that is filled with one electron [Fig. 3.1(c)]. To obtain the minimum repulsion, all four hybrid orbitals are optimized in their position in space. Such optimization of the orbitals leads the tetrahedral geometry in which four σ bonds can be formed with the carbon neighbors at an individual angle 109.5° from each other. As a typical example methane (CH_4) molecule may satisfy such a specific bonding arrangement. Since diamond is the three-dimensional carbon allotropic with the atoms arrangement in order to form face-centered cubic crystal structure, it is also called a diamond lattice [Fig. 3.1(a)].

Note that in diamond existing carbon atoms are usually in sp³ hybridization and connect to σ bonds from the four nearest neighbors having a bond length of 1.56 Å. The σ bonding concavity occurs due to the overlapping of the two hybrid orbitals in which each orbital contains one electron. However, diamond is known as the material that can have extreme mechanical properties which originate from the strong sp³ covalent bonding between its atoms, its literal meaning in ancient Greek αδαμασ – adamas 'unbreakable'. A diamond can have the highest hardness values as well as extremely high thermal conductivity among known materials. It is an electrical insulator material with a band gap of ~5.5 eV, and is also transparent for visible light [1, 2].

The second possible hybridization schematic is represented in Fig. 3.1(c), according to this, such an atomic orbital can consist by mixing three and four orbitals electrons. The mixing of one 2s orbital and two 2p orbitals can lead to the formation of three sp² hybrid orbitals, in which each orbital is filled from only one electron. Moreover, the three sp² hybrid orbitals can rearrange themselves to make them far apart, as a consequence, they can form a trigonal planar geometry in which the angle between the individual orbital is around 120°. The other p-type orbitals usually do not mix and are perpendicular to the corresponding plane. Under such a configuration three sp² hybrid orbitals may form σ bonds with the three nearest neighbors as well as side-by-side overlap of the unmixed pure p orbitals which allows to form π bonds between the carbon atoms, this attributes for the carbon–carbon double bond. Typical examples include ethylene (C_2H_4) and benzene (C_6H_6) aromatic molecules sp² hybridization.

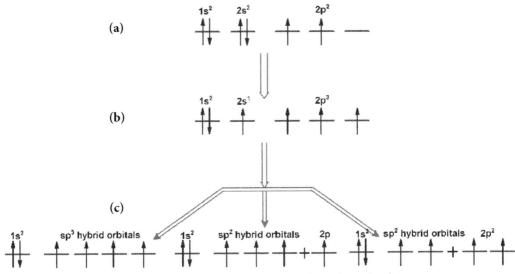

Figure 3.1   Schematic for electronic configurations of carbon:
(a) ground state; **(b)** excited state; **(c)** sp³, sp² and sp hybridized states.

In the three-dimensional crystal, the sp² stacked layer hybridization of the carbon atoms makes the graphite structure. In such hybridized configuration every carbon atom is usually connected to three others making an angle of 120° and bond length 1.42 Å. In such anisotropic structure, it was demonstrated that the presence of strong σ covalent bonds between carbon atoms stay in a plane, while the π bonds can provide weak interaction between adjacent layers. This kind of atomic armament within the crystals can be of good use as pencils owe to their ability to mark surfaces as a writing material due to formation of nearly perfect cleavage between basal planes related to the anisotropy of bonding.

The third and last possibility of hybridization, in which mixing one 2s orbital and one 2p orbital atomic orbitals with four can lead to the formation of two sp hybrid orbitals, where each filled orbital can have one electron [Fig. 3.1(c)]. Their geometry could be in a linear form with the sp orbitals angle of 180°. And the remaining two p-type orbitals cannot mix perpendicularly to each other. Under such a configuration the two sp hybrid orbitals can form σ bonds with the two nearest neighbors, while the side-by-side overlap of the two unmixed pure p orbitals can form π bonds between the carbon atoms. Such configurations are accounted for the carbon–carbon triple bond formation (one σ bond and two π bonds). The most typical example is acetylene (H−C≡C−H) that the linear molecule can satisfy this specific bonding condition. The carbon based materials also have the ability to form one-dimensional chains, typically known as carbynes. They are traditionally classified as cumulene that have monoatomic chains with double bonds, =C=C=. A similar form of polyyne are those that have dimerized chains with alternating single and triple bonds, −C≡C−. Although sp^2 and sp^3 carbon-based materials structures have been extensively synthesized and characterized, but precise synthesis of the carbynes is still challenging due to their high reactivity of chain ends as well as a strong tendency to form interchain crosslinking [3]. Cataldo was the first to synthesize the linear carbon chains up to a few tens of atoms adopting the chemical route, the chain ends stabilization with the nonreactive terminal groups were provided by Kavan and Lagow et al. [4–6]. Though, all the demonstrated systems consisted of a mixture of carbons and other chemical elements. Moreover, a pure carbon environment was also achieved using the supersonic cluster beam deposition as well as the electronic irradiation of a single graphite basal plane (graphene) inside a transmission electron microscope [7–11].

GRAPHITE

Graphite has a thermodynamically stable phase of carbon under the ambient conditions that have been used by mankind for centuries. As an example, graphite was deposited near Borrowdale in the English Lake District as a material for lining the molds of cannonballs. Thus, graphite can be used for an important and very diverse range of applications such as nuclear reactor moderators, pencils, electric motor brushes and addition of carbon to steel making.

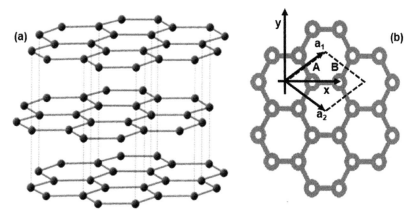

Figure 3.2 **(a)** ABAB stacking atomic structure of graphite, for the most common and thermodynamically stable form; **(b)** Sketch of primitive unit cell (dotted region) and a_1 and a_2 lattice vectors for the graphene.

Graphite atomic structure is the layered material in which each layer consists of a hexagonal lattice of carbon atoms joined through the strong covalent bonds. On the other hand, few bonds between the layers may connect through the weak van der Waals bonds as shown in Fig. 3.2(a). When graphite has a single atomic layer then it is called graphene. The graphite primitive unit cell size depends on how the individual layers stack to form the graphite crystal. Graphite can exist in nature with various stacking arrangements, their most common and thermodynamically stable stacking is known as Bernal (or ABAB) stacking. In such a stacking the B atom in the second layer should be directly above A atom in the first layer, then in the third layer there is an A atom at the location, similar to the first layer. Therefore, the primitive unit cell of Bernal-stacked graphite consists of four atoms in two adjacent layers. The typical graphite crystal structure is exhibited in Fig. 3.2(b).

Since graphites exhibit high anisotropic behavior, the electronic properties are also greatly influenced. In a specific individual layer graphite can have very high conductivity, while in a normal direction of this plane conductivity may be somewhat lower. Theoretically graphite electronic dispersion has been studied for many years prior to the investigation of graphene [12–14]. The tight-binding approximation was considered the most appropriate method to explore the electronic properties of the graphite.

CARBON NANOSTRUCTURES

Carbon nanomaterials can also have a rich polymorphism in their various allotropes by showing the possible dimensionality such as fullerene molecule (0D), nanotubes (1D), graphite platelets and graphene ribbons (2D), nano-diamond (3D), etc. Due to their extraordinary versatility, nanomaterials can have distinct physical, chemical and biological properties, therefore, carbon nanostructures can play an important role in nanoscience and nanotechnology. Specifically, carbon nanostructures are recognized for their excellent electrical conductivity, supreme mechanical strength, high thermal conductivity, extraordinarily high surface area, excellent photoluminescent properties [15], high transparency and structural stability [16, 17]. Such unique properties make carbon nanoarchitectures promising for their applications as thin film transistors [18], transparent conducting electrodes [19], photovoltaics [20], supercapacitors [21], biosensors [22], drug delivery [23], tissue engineering [24], photothermal therapy [25] and biological molecules [26]. In addition to these, a number of carbon-based nanomaterials can possess powerful bactericidal properties toward pathogenic microorganisms. In which the complex bacterial mechanism inactivates due to the intrinsic properties of the nanostructured material. This can directly correlate to their composition and surface modification as well as the nature of the target microorganisms. Thus characteristics environment of the cells and nanostructure material can have an interaction [27].

Therefore, carbon based nanoscience started from the discovery of C_{60} when Kroto et al. introduced Buckminsterfullerene [28]. The Buckminsterfullerene has a cage-like molecule of 7Å in that diameter contains 60 carbons atoms which lay out on a sphere as shown in Fig. 3.3(a). The Buckminsterfullerene C_{60} structure consists in a truncated icosahedron associating with 60 vertices and 32 faces (in which 20 hexagons and 12 pentagons under the absence of the pentagons vertex share), accompanied by a carbon atom at the vertices of each polygon and a bond along each polygon edge. Every carbon atom in this structure is covalently bonded with three others owing to $sp^{2+\delta}$ hybridization (here δ is due to the curvature). Their average bond length around 1.46Å in five-member rings (single bond) and the bond connecting five-member rings is 1.4Å (the bond corresponds to fusing six-member rings). This number of carbon atoms in a distinct fullerene cage can be different. Therefore, the fullerene molecules can be represented by the well-established formula C_n, here n represents the number of carbon atoms existing in a cage. Under this configuration the C_{60} nano-soccer ball (or buckyball) is the most stable, and the member of the fullerene family are well characterized. The name of these molecules C_n was taken from the name of the inventor and architect Buckminster Fuller. The C_{60} molecule still dominates research on fullerene, a large number of stimulation creativities as well as the interest of scientists have paved the way for a whole new chemistry and physics of nanocarbons [29].

Graphitic onions were introduced in 1988 and their first electron microscope images reported by Sumio Iijima in year 1980. In 1999 Harris [30] had claimed that only nest icosahedral fullerenes ($C_{60}@C_{240}@C_{540}@C_{960}...$) [31] can contain the pentagonal and hexagonal carbon rings, this schematic representation is shown in Fig. 3.3(b). In this order the reconstruction of polyhedral graphitic particles into almost spherical carbon onions (nested giant fullerenes) was explored using high-energy electron irradiation inside a high-resolution transmission electron microscope (HRTEM) [32]. Moreover, the formation of C_{60} was also reported *in situ* by the creation of local defects in graphene through the electron irradiation in a HRTEM [33]. According to this interpretation such carbon onions can form quasi-spherical nanoparticles consisting of fullerene-like carbon layers which enclosed the concentric graphitic shells. As a consequence, their electronic and mechanical properties can differ from the other carbon nanostructures owing to its high order structural symmetry.

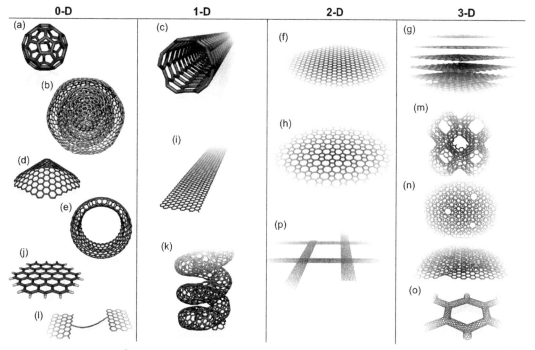

Figure 3.3 Various sp^2-like hybridized carbon nanostructures with different dimensions, like 0D, 1D, 2D and 3D: **(a)** C_{60}: Buckminsterfullerene; **(b)** nested giant fullerenes or graphitic onions; **(c)** carbon nanotube; **(d)** nanocones or nanohorns; **(e)** nanotoroids; **(f)** graphene surface; **(g)** 3D graphite crystal; **(h)** Haeckelite surface; **(i)** graphene nanoribbons; **(j)** graphene clusters; **(k)** helicoidal carbon nanotube; **(l)** short carbon chains; **(m)** 3D Schwarzite crystals; **(n)** carbon nanofoams (interconnected graphene surfaces with channels); **(o)** 3D nanotube networks, and **(p)** nanoribbon 2D networks.
(Reproduce from permission, Terrones et al., 2010, Nano Today 5, 351, copyright @ Elsevier).

In the order of carbon nanostructured materials investigations, one of the important forms of these materials, i.e. carbon fibers or multiwall carbon nanotubes were introduced in 1976, using the modified Chemical Vapor Deposition (CVD) method. Later this method was extensively used for the production of the conventional carbon fibers and their structures usually characterized from TEM analysis [34]. Carbon nanotubes (CNTs) have been present since 1991 after the discovery of C_{60}. Graphite microtubules multiwall nanotubes (MWNTs) were produced by the arc discharge between two graphite electrodes under the inert atmosphere and their characterization from using HRTEM by Iijima [35]. This experimental interpretation had confirmed that MWNTs atomic structures consist of the nested graphene nanotubes which were terminated by fullerene-like caps. After this discovery the single wall carbon nanotubes (SWNTs) were synthesized using the same carbon arc technique in conjunction with metal catalysts by Iijima, Ichihashi and Bethune in 1993, the SWNTs photographs shown in Fig. 3.3(c) [36, 37]. Thus, CNTs are allotropic forms of carbon that can have a long and hollow cylindrical-shaped nanostructure with a length-to-diameter ratio around 10^8 [38]. This length-to-diameter ratio is significantly higher in one-dimensional material.

Therefore, carbon nanotubes can be frequently considered members of the fullerene family in which their ends may be capped from a buckyball hemisphere. Their cylinder walls formed from the carbon rolled up of one-atom-thick sheets with a specific or discrete chiral angle. The nanotube diameter as well as rolling angle leads to specific properties, such as, a SWNT acting as a metal or a semiconductor depending on its geometry [39]. The occurrence of the long-range van der Waals and π-stacking like weak interactions allows the individual nanotubes to naturally align into ropes or bundles [40]. Such carbon nanotubes usually can have valuable unusual nanotechnology properties, such as electronic, mechanical, optical as well as in other areas of materials science. Specifically, the extraordinary mechanical properties, electrical and thermal conductivity of carbon nanotubes have already found various applications with their additives (primarily carbon fiber) in the area of the composite materials, as for the instance, baseball bats, golf clubs, or car parts, etc. [41].

Considering the initial consecutive discoveries of fullerenes and carbon nanotubes several investigators have been made the effort to synthesize and characterize the different kinds of nanostructures [Fig. 3.3 (d-o)] including graphitic-like nanostructures, nanocones [42], nanopeapods [43], nanohorns [Fig. 3.3(d)] [44], carbon rings or toroids [Fig. 3.3(e)] [45]. In this order in 2015 Georgakilas et al. presented a general classification for the different kinds of nanostructured materials based on various recent investigations [46]. This nanostructured materials classification is also based on their diverse crystallographic structures, shapes, dimensions, geometries and chemical bonds. According to this classification carbon based nanostructured materials can be divided into two general groups considering their types of covalent bonds connected with the carbon atoms. In the first category they comprised all the graphenic nanostructures for those predominantly composed sp^2 carbon atoms and compactly crammed or arranged in a hexagonal honeycomb crystal lattice. Additionally, in some instances they could also incorporate some C–C (sp^3 carbon atoms) at defect sites. Several nanostructures have been classified in this category such as, graphite, graphene, other graphene-related and derived materials, carbon nanotubes, nanohorns, onion-like carbon nanospheres, C-dots, carbon-based aerogels, carbon fibres and their composites [46]. Such materials are included in this category due to reasons that all these graphenic nanoallotropes have versatility and the utility of carbon to catenate by forming three indistinguishable covalent bonds with other carbon atoms with the sp^2 orbitals [47]. This leads a common view, in these materials a two-dimensional lattice can formed tightly compact hexagons. As an example, a one-atom-thick sheet of sp^2-hybridized carbon with two-dimensional hexagons, such an arrangement in a hexagonal lattice usually appears in graphene. Here it should be noted that the dimensions of the layer are usually not predetermined but they are usually in the excess of 500 nm.

According to their classification, the second group of carbon nanostructured materials can be made up from the both C–C and C=C (i.e., sp^3 and sp^2 carbon atoms) bonds from the different ratios and contains combinations of amorphous and graphitic regions (or involves primarily sp^3 carbon atoms). Beside a few carbon dots can containing agraphitic (non-graphitic) structures have also be considered in this group. As an instance, nanodiamonds with prominent carbon nanostructures belong to this group. There is a clear distinction from the first category of nanostructured materials with the second category nanoforms, which are usually not fabricated from graphene parts or monolayers, like carbon nanotubes [46]. This distinction between the described two category nanostructured materials can be treated as main regularity rule for them.

Additionally, the carbon nanoallotropes can also be further classified based on their morphological features. For example carbon nanostructures with the internal void spaces; such as, carbon nanotubes, nanohorns, fullerene, could fall into this category. These nanostructured materials empty internal spaces may allow the hosting of foreign or guest molecules, metals, atoms or other nanostructures. Thus, they provide an extensive opportunity to create a new nanoenvironment with certain possible chemical reactions. However, a further second classification of the nanostructured materials is entirely opposite to first group classification. According to this classification approach, not all the carbon nanostructured materials could have internal void spaces, such as, carbon dots, graphene, graphene quantum dots, and nanodiamonds [46]. Moreover, based on these classifications of the nanostructures/ or dimensionality they present a common schematic as shown in Fig. 3.4. with clear distinguishing in zero-dimensional nanoallotropes (like, carbon dots, fullerene and nanodiamonds) and one-dimensional (carbon nanofiber, carbon nanotubes, etc.) as well as two-dimensional nanostructures few-layer graphenes, graphene and graphene nanoribbons) [46].

Hence, in the recent years predominantly controlled manipulation, reduction and modification of sample dimensions into a small number of nanometers has been attracting a lot of attention. Due to the fact that the physicochemical properties of materials on the nanoscale changes considerably from those at a larger scale or in bulk. Thus, nanostructured materials could be designed and developed by using the modified, controlled and size-selective production of nanoscale building blocks under the tuneable and enhanced physical and chemical properties [48].

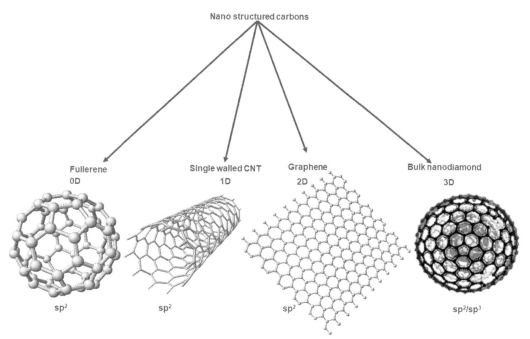

Figure 3.4 Various forms of the potential carbon nanostructures.

Fullerene

A discovery in 1985 that revolutionized the area of the advance materials science changing the carbon face from the synthesis of Buckminsterfullerene (C_{60}) using pulsed laser vaporization of graphite [28]. It demonstrated that carbon fibers at an atomic scale composed of graphite that are thermodynamically stable three-dimensional allotrope of carbon. Usually electrons and phonons of the graphite are associated with three degrees of freedom, while, C_{60} is the zero-dimensional allotrope of carbon. In which electrons and phonons are confined on a microscopic level in all three dimensions. Their lattice is formed from 60 carbon atoms in which some of the hexagons are replaced by pentagons, therefore, the lattice forms a sphere instead of a flat sheet as shown in Fig. 3.5. The smallest possible fullerene diameter is accessed about 0.7 nm and larger diameters of fullerenes were also synthesized by several investigators [49].

Figure 3.5 Buckminsterfullerene (C_{60}), zero-dimensional allotrope of carbon that are made from a graphene lattice in a spherical shape by the replacement of some hexagons with pentagons.

This C_{60} innovation hypothesized [39] that an extremely small carbon fiber can be synthesized with each single pristine layer of graphene and could extend around the fiber and join back on itself. The potential thickness of such a fiber could be just a single layer of graphene. Usually these structures are called multi-walled carbon nanotube (MWCNT), the typical demonstrated diagram is represented in Fig. 3.6 as well as single-walled carbon nanotube (SWCNT) is given in Fig. 3.7(a, b, c). The MWCNTs was introduced and extensively studied by Iijima in 1991 using a laser ablation method that was similar to the synthesis of fullerenes [35], this discovery was the foundation of the synthesis of SWCNTs in 1993 [36, 37].

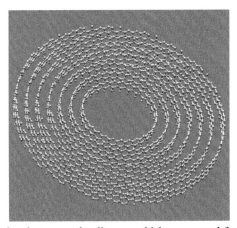

Figure 3.6 Multi-walled carbon nanotube diagram which are created from continuous concentric graphene tubes. The innermost tube is 1.42 nm diameter.

Later it was recognized that MWCNTs had been synthesized several decades prior to Iijima's report in 1991, as well as the discovery of C_{60} in 1985 [28, 35]. This work was done by researchers in the former USSR, but their importance was not recognized due to the non-existence of the term nanotechnology in 1950. It did not find worldwide readership as it was published in the Russian language. However, in 1952 it was recognized after the synthesis of MWCNT by Radushkevich and Lukyanovich, researchers at the Institute of Physical Chemistry, USSR Academy of Science [50, 51]. However, some details may have been lost, as it is known that Radushkevich and Lukyanovich tworked together with Dubinin on the adsorption of carbonaceous materials (materials produced by combustion of coal) on transition metals. They had noticed that in some cases, quite unusual structures were formed, based on a study they published in 1952 [50], they unambiguously identified them (using transmission electron microscopy), now known as MWCNT. At present transition metals are widely used as the catalyst to grow carbon nanotubes (both multi-walled and single-walled) and transmission electron microscopy is still the most convenient direct method to counting the number of graphitic walls that make up in a nanotube.

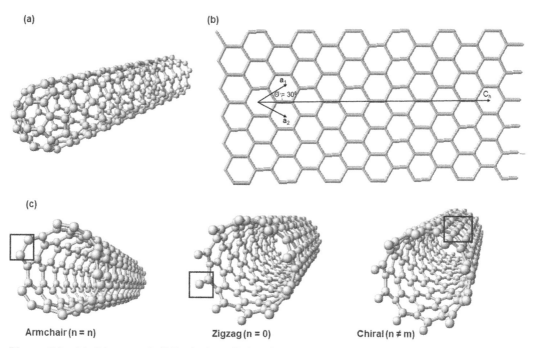

Figure 3.7 **(a)** Diagram of (5,5) single-walled carbon nanotube capped by half of a C_{60} molecule, **(b)** The (5,5) single-walled carbon nanotube section of graphene lattice formation with their lattice vectors (a_1, a_2) and chiral vector C_h and chiral angle θ, **(c)** Schematic diagrams of the achiral armchair $(n = m)$, achiral zigzag $(n, 0)$ and chiral $(n \neq m)$ for the single-walled carbon nanotubes.

Later several published reports also demonstrated the biological activity of fullerene-caged particles [52–54]. To obtain this, various mechanisms were proposed for the bactericidal action of fullerene materials. Specifically, fullerenes and their derivatives were demonstrated as powerful antibacterial activity against a wide spectrum of microorganisms under light exposure [55, 56]. Subsequently, it was stated that bactericidal behavior can be related to the unique structure of the fullerene particle. As fullerene has a closed-cage nanoparticle structural design, in which the conjugation can be extended through-electrons. Such a structural configuration of fullerene perhaps is the main reason to absorb light as well as generate reactive oxygen species [57]. According to this conceptual statement, with increasing time of light exposure on fullerene (C_{60}) it can illuminate the photons, therefore, C_{60} will go under excitation from the ground state to an extremely short-lived (~1.3 ns) excited state. Subsequently, the excited state deteriorates to a lower triplet state that has a longer lifetime (50–100 μs) [45]. Due to presence of molecular oxygen ($^3O_2$) fullerene can produce the Reactive Oxygen Species (ROS), along with the singlet oxygen ($^1O_2$), however, their energy transfer can be from the photochemical pathway, while superoxide anion (O_2^-) may go through electron transfer pathway, as illustrated in Fig. 3.8 [58]. These radicals are short-lived oxidants containing one or more unpaired electrons excited in their highest occupied atomic/molecular level [59, 60]. In such a case, the ROS it is usually accepted to be responsible for eukaryotic lipid peroxidation and eukaryotic cell membrane interruption [55, 61–63]. On the other hand, the high level of ROS can be lethal to microorganisms [62], triggering damage to cellular molecules such as, lipids, proteins and nucleic acids [64]. Surprisingly there are some exceptions like fullerenes particles in dark sites may act as antioxidants to avoid the lipid peroxidation induced by hydroxyl and superoxide radicals [65].

Another antibacterial mechanism is described in terms of the physical interaction between fullerenes and outer microbial membrane, in which the fullerene NPs induce cell membrane disruption and/or DNA cleavage due to high surface hydrophobicity of the particle, that can simply interact with the lipids membrane [56]. As we know that different bacterial species can have dissimilar cell wall components, this may be the reason of dissimilar fullerene–cell interactions. Usually, fullerene particles are more biologically active for the gram positive bacterial species rather than gram negative microorganisms, therefore, the bactericidal success depends on the fullerene insertion into the bacterial cell wall [66, 67].

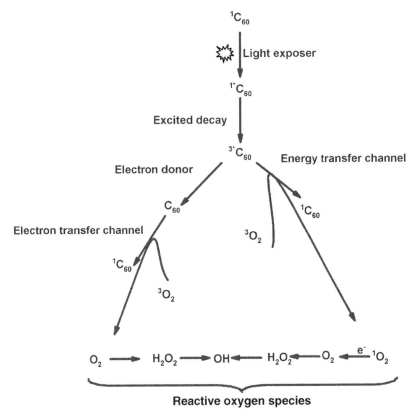

Figure 3.8 Fullerene (C_{60}) schematic representation for their photochemical changes in reactive oxygen species generation.

In terms of physical characteristics, the electrostatic forces between fullerenes and a bacterial surface may play an important role during their interactions. To verify the electrostatic relations in fullerenes *E. coli* and *S. oneidensis* were performed experimentally with four forms of fullerene compounds (C_{60}, C_{60}–OH, C_{60}–COOH and C_{60}–NH$_2$). They demonstrated the positively charged C_{60}–NH$_2$, at low (10 mg/L) concentrations, an acute effect on cellular integrity as well as a reduced substrate uptake for both microorganisms [68]. The neutrally charged (C_{60} and C_{60}–OH) may have a mild antibacterial influence on *S. oneidensis*, while the negatively charged C_{60}–COOH did not have any impact on the growth of either microorganism. This fact established the interaction of positively charged fullerenes with the negatively charged bacterial membranes may be more effective than neutrally and negatively charged fullerene particles [68]. Similarly, several investigators demonstrated that electrostatic attraction can play a major role in the cytotoxic action of fullerene derivatives, owing to this membrane stress mediated by direct physical contacts, however the role of oxidative stress was considered small [55].

The respiratory bacterial chain (located in the membrane) can also be affected from fullerene particles, this could also be considered as a significant bacteriostatic mechanism. It is possible to obstruct fullerene nanoparticles with the cellular metabolism chain to oppose the physical disruption in the bacterial membrane [69]. The higher concentrations of fullerene derivatives probably increase the uptake of O_2, this may trigger an increase in its conversion for the H_2O_2, therefore this interferes with the respiratory chain [70].

In order to investigate the various aspects of such a system, several theoretical approaches (simulations) were proposed to predict the mechanisms for the fullerenes penetrating into the microbial membranes. The molecular dynamic simulations outcomes demonstrated that C_{60} translocation largely depends on the specific lipid structures of the target pathogen. As C_{60} has a minimal tendency to enter homogeneous bilayers of incomplete core lipopolysaccharides, the translocation of C_{60} into bilayers of complete core lipopolysaccharide may not have a thermodynamically favored process. The same simulation revealed that small changes in temperature can also reflect in the ambient ion concentrations, lipopolysaccharide core sugar length or the incidence of phospholipid defects makes a large difference in the interactions between the C_{60} and the surface membranes [71]. It is also important to recognize that the bio-reactivity of nanomaterials with biological targets not only depends on cell wall structure, but is also subjected to cellular enzymes and metabolic activities of the microorganism [72]. Thus, the influence in ambient conditions and microorganism cellular activities may help to understand the inconsistent toxicology.

There are different types of functionalization that are considered for fullerene compounds to control the interactions through biological molecules. The carboxy-functionalized fullerene into the microbial wall has been extensively examined, in which the antibacterial performance initiated through the insertion into the cell wall and led to damaging the membrane's integrity [73]. Subsequently, the antimicrobial activity of fullerene with different functionalities like two C_{70}-derivatives have been created [74], particularly, one with a decacationic side chain (LC17) and another with the same decacationic side chain with an extra deca-tertiary-amine side chain (LC18) [74].

The C_{70} is also highly efficient as a broad-spectrum antimicrobial photosensitizer capable of eradicating six logs of both gram-positive and gram-negative microorganisms. Surprisingly, the attachment of an additional arm can allow moiety to act as an effective electron donor to improve the generation yield of hydroxyl radicals under normal light illumination. It has been noted that the white light can be more bio-active for LC17, while ultra-violate light can be more bio-active for LC18 [74].

Modification in cationic C_{60} from the iodide may develop a powerful bactericidal fullerene. The cationic C_{60}/iodide antimicrobial mechanism can involve the photo-induced electron reduction as $^1(C_{60}>)^*$ or $^3(C_{60}>)^*$ through iodide generation I or I_2, with successive intermolecular electron-transfer actions of $(C_{60}0>)^-$ to get a yield of reactive radicals [75]. Therefore, it is significant to monitor the ability of fullerene materials to generate ROS that may be strongly influenced from the chemical modification of the cage [76]. As an example, usually the rate of ROS (singlet-oxygen) production is slower for functionalized C_{80} compared to unfunctionalized fullerenes [76]. Thus the chemical fictionalization of the fullerene particles can reduce bond angles from 120° in sp^2 down to 109.5° in sp^3, this reduction makes the molecule more stable [77].

Fullerenes also can have a high order of insolubility in water; however, uniformly distributed aqueous suspension may be prepared using fullerene derivatives [78]. It has also been demonstrated the fullerene suspensions (e.g., aggregations of C_{60}) would be able to possess biological activities against microorganisms; it may be distinct from bulk solid fullerene [79]. In the case of the aquatic systems, it was argued that the fullerene particles may not necessarily puncture the microbial cell and cannot generate ROS, but instead exert toxicity as a particle via chemical interactions on direct contact [80, 81]. Regarding this, it was demonstrated that fullerene aqueous solution can be an efficient photo-induced antibacterial agent. Even at a low concentration ($C_{60} = 1.8 \times 10^{-2}$ mM), they are able to effectively suppress the growth of gram-positive microorganisms [82]. Subsequently, it was also demonstrated that C_{70} suspension can be more photoactive than nC_{60} (forming more $^1O_2$

than nC_{60} in the wavelength range 300–650 nm), therefore a consistent ratio of 1.69 ± 0.05 times can create the $^1O_2$ of nC_{60} [83]. In a similar way, the suspension of C_{60}/pyrrolidinium may provide extremely an active broad-spectrum bactericidal photosensitizer. This may be capable to eradicate more than 99.99% of bacterial and fungal cells in vitro once irradiates with white light on it [84]. Therefore, irradiation of dissolved polyhydroxylated fullerene (fullerol) by UV radiation (310 to 400 nm) can significantly increase the inactivation of bacteriophage MS_2 (up to 4 log) owing to ROS generation. While in the absence of UV, fullerol may have limited biological activity due to negligible ROS production [85]. Although, usually soluble functionalized fullerenes has one shortcoming in their absorption range, that they are normally inclined toward the blue to green visible spectrum rather than red/or far red band which can have a better tissue penetration [86]. Similarly, unmodified fullerenes like C_{60} can have high hydrophobicity and innate tendency to aggregate and prevent efficiently photo-activity [87]. Therefore, it is significant to demonstrate that the antibacterial performance of fullerene suspensions may also be influenced from the used preparation methods. As an example, the aqueous fullerene suspensions have been prepared by adopting four different procedures, using tetrahydrofuran (THF) as a solvent (THF/nC_{60}), sonicated C_{60} dissolved in toluene with water (son/nC_{60}), C_{60} powder stirring in water (aq/nC_{60}) and through a solubilizing agent (PVP/C_{60}). All these fullerene derivatives have revealed antibacterial activity toward B. subtilis. This outcome also demonstrated that the THF/nC_{60} could have more effective antimicrobial activities than other preparations owing to their variability in the extraction method [88]. The key fullerene antibacterial activities can be summarize as [87]:

- Fullerene has the capability to induce cell membrane disruption and/or DNA cleavage in microorganisms.
- Fullerenes have the ability to inactivate microorganisms through influencing their cellular energy metabolism chain.
- Under the light illumination, fullerenes usually yields a high rate of ROS, this could increase the antibacterial performance.

Multi-walled Carbon Nanotubes (MWCNTs)

A new class of hollow carbon structured materials were introduced based on the discovery of fullerenes or buckyballs (C_{60}) by Kroto et al. [89]. This new class of materials are composed of 60 C atoms forming a structure resembling a football, this innovation boosted the subsequent nanomaterials revolution. In 1991 Iijima et al. [35] discovered hollow tubes with similar crystalline structures as buckyballs containing multiple shells that provided more ethnicity to these kinds of materials. Later these materials were classified as the multi-walled carbon nanotubes (MWCNTs) [90]. However, according to available literature while it was Iijima's discovery, it was actually Bacon et al. [91] who first reported nanotubes in 1959. But at that time they could not recognize the tubes internal structure and reported it as "filament like graphite with a scroll structure" [92, 93].

Similar work was done by researchers from the former USSR, however they did not receive any significance as at that time (the 1950s) because the concept of nanotechnology did not exist. Therefore, their work did not receive worldwide attention due to above described reasons,however, later it was recognized [94, 95] when different institutions did work together. They had noticed that nanostructures formed some cases are not in usual way, and published different research articles on this topic [94].

Usually internal diameter of a SWCNT is in the range of 0.4–3 nm and the thickness of the wall can be considered to the same as a graphene sheet. However, if the nanotube is comprised of two to 50 coaxial sheets or walls it is defined as a multiwalled carbon nanotube. The MWCNT chirality or helicity is also an important characteristic. The hexagonal arrangement of atoms in CNTs as well as in graphene sheets can be described using the chiral vector, ($\vec{C}_h$), or the chiral

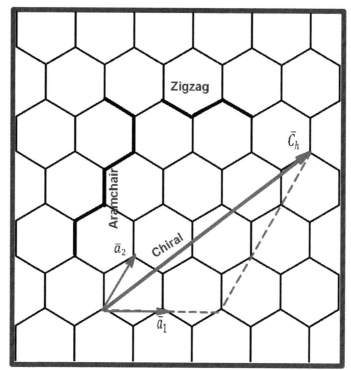

Figure 3.9(a) Schematic representation of a graphene sheet structure with respective to single-walled carbon nanotubes chiralities.

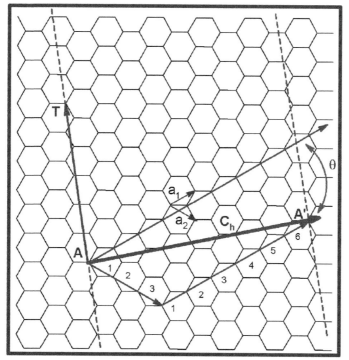

Figure 3.9(b) Schematic for the graphene network, in which lattice vectors are indicated by a_1 and a_2 for the chiral vector $C_h = 6a_1 + 3a_2$. The translational vector T specifies the direction perpendicular of C_h (tube axis). The chiral angle θ in between the C_h and a_1 'zigzag' graphene lattice is indicated.

angle θ, as illustrated in Fig. 3.9. The chiral vector, $(\vec{C}_h)$, can be defined with help of the following equation:

$$\vec{C}_h = m\vec{a}_1 + n\vec{a}_2 \tag{3.1}$$

Here integers n and m are the number of steps along the unit vectors. Fundamentally CNTs are classified into three categories chiral, armchair and zigzag as given in Table 3.1. (a_1 and a_2 are vectors)

Tables 3.1 Carbon nanotubes chirality along with their integers and chiral angle

	Integers (m, n)	Chair angle (θ)
ZigZag	$m \neq n, \quad n = 0$	$\theta = 0°$
Armchair	$m = n \neq 0$	$\theta = 30°$
Chiral	$m \neq n \neq 0$	$0° < \theta < 30°$

The quality of the MWCNTs is of major importance in order to harness their outstanding thermal conductivity. In MWCNTs structural defects such as pentagons [96–99], pentagonheptagon pairs [97–99], vacancies [98, 99], interstitials [97–99], edges [97–99], disorder (e.g. a-C, fullerenes and distorted or incomplete mesoscopic graphite shells) [97–99] and impurities such as metal catalysts are also present, as illustrated in Fig. 3.9. These features have a strong detrimental effect on MWCNT properties in general [97–103].

MWCNTs can be described as, an intermediate material between graphite and SWCNTs. On closer inspection, though, MWCNTs are more complex than either crystalline graphite or SWCNTs, as subtle forms of disorder exist. MWCNTs are composed of both scrolls and cylinders, rather than merely the purely cylindrical structure [97, 104]. However, the most important contributions to MWCNT disorder include tapering cylinders, variable numbers of carbon layers and partial interior filling. The cylindrical crystalline structure can also be strongly compromised by certain additives [97].

Single-walled Carbon Nanotubes (SWCNTs)

Contrast to MWCNTs in history and discovery there is no controversy regarding the discovery of SWCNTs. SWCNTs were discovered independently by two different groups of researchers (Iijima and Ichihashi at NEC Corporation in Japan, Bethune and co-workers at IBM in the United States). Their findings were published simultaneously in Nature in 1993 [36, 37].

Usually the SWCNT, is the tube which is formed from a single graphitic sheet [Fig. 3.7(a)] rolled into a tube and joined back on it. These tubes can be (and normally are) terminated by half of a fullerene molecule. The smallest SWCNT may be capped by half of the smallest possible fullerene molecule, C_{60} (the nanotube having a diameter of 0.7 nm) [39]. However, larger SWCNTs can be capped by larger fullerene molecules.

The key properties of SWCNT can also be defined in terms of the graphene lattice vectors with their structural descriptions. The SWCNT chiral vector circumnavigates the nanotube perpendicular to the tube axis. To form the continuous graphene lattice of the tubes, the chiral vector $(\vec{C}_h)$ must consist of integer (n, m) multiples of the graphene lattice vectors $\vec{a}_1$ and $\vec{a}_2$ and their diameter d_t can be obtained from the chiral vector using Eq. (3.1). Further diameter d_t can be obtained by the relationship:

$$d_t = \frac{|\vec{C}_h|}{\pi} \frac{a}{\pi} \sqrt{n^2 + m^2 + nm} \tag{3.2}$$

The unrolled graphene sheet is exhibited in Fig. 3.7(b). The graphitic sheet lattice vectors and the chiral vector $\vec{C}_h$ are marked, along with the chiral angle θ. Here θ is the angle between the graphene lattice vector $\vec{a}_1$ and the chiral vector. The hexagonal symmetry of the graphene lattice allows the SWCNT to be expressed from a chiral angle in the range $0 \leq \theta \leq 30°$, their empirical values can be obtained from the following relationship:

$$\cos\theta = \frac{\vec{C}_h \cdot \vec{a}_1}{|\vec{C}_h||\vec{a}_1|} = \frac{2n+m}{2\sqrt{n^2+m^2+nm}} \tag{3.3}$$

These nomenclatures were developed to classify the SWCNTs based on their chiral angle θ and chiral indices (n, m). A SWCNT having an angle $\theta = 0°$, $30°$ and corresponding vectors $(n, 0)$, (n, n) are designated as zigzag and armchair. While, a SWCNT having an angle between $0 \leq \theta \leq 30°$ with the vector $(n, m$, where $n \neq m)$ is called chiral. Hence, the names armchair and zigag refer to the shape and cross section of the graphitic lattice circumnavigating of the nanotube. The term chiral represents the SWCNTs category having a mirror image for those are not identical to the original SWCNTs. Therefore, a chiral SWCNTs possessing vector (n, m) can exist in two different forms with the equal appearance of left-handed or right-handed helicity. Such helicity can produce the mirror image of an armchair or zigzag SWCNTs.

SWCNTs structure can only exist on an extremely small scale, because SWCNTs possessing diameter more than ~2.5 nm spontaneously collapse [105, 106]. This is due to a larger SWCNT; the collapse energy saving (under the adjacent walls formation with the flat nanotubes from the van der Waals bonds) exceeds the strain energy under the deformation of the graphitic lattice at the edges of the collapsed tubes. Therefore, only a small diameter (~1 nm) SWCNT (flatter) structure can exist, this makes them an appropiate candidate for electronic use. However, the length of SWCNT can be up to microscopic dimensions. Typically, SWCNTs lengths around 1 μm and it can be in excess of 10 cm [39, 107, 108]. Thus, SWCNTs have an externally high aspect ratio (at least 10^3), therefore, it can be considered as a genuine one-dimensional allotrope of the carbon, under the negligible effect of the fullerene end caps and the effect of the quantum confinement of the electrons in the axial direction.

ELECTRONIC PROPERTIES OF CARBON NANOTUBES

Structural Parameters of the Carbon Nanotubes

The structural parameters of the carbon nanotubes were first introduced in detail by Sumio Iijima describing the "Helical microtubules of graphitic carbon" [35]. He made the concentric cylindrical shells microtubules having a space of about 3.4 Å. This is identical to conventional graphite materials. The synthesized microtubules diameter is in the range from a few nanometers for the inner shells to several hundred nanometers for the outer shells, these days they are known as carbon nanotubes. In the past decade the arc discharge methods with transition metal catalysts was used to synthesize the carbon nanotubes and make a single graphene layer rolled into a hollow cylinder [36, 37]. These newly invented structures are called single wall carbon nanotubes (SWCNTs) having diameters of about 1 nm with an impassive perfect crystalline structure. It was considered as the ultimate carbon-based one-dimensional systems.

The rolled graphene strip single wall carbon nanotube structure can be identified from the concepts of the two equivalent sites on a graphene sheet that contains a chiral vector, as illustrated in Fig. 3.9(a, b). Here the chiral vector can be defined by two integer numbers (n and m) through the relative position of the pair of atoms on the graphene network that form a tube when rolled. By using Eqs. (3.2) and (3.3) the zigzag, armchair nanotubes and achiral tubes structures can described. The structural schematic of the different kinds of nanotubes are illustrated in Figs. 3.10(a, b, c).

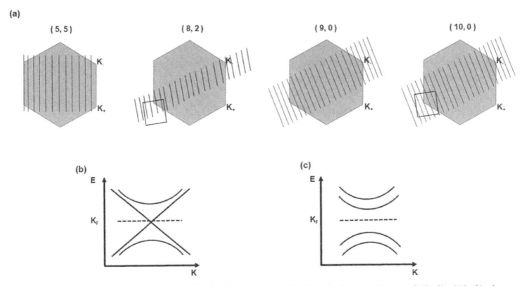

Figure 3.10 **(a)** Schematic structures for (5, 5) armchair, (8, 2) chiral nanotubes, and (9, 0), (10, 0) zigzag. The allowed k-vectors for the same nanotubes mapped onto the graphene Brillouin zone. **(b)** the allowed k-vectors including K point of the graphene Brillouin zone for the metallic nanotubes. **(c)** the semiconducting nanotubes K point with no allowed vector in the energy band gap.

In the structural determination chiral vector also provides the empirical information about the unit cell. Like the translation period along the tube axis can describe the smallest graphene lattice vector $\vec{T}$ on the $\vec{C}_h$. This translation vector linear combination can be shown from the expression:

$$\vec{T} = t_1 a_1 + t_2 a_2 \tag{3.4}$$

By using the orthogonality condition expression $\vec{C}_h \cdot \vec{T} = 0$ can be obtained, therefore, values of t_1 and t_2 can be defined. The length of the translation vector can be expressed as:

$$t = \vec{T} = \sqrt{3a}\sqrt{n^2 + nm + m^2 / N_R} \tag{3.5}$$

Here N_R is the greatest common division factor. The nanotubes unit cell of the cylindrical surface have height t and diameter d_t. The number of carbon atom per unit cell can be defined as:

$$N_C = 4\frac{n^2 + nm + m^2}{N_R} \tag{3.6}$$

Electronic Structure of CNTs

The electronic structure of the CNTs could be described with the help of the single-band tight-binding model of graphene based on the nearest-neighbor approximation as well as other recently developed models like the zone folding approach, etc. Single-band tight-binding model of graphene has received much attention with the zone folding approach owing to their advantages in term of ordered nanolevel electronic transport. To put it simply, these models collectively could demonstrate the prediction of the electronic properties [109]. The concept of the zone folding approach allows to ignore the curvature effects, thereby it could be a good approximation for the large radii ($d_t > 1$ nm) carbon nanotubes. In the zone-folding approach a nanotube is considered as a piece of graphene sheet having periodic boundary conditions along the circum-ferential direction. The mathematic expression can be conveyed as:

$$\psi_k(\vec{r} + \vec{C}_h) = e^{ik\vec{C}_h}\psi_k(\vec{r}) = \psi_k(\vec{r}) \tag{3.7}$$

The vectors $\vec{r}$ and $\vec{C}_h$ should lie on the nanotubes surface. This equation represents equivalence of applying Bloch's theorem. Here boundary conditions describe a quantization of permissible wavevectors around the nanotube circumference, $k \cdot \vec{C}_h = 2\pi q$ (here q is an integer). However, the wavevectors along the axis of the nanotube remain continuous. This allows the wavectors plots in a reciprocal space, thus a set of parallel lines in those directions and spacings that depend on the indices can be left (n, m) as shown in Fig. 3.10(b). Dispersion of each accepted wavevector circumference direction through the cutting dispersion relation of two-dimensional structure along the cutting lines superposition curves gives the electronic of the (n, m) nanotube. A nanotube is called metallic whenever one of the cutting lines crosses the wave vector (K) (either K^+ or K^-) point. Depending on the rule of metallicity by imposing the K allowed wavevector for the given (n, m) nanotube, i.e. $\exp^{i\vec{K}\vec{C}_h} = 1$. Using the boundary condition $K = |K|a_2/a$ with $|K| = 4\pi/3a$, the integer $n + 2m$ should be the multiple of 3, this means $n + 2m \equiv 0 \pmod{3}$. Hence, $3m \equiv 0 \pmod{3}$ for any m, similarly $n - m \equiv 0 \pmod{3}$. Thus, a nanotube can be defined by the (n, m) indices and it will be metallic if $n - m = 3l$, (here l is an integer), while the nanotube is semiconductor when $n - m = 3l \pm 1$.

Usually nanotubes are semiconductors and only a fraction (1/3) of metallic behavior. Their (n, n) armchair nanotubes are always metallic while $(n, 0)$ zigzag nanotubes are metallic when n is multiple of 3. The metallic nanotubes can be in the vicinity of E_F $(k = K + \delta k)$ with the dispersion relation:

$$E_\pm(\delta k) \approx \pm \frac{\sqrt{3}a}{2}\gamma_0|\delta k| \tag{3.8}$$

This equation represents the energy-momentum liner relationship, as shown in Fig. 3.10(c).

Whereas the semiconducting nanotubes conduction and valence bands emerge states with k vectors which located on the allowed lines closest to the K point [see Fig. 3.10(c)]. If the dimensional parameters are selected $n - m = 3l + 1$, this can provide a gap opening at the Fermi level with a magnitude according to the following relationship:

$$\Delta E_g^1 = \frac{2\pi a\gamma_0}{\sqrt{3}|\vec{C}_h|} = \frac{2a_{cc}\gamma_0}{d_t} \tag{3.9}$$

This expression clearly indicates that with the increasing nanotube diameter (d_t), the energy band gap is in a decreasing trend [110]. In case of a large diameter limit one can get a zero band gap semiconductor, as an example nanotube (17, 0) with a diameter 1.4 nm can have $\Delta E_g^1 \approx 0.59$ eV.

Hence, the usual one-dimensional nanotubes zone edges can be denoted from X and X′, their time reversal symmetry X′ = X. As an example band folding of (5, 5) armchair nanotube schematic is given in Fig. 3.11. It should be noted that the K points fold at a distance of $\pm 2\pi/(3a)$ from the point while the zigzag nanotubes can be folded onto the point itself. Therefore, with the help of the described dispersion relationship different kinds of nanotubes (such as (5, 5), (8, 2), (9, 0), etc.) electronic structure can be built.

Curvature Effects

The described electronic structure based CNTs properties can be directly obtained from the confinement of the electrons around the tube circumference by imposing the restriction of the allowed k Bloch vectors. This means that it disregards the curvature effects, however, the curvature is significantly important to further reduce the nanotube diameter. Specifically, in the case of the cylindrical geometry, the carbon atoms may be placed onto a cylindrical wall. This can be produced in the following consequences:

(i)   The perpendicular and parallel C–C bonds of the axis become different; therefore, lengths of the lattice parameters a_1 and a_2 may be different.

(ii) This facilitates the formation of an angle of two p_z orbitals located on bonds renormalization, therefore, the hopping terms γ_0 can exist between a given carbon atom and its three neighbors.

(iii) This creates the broken planar symmetry by mixing π and σ bonds, as a result, it forms hybrid orbitals with those exhibiting partial sp^2 and sp^3 characters.

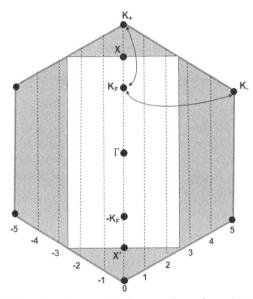

Figure 3.11 Schematic of the Brillouin zone for hexagonal graphene (gray colour) together with the rectangle Brillouin zone (white color) for a four-atom unit cell along. The allowed k vectors for a (5, 5) nanotube lie on the black lines depicted in figure. The (5, 5) nanotube band structure can be computed by folding the corners of the hexagonal Brillouin zone onto the rectangular cell.

Therefore, the effect of finite curvature on the electronic properties on nanotubes can be briefly described as: modifications (i) and (ii) are the conditions at which occupied and unoccupied bands crossing (i.e. at k_F). Owing to this shift, the Fermi vector k_F moves away from the Brillouin zone corners (K point) [111]. The change in curvature can be taken into account for armchair nanotubes shifts. Though, due to symmetry reasons, the metallic nature of armchair tubes remains insensitive to finite curvature. On the other hand, the non-armchair metallic nanotubes, k_F can have a shift away from the K point perpendicularly to the allowed k-lines, due to this the system allows the formation of a small band gap at E_F.

Hence the presence of curvature effects accounted for the sole zero-band gap in the (n, n) armchair nanotubes, when (n, m) tubes $n - m = 3l$ (here l is a non-zero integer) lies into the category of tiny-gap semiconductors. Usually, the armchair tubes are characterized as type I metallic tubes and others type II. Other than these two, the remaining nanotubes belong to the intermediate-gap semiconductors (with gaps a few tenths of an eV). The tiny-gap semiconducting nanotubes may also exist due to the induced secondary gap by curvature. This depends on the tube diameter (according to $\frac{1}{d_t^2}$) and chiral angle [111]. While secondary gap in quasi-metallic zigzag nanotubes (chiral angle = 0) can be assessed with the help of the following relationship:

$$\Delta E_g^2 = \frac{3\gamma_0 a_{cc}^2}{4d_t^2} \qquad (3.10)$$

In the case of metallic tubes at room temperature, the vanishing effect of the curvature is small, therefore, it is usually considered as $n - m = 3l$ tubes. The experimental verification of curvature effect can be directly confirmed from the scanning tunneling spectroscopy density measurement that depends on the zigzag nanotubes diameter $\frac{1}{d_t^2}$, together with the true metallic nature of armchair nanotubes. Their band folding depends on the single-band tight binding approach [109, 112, 113].

Nanotubes in Bundles

Only a special achiral subset of carbon tubes known as armchair nanotubes and their true metallic behavior was predicted. Such single-wall (n, n) nanotubes can occur only in real one-dimensional cylindrical conductors, those having only two open conduction channels. This can be correlated to their energy sub bands laterally confined in a system across the Fermi level. Therefore, with the tubes increasing length, their conduction electrons ultimately become localized owing to the residual disorder in the structure, which produces the interactions between the tube and its surrounding environment. However, it has been theoretically demonstrated that unlike the normal metallic wires, their conduction electrons in armchair nanotubes experience an effective disorder averaged over the tube's circumference. This may lead to electron mean free paths getting larger with the increasing nanotube diameter. Such an increase accounted for the exceptional ballistic transport properties and localization lengths of 10 μm or more for tubes, the corresponding tubes diameters have been typically produced experimentally.

The close-packing of individual nanotubes into ropes cannot significantly change the electronic properties, while theoretical abinitio calculations predicted that broken symmetry of the (10, 10) tube owing to interactions between tubes in a rope induces a pseudo gap about 0.1 eV at the Fermi level [114]. Consequently, the pseudo gap is modified extensively with the several fundamental electronic properties of the armchair tubes. As an example, a semi-metallic-like temperature dependent electrical conductivity and the presence of the finite gap in the infrared absorption spectrum nanotubes bundles.

As discussed above, the isolated (n, n) armchair nanotubes electronic properties can be obtained from their geometrical structure by imposing the two linear $\pi-\pi*$ bands across at the Fermi energy [Fig. 3.12(a)]. Such linear bands can provide the rise in constant density of states near the Fermi level with true metallic behavior. Since atomic structure of an isolated (n, n) nanotube have n mirror planes including the tube axis, the existing π-bonding and π-antibonding states are even and odd under the symmetry operations. This means band crossing is allowed, therefore, the armchair nanotube has a metallic behavior, as represented in Fig. 3.12(a). It should be noted that the symmetry of the isolated (n, n) tube which induces the intrinsic metallic behavior of the tube and their extraordinary ballistic conduction [110].

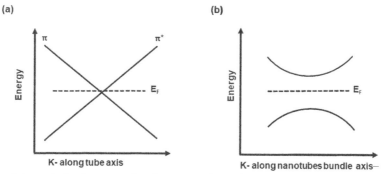

Figure 3.12 Schematic for the energy band structures for an isolated and a bundle *(10, 10)* nanotubes. **(a)** the crossing of the two linear $\pi-\pi*$ bands for the isolated tube, **(b)** the band gap opening due to the breaking of the mirror symmetry.

Breaking this kind symmetry completely alters the structural geometrical picture of nanotube. However, if the tubes in the rope are separated enough to eliminate the individual nanotube interactions, in which case the band structure of nanotubes remains identical. However, if the inter-tube distances are small enough in the bundle, then each nanotube may experience the potential owing to the neighboring tubes [114]. Hence, when perturbation occurs, the corresponding Hamiltonian at point k where the two $\pi-\pi*$ bands crossing, and empirically it can be described as:

$$H_k = \begin{pmatrix} \varepsilon_0 + \delta_{11} & \delta_{12} \\ \delta_{12} & \varepsilon_0 + \delta_{22} \end{pmatrix} \tag{3.11}$$

Here ε_0 is the unperturbed energy and δ_{11} and δ_{22} are the diagonal matrix elements that reflect energy and location shift in k-space at band crossing, while, δ_{12} and δ_{21} are the off-diagonal elements that represent quantum-mechanical level repulsion. The opening of a band gap can be schematically represented as shown in Fig. 3.12(b). In the case of high symmetry vertical line through k, the off-diagonal matrix elements may be zero, as a consequence a crossing may persist. Although the inter-tube interactions may dramatically change the physics of the ropes at a general k point, however, crossing is persevered under the unchanged symmetry of the nanotube bundle (i.e. for (6, 6) armchair nanotubes) [115]. Such inter-tube interactions may break the rotational symmetry of armchair (n, n) tubes due to the local configuration, that can be measured experimentally using low-temperature scanning tunneling spectroscopy [116]. Hence, it reflects the magnitude of the pseudo gap inverse dependence on nanotube radius.

DEFECTS IN CARBON NANOTUBES

The physical properties of carbon nanotubes have been described only based on their geometrical aspect and local environment investigations. However, carbon nanotubes network is not perfect and ideal as was initially considered. This may due to the existence of defects such as pentagons, heptagons, vacancies or dopants. These factors can dramatically modify the electronic properties of the one-dimensional nanosystems. Therefore, the existence of defects in the carbon network provides a way to tailor the intrinsic properties of the tube, and to involve the carbon nanotubes for the novel potential applications purpose in nanoelectronics.

Since carbon nanotubes metallic/semiconducting character is due to their chirality, this sensitive property can be used to form all-carbon metal–semiconductor, semiconductor–semiconductor or metal–metal junctions. These kinds of junctions may have great potential for their applications at the nanoscale dimensions that are made entirely from carbon atoms. The key issue to construct this kind of on-tube junction is the joining of two half-tubes with different helicity seamlessly to each other, without significant cost of energy or disruption in structure. In the several investigations, it was pointed out that the introduction of pentagon–heptagon pair defects into the hexagonal network of a single wall carbon nanotube changed the helicity of the carbon nanotube and altered their fundamental electronic structure [114–118].

Defects can play a significant role when they induce with zero net curvature to prevent the tube from flaring or closing. Due to the fact that the smallest topological defect with minimal local curvature (less energy cost) and zero net curvature exists in a pentagon–heptagon pair. Under this condition when the pentagon attaches to the heptagon, their structure is aniline and creates only topological changes (without net disclination), therefore, it can be treated as a single local defect. This kind of pair creates only a small local deformation in the width of the nanotube, it can also make a small change in the helicity that depends on their orientation in the hexagonal network. As a typical example two nanotubes show the different electronic properties for a single 5–7 pair (Fig. 3.13).

Similarly, nanotubes with the structural parameter (8, 0) can have a band gap 1.2 eV in the tight-binding approximation and (7, 1) tube metallic character (however, a small curvature – induced gap can exist close to Fermi level) [1].

Such joining of the semiconducting and metallic nanotubes through the pentagon–heptagon 5–7 pair can be incorporated in a hexagonal network. This could be the basis of a nanodiode (or molecular diode) for nanoelectronics [1]. A quasi -1D semiconductor -metallic junction schematic is illustrated in (Fig. 3.13). According to this observation within the band folding (7, 1) the half-tube is metallic, while (8, 0) half-tube falls in the semiconducting category. Therefore, such defective nanotubes can fulfill the desired nanoscale metal–semiconductor Schottky barriers, semiconductor heterojunctions, metal–metal junctions for modern nanotechnology with the novel properties. This could also act as building blocks in future nanoelectronic devices.

Investigators have experimentally fabricated some kinds of nanoscale junctions, such as using the beam of a transmission electron microscope illuminate on the local nanostructures. The covalently connected crossed single-wall carbon nanotubes can be created by using the electron beam welding at elevated temperatures [86, 118]. Usually such molecular junction geometries (X, Y and T) are stable after completion of the irradiation process. To explore the relevance of such kinds of created nanostructures, various models have been used to generate molecular junctions. It has been demonstrated that the presence of heptagons can play a key role in the topology of nanotube-based molecular junctions. An ideal X nanotube (5, 5) armchair nanotube intersects an (11, 0) zigzag tube schematic is depicted in Fig. 3.14. To create a smooth topology at the molecular junctions, at each crossing point six heptagons can be introduced [119].

In order to define the local densities of states of the metallic (5, 5) nanotube and the (semi-conducting) (11, 0) nanotube, specific regions can be considered where two nanotubes cross each other to enhance the electronic states at the Fermi level. The presence of localized donor states in the conduction band can be correlated to the existed heptagons. However, a signature of high curvature of the graphitic system in the valence band close to Fermi energy can be occur [119]. Owing to the existence of van Hove singularities for the two achiral nanotubes can be less pronounced in the junction region (Fig. 3.14). Thus, the local density of states of CNT-based junction models reveal their importance in electronic device applications and provide a path towards the controlled fabrication of nanotube-based molecular junctions and the network architectures for the exciting electronic and mechanical behavior.

Figure 3.13   Atomic structure of an (8, 0)/(7, 1) intramolecular carbon nanotube junction. The large light-gray balls denote the atoms forming the heptagon–pentagon pair (Reprinted with permission from L. Chico et al., 1996 Pure carbon nanoscale devices: nanotube heterojunctions, 76, 971, copyright @ American Physical Society).

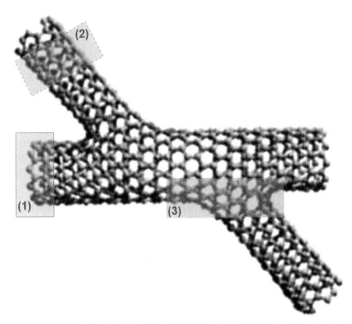

Figure 3.14   Atomic structure of an ideal X-junction, created by intersecting an (5, 5) tube with an (11, 0) tube (Reproduced with permission, Terrones, M. 2002 Molecular junctions by joining single-walled carbon nanotubes, 89, 075505, copyright @ The American Physical Society).

Thus, the ideal defective tubes can have intriguing electronic properties like local density of states with an important enhancement of electronic states close to the charge neutrality point, independent of orientation, tube diameter and chirality. These new nanostructures could offer different advantages compared to the conventional carbon nanotubes in many applications. The defective carbon nanotubes could be more stable than C_{60} with an enhanced cohesive energy of 0.3–0.4 eV/atom, this allows the potential synthesis of this new class of nanotubes. However it remains challenging to synthesize such ideal topologies under controlled nanostructures [120].

PROPERTIES OF CNTs

Mechanical Properties

Mechanical properties are strongly dependent on the structure of the nanotubes. There are three kinds of nanotubes such as bundles of single-wall nanotubes (SWNTs) (sometimes called ropes) arc-grown multi-walled nanotubes (MWNTs) and catalytic multi-walled tubes that always contain structural defects. Usually the measured specific tensile strength of a single layer of a multi-walled carbon nanotube can be as high as 100 times that of steel.

Elastic Properties of CNT

Young's modulus (E) of a material is the primary step to know their structural element for various applications. Usually the structural engineering depends on the theory of elasticity, that is concerned with the design and stresses below the elastic limit, the safe loads being half that of the elastic limit or less [121].

Young's modulus of a solid material can be directly related to the cohesion, hence, the chemical bonding of the constituent atoms. More accurately, a covalently bonded solid potential energy shape of a pair of atoms as a function of inter-particle separation determines the elastic properties in an ideal crystalline solid (i.e. a perfect solid free of point defects, dislocations and grain boundaries). Such crystal associate energy Taylor function expansion $U(x)$ around the equilibrium position r_0 fulfill Hooke's law of conditions for small strains (i.e. force is proportional to displacement) along with the second derivative of the energy at proportionality term r_0. Like an isotropic material thin rod length l_0 and their cross-sectional area A_0, then Young's modulus can be defined from the following relationship

$$E = \text{stree/strain} = (F/A_0)/\delta l/l_0 \qquad (3.12)$$

A molecular solid usually has a low modulus (usually less than 10 GPa) due to the occurrence of weak (typically 0.1 eV) van der Waals bonds, while a covalently bonded solid (like graphite, diamond, SiC, BN...) has a high modulus (higher than 100 GPa). Generally, in each class of solids (defined by the nature of the bonding) the experimental findings have demonstrated that the elastic constants follow a simple inverse fourth power law with their lattice parameter. A small variation in the lattice parameter of a crystal may induce important fluctuations in its elastic constants. As an example, C_{33} graphite (corresponding to the Young's modulus parallel to the hexagonal c-axis) extensively depends on the temperature owing to interlayer thermal expansion [122]. Thus, Young's modulus of a CNT can be related to the sp^2 bond strength that should be equal to a graphene sheet when the diameter is not too small to distort the C–C bonds.

Theoretical Aspect

Several theoretical models have been proposed for the evaluation of Young's modulus of the solid. Therefore, it is important to a view Young's modulus and its dependence on the nanotube diameter and helicity. The different theoretical model's outcome depends on the type of method and the potentials used to describe the interatomic bonding. According to the classical theoretical model of the CNTs, Young's modulus can be written as the second derivative of the strain energy divided by the equilibrium volume [123]. The continuum elastic theory predicts that $1/R^2$ variation of strain energy with an elastic constant is equal to C_{11} of graphite (this corresponds to Young's modulus parallel to the basal plane) and independent of the tube diameter [124]. Thus, according to classical approximation of Young's modulus is not expected to vary when wrapping a graphene sheet into a cylinder. This result was obtained owing to not considering the atomic structure; therefore, the elastic constants are the same in a planar geometry. Though, the classical approximation is more appropriate for the larger diameter CNTs. On the other hand, in the case of the very small diameter tubes the atomic structure and bonding arrangement must be included in a realistic model [125–130]. In these models both ab initio and empirical potential-based methods have been used to calculate the strain energy as a function of the tube diameter (and helicity). A common conclusion was drawn from all models, that only a few corrections in $1/R^2$ behavior are desired. Owing to this corrective function, a small deviation of the elastic constant along the C_{33} (in standard notation) was observed. Thus, it is worth noting that elastic constants of the nanotube dependence on diameter were found to be different from various models. As an example, two contrasting empirical potentials can give divergent values of the elastic constant with different trends of diameter functions. Therefore in standard C_{33}, sometimes a decrease can be experienced, otherwise it shows inverse behavior [131]. Moreover, a collective direct-spun mats model was introduced to overcome the major deficit in stiffness and strength problem of the individual CNTs with bulk CNT materials, considering the key issue that a random, interconnected network of CNT bundles possesses inferior tensile properties to those of individual CNTs [132].

Experimental Interpretation

Experimentally Young's modulus of any material can be measured, using large quantities of material to shape it into a rod. Typical examples include bar, beam and shell models. The precision of the kind of method is high, and experimental findings may be beyond theoretical predictions. The opposite situation may be true for CNTs due to a number of reasons. As the nanometer size of CNTs allows molecular dynamic simulations to be performed and compared directly with experimental data (present and future). Since nanometer size increases the difficulty of experiments and also their precision, usually second-order effects are observed, such as the curvature and the helicity. Lourie and Wagner used the bar model to demonstrate the compressive response by using the micro-Raman spectroscopy [133]. By using this model, they found SWCNT and MWCNT Young's modulus in the range of 2.8–3.6 TPa and 1.7–2.4 TPa, respectively. Yu et al. performed the direct tensile loading tests for the SWCNT and MWCNT [134, 135]. They obtained Young's modulus for the SWCNT and MWCNT in the range of 320 to 1470 GPa and 270 to 950 GPa, respectively.

The cantilevered beam model was experimentally demonstrated by Wong et al. [136], by using the bent individual MWCNT from an atomic force microscope tip. They measured the static response for the analytical solution from a cantilevered beam, and obtained Young's modulus value at 1.28 ± 0.59 TPa. Salvetat et al. used the simple-supported beam model for the deflections of individual MWCNTs and SWCNT ropes [137,138]. They measured the arc discharge grown MWCNTs Young's modulus ~1 TPa, while CNT grown from the catalytic decomposition of hydrocarbons, the modulus is 1–2 order of magnitude smaller. The shear modulus for SWCNTs has also been reported.

Treacy et al. [139] first reported on Young's modulus fitting MWCNT with experimental data. They analyzed the thermal vibration of MWCNT which modeled as a continuous beam. A total of 11 MWCNT's Young's modulus values t were reported in the range of 0.4 to 4.15 TPa with a mean of 1.8 TPa. Krishnan et al. [140] also conducted a similar experimental study on SWCNT, they analyzed the average Young's modulus of 1.3–0.4/+0.6 TPa for the measured 27 SWCNTs. By using the same structural model Poncharal et al. measured the resonance frequency of MWCNTs by driving the resonance of a counter electrode and RF excitation [141]. They obtained Young's modulus approximately at 1 TPa for MWCNT with a radius smaller than 12 nm. Moreover, they also commented that when the resonance response fits with the assumption of a homogeneous resonating beam for larger diameter MWCNTs, a sharp drop in Young's modulus may occur. With this experimental finding, they were able to demonstrate the occurrence of the rippling pattern and its influence on the resonance behavior for the larger diameter MWCNTs. By using this concept, other authors have also examined the ring-pattern buckling in compressively loaded MWCNTs [142, 143].

Structural Instability of CNT

High aspect ratio of the CNT structure can be directly connected to their susceptibility for the structural instability. Yakobson et al. [144, 145] presented a numerical study on the Tersoff–Brenner potential. In the case of compressive loading, a buckling strain of 0.05 was reported, followed by three subsequent buckles on further loading. The bending and torsion governing buckling was demonstrated, that is characterized by a collapse in the cross-section, as a consequence, a kink or ribbon-like structure was observed [144, 146–149].

Despres et al., Iijima et al. and Ruoff et al. presented the experimental study of the buckling for CNT using high resolution TEM (HRTEM), while Wong et al. used AFM instead [136, 150–153]. Falvo et al. and Hertel et al. used AFM as a loading tool to bend MWCNT in the experiment [154, 155]. Lourie et al. presented the embedded CNT into a polymer film to apply both compression and bending. They observed the local rippling in the buckled regions [156].

In the buckling mode, the major factor of contribution is the radial deformability of the tubes. Experimentally Ruoff et al. [157] studied two MWCNT systems adjacent to each other, they concluded that CNTs in anisotropic physical environments (on a surface, near other objects such as other CNTs) are not perfectly cylindrical, this is due to the van der Waals attraction which develops an interfacial region like adhesion layer in biological vesicles. Further, a closest-packed SWCNT crystal was studied by Tersoff and Ruoff [158]. They demonstrated that tubes with diameters smaller than 1 nm are little affected (in their geometry) by the inter-tube interaction. But in the case of tubes with a diameter exceeding roughly 2.5 nm are differently behaves. Moreover, Lopez et al. reported the polygonized SWCNTs in contact [159].

Chopra et al. studied the fully collapsed MWCNTs using HRTEM, while Benedicts et al., Hertel et al. and Avouris et al. explored the radial deformation of MWCNTs on a substrate by both experiment (using AFM) and simulation [158, 160–163]. Yu et al. also reported the fully or partially collapsed MWCNTs on surfaces, including an energetic analysis of the contact [164, 165]. Lordi and Yao presented the simulation and indicated that the radial deformation can be reversible and elastic, depending on the type of CNT [166]. Gao et al. studied the dependence of the cross-sectional shape on the isolated SWCNT diameter [167]. They demonstrated that an essentially circular shape is the stable cross-sectional shape if the radius of CNT is less than 1 nm, between 1 and 2 nm, both near-circular and collapsed shapes may be favored, if the radius larger than 3 nm SWCNT should collapse to a ribbon.

The radial deformation characterization can also be performed from the nano-indentation test using contact mode AFM [168]. The deformability (up to 46%) of the tube and resilience to a significant compressive load (20 µN) has been reported. Yu et al. conducted a similar experiment using AFM in tapping-mode [169]. They found that the elastic constant corresponding to the radial deformation is in the range 0.3 GPa to 4 GPa, by using the Hertzian contact model. The radial deformability of CNT was studied by Chesnokov et al., they measured the volume compressibility (0.0277 GPa−1) and found it to be smaller than graphite (0.028 GPa−1) [170, 171].

In case the CNT completely collapses to a ribbon, then the interlayer interaction is analogous to two stacked graphene sheets. Due to a slight change in the interlayer registration between two surfaces, a metastable configuration might exist. This was demonstrated from Yu et al. [172], according to them a simple energetics analysis reveals the presence of an energy barrier that prevents the twist in MWCNT ribbon from untwisting. Hence in brief, the low-dimensional geometry makes structural instability an important issue for mechanical application of CNTs. Further innovation is needed for understanding the mechanics in partially or fully collapsed CNT and CNT bundles

Strength of CNTs

The strength of a material is not as well defined as that of Young's modulus, as it depends not only on the type of material, but also on its history, the atmosphere, the pressure, temperature and the measuring system (fluctuations in load can modify the strength). It is also closely linked to structural defects and imperfections that are present in the solid, with only a few cases in which materials strengths approach the theoretical limit. Usually two kinds of solids associated with two different types of stress/strain curves can be distinguished. Typically, the first one has brittle conditions which are characterized from the absence of plasticity, i.e. the rupture occurs in the elastic regime. Usually fracture stress of ceramics and glasses exhibits the strength under a brittle breaking mechanism. The second one is the ductile conditions which can be encountered in metals and simple ionic solids. Such materials strength can be associated with the yield stress when the material ceases to act elastically. Therefore, the strength and the breaking mechanisms of a material depend widely on the mobility of dislocations [121].

It is often noticed that the strength of a brittle solid depends on the size of the sample. The graphite whiskers generally exhibit high strength (20 GPa), however, the typical strength for larger fibers is around 1 GPa [173]. This behavior occurs due to the fact that a number of flaws can considerably be reduced in whiskers. According to Griffith's description in brittle solids, fracture occurs through local decohesion at the tip of an extending sharp crack instead of simultaneous bond breaking across the whole fracture plane. This gives the answer why real strengths order of magnitude are lower than theoretical ones.

Temperature is an important parameter in the strength of a material owing to the motion of the dislocations being thermally activated. All covalent materials and CNTs are brittle at low temperature, with any diameter and helicity. It could be demonstrated as, room temperature flexibility of CNTs does not undergo any plastic deformation, however, their high strength and unique capability of the hexagonal network can distort under relaxing stress. This has been well described theoretically as well as experimentally. But the problem remains whether or not plastic flow can occur at high temperature and what kinds of dislocations are involved. Many-body inter-atomic potentials were studied by Tersoff and Brenner, who performed the molecular dynamics simulations on a large number of atoms at high temperature. Their study provided an interesting outcome on the mechanism of strain release under tension [174]. This leads to the spontaneous formation of double pentagon–heptagon pairs in strained nanotubes at high temperatures. Such defects can act as nucleation centers for dislocations which are formed by a single pentagon–heptagon pair whereas the Burgers vector is the primitive vector of the Bravais lattice [175, 176]. The orientation of the easy-gliding line depends on the tube helicity, therefore, it permits the distinct behavior of the high temperature when a tensile strain is applied along the tube axis. Further, Nardelli and co-workers explicitedly demonstrated the behaviour of CNTs under tensile strain at high temperature that depends on their symmetry and diameter [177].

Thermal Properties

Nanoscale graphitic structures such as carbon nanotubes are of great interest not only for their electronic and mechanical properties, but also for their thermal properties. Due to their small size, the quantum effects are important, their low-temperature specific heat and thermal conductivity can be directly related to one-dimensional quantization of the phonon band structure. Theoretical interpretation of low-temperature specific heat allows the determination of the on-tube phonon velocity and the splitting of phonon sub-bands on a single tube with their interaction between neighboring tubes in a bundle. The thermal conductivity of nanotubes can be examined both theoretically and experimentally. Theoretically it was predicted at room-temperature, their thermal conductivity may be larger than graphite or diamond. The experimental evidence has demonstrated that room-temperature thermal conductivity over 200 W/m K for bulk single-walled nanotubes (SWNTs), and over 3000 W/m K for individual multiwalled nanotubes (MWNTs). Moreover, the addition of nanotubes to epoxy resin can double the thermal conductivity for a loading of only 1%, this may directly relate nanotube composite materials to the thermal management and their applications.

Specific Heat of the CNTs

Specific heat of the material is a sensitive probe of the low-energy excitations. In case of the three-dimensional graphite, two-dimensional graphene and nanotubes, phonons are under the dominant excitations, therefore, the phonon specific heat C_{ph} dominates over all the temperatures. The C_{ph} depends on the phonon density of states $\rho(\omega)$, it can be obtained by integrating $\rho(\omega)$ together with the temperature-dependent convolution factor accounting for the temperature-dependent occupation of each phonon state [178]

$$C_{ph} = \int k_B \left(\frac{\hbar\omega}{k_B T} \right)^2 \frac{e^{\left(\frac{\hbar\omega}{k_B T} \right)}}{\left(e^{\frac{\hbar\omega}{k_B T}} - 1 \right)} \rho(\omega) d\omega \qquad (3.13)$$

Here symbols are in well defined manner. According to the relationship at a given temperature T, the convolution factor decreases. At a low temperature, C_{ph} can probe only the lowest energy phonons (i.e. the acoustic modes). Therefore, such dispersion can be expressed in terms of powder law $\rho(\omega) \propto k^\alpha$. Hence for a single mode, Eq. (3.13) can be modified as:

$$C_{ph} \propto T^{(d/\alpha)} \qquad (3.14)$$

here d is the dimensionality of the system, in case of the linear dispersing mode coefficient ($\alpha = 1$). Therefore, the specific heat is linear in T for a one-dimensional system, which follows the Debye T^3 characteristic for a three-dimensional system.

It is well established that isolated nanotubes band structure can be related to phonon density of states [179, 180]. Usually (10, 10) phonons density of states have been demonstrated that is similar to the density of states of a single two-dimensional sheet of graphene. When it is rolled in the form of a nanotube, the two-dimensional bandstructure folds into a large number of one-dimensional sub-bands. The (10, 10) nanotubes three acoustic and three optical phonon bands become 66 separate one-dimensional sub-bands. The experimental evidence can be visualized from the one-dimensional van Hove singularities. However, the overall density of states may be similar at high energies. Therefore, the high temperature specific heat should be roughly equal as well. This is due to high-energy phonons being more reflective of carbon–carbon bonding than the geometry of the graphene sheet. However, at low energies, the geometry of the nanotube substantially differs from the parent graphene sheet owing to the phonon structure. Theoretically the low-energy phonon band structure of an isolated (10, 10) nanotube is interpreted as: four acoustic modes existing. All four can have a linear dispersion, $\omega = vk$, near the zone center. Their longitudinal (LA), transverse (TA) and twist modes can be exist at velocity (v) = 24, 9 and 15, km/sec. Additionally, only the four acoustic modes may be present below 2.7 meV, that can produce a constant density of states. Moreover, comparing to two-dimensional graphene and three-dimensional graphite can show very different low-energy phonon structure [39]. This is due to the fact that a single graphene sheet or rolled sheet has quadratic expansion of energy. Further, the intertube interaction in bundles depresses the low-energy density of states in SWNT bundles.

Theoretically the specific heat of the nanotubes can be calculated using the $C(T)$ equation. The contribution of each acoustic mode to $C(T)$ is can be expressed as [181]:

$$C_{ph} = \frac{\pi k_B^2 T}{\hbar v \rho_m} \tag{3.15}$$

Here ρ_m is the linear mass density. The linear behavior at low T can be directly related to one-dimensional quantized nature of the nanotube phonon band structure. The interlayer coupling (in graphite and intertube coupling (in strongly coupled bundles) depresses the $C(T)$ at low T (around below 8 K). While in the actual samples, the temperature at which the measured $C(T)$ diverges from the single-tube curve gives the intertube coupling.

On the other hand, the electronic specific heat of a one-dimensional metal nanotube possesses constant density of states near the Fermi level. At low temperature the electronic heat capacity is linear with the temperature, it can be obtained from the following relationship:

$$C_{el} = \frac{4\pi k_B^2 T}{3\hbar v_F \rho_m} \tag{3.16}$$

Here v_F is the Fermi velocity. Therefore, the ratio between the phonon and electron specific heat can be expressed as:

$$\frac{C_{ph}}{C_{el}} \approx \frac{v_F}{v_{ph}} \approx 100 \tag{3.17}$$

Thermal Conductivity

Since diamond and graphite exhibit the highest known thermal conductivity at moderate temperatures, similarly nanotubes could also have good in this regard as well. The room temperature thermal conductivity (6600 W/m K) of nanotubes is high. While it's low temperature thermal conductivity should have the effects of one-dimensional quantization probably as specific heat. Therefore, thermal conductivity in a highly anisotropic material is most sensitive to the high velocity and high-scattering-length phonons. Thus, in nanotube bundles the thermal conductivity should directly probe tube phonons and be insensitive to inter-tube coupling [182].

Usually thermal conductivity K is a tensor quality, therefore, it is important to consider the diagonal elements (i.e. in Z-direction):

$$K_{ZZ} = \Sigma C v_z^2 \tau \tag{3.18}$$

Here C represents specific heat and v_z and τ are the group velocity and relaxation time of the considered phonon state. At a low temperature ($T \ll \Theta_D$) a roughly constant relaxation time can be determined from scattering of fixed impurities, defects and sample boundaries. Thus, in simple materials, the low-temperature thermal conductivity has the same temperature dependence as specific heat. While in anisotropic materials, this relationship does not hold strictly. Owing to the contribution of each state being weighed by the scattering time and the square of the velocity, the thermal conductivity preferably samples states with high v and t. However, graphite thermal conductivity parallel to the basal planes that are only weakly dependent on the interlayer phonons. Similarly, nanotube bundles (SWCNT) has $K(T)$ dependence only on the tube phonons, rather than the intertube modes.

Thermal conductivity particularly in low dimensional systems such as one-dimensional system, the ballistic electronic channel due to their quantized electronic conductance can be expressed as:

$$G_0 = \frac{2e^2}{h} \tag{3.19}$$

In the same way, the thermal conductance is independent of materials parameters, existing as a quantum of thermal conductance in a single ballistic one-dimensional channel, their linear relationship with the temperature can be define as:

$$G_{th} = \frac{\pi^2 k_B^2 T}{3h} \tag{3.20}$$

Using this condition Rego and Kirczenow first examined the quantum in detail [183]. Schwab et al. provided the experimental confirmation of this value with the interpretation of the lithographically defined nanostructures [184].

Hence at high temperatures the three-phonon Umklapp scattering begins to limit the phonon relaxation time. It can be recognized from the decreasing phonon thermal conductivity with increasing temperature. Mainly due to the requirement of production of a phonon beyond the Brillouin zone in Umklapp scattering because of the high Debye temperature of diamond and graphite. Therefore, the maxima of the thermal conductivity in these materials can be near 100 K; which is significantly higher than other materials. However, in the less crystalline graphite like carbon fibers, the maximal $K(T)$ can occur at higher temperatures, owing to dominant defect scattering over the Umklapp scattering [185]. Though it is a fact that in low-dimensional systems, it is difficult to conserve both energy and momentum in Umklapp processes [186], it might be possible to suppress Umklapp scattering in nanotubes relative to two-dimensional or three-dimensional forms of carbon.

The yields of the combined contribution of the electrons and phonons can be obtained from the Wiedemann–Franz law relationship [178]:

$$\frac{K_e}{\sigma T} \approx L_0 = 2.45 \times 10^{-8} \left(\frac{V}{K} \right)^2 \tag{3.21}$$

Here σ is the electrical conductivity and K_e is the electron thermal conductivity. In this way, the total thermal conductivity can be measured.

The room temperature thermal conductivity of SWCNT reportedly grows with increasing temperature and a decreasing trend can be observed at high temperature [182]. It also detects the intrinsic thermal conductivity of an individual tube from the measurements. The disordered samples room-temperature thermal conductivity has been measured ~35 W/m K, while in the aligned nanotubes, the thermal conductivity was significantly higher, above 200 W/m K [187, 188].

On other hand, Yi et al. measured $K(T)$ for the bulk samples of MWNTs. They found a roughly T^2 temperature dependence up to 100 K [189]. They obtained room temperature thermal of MWNTs around 25 W/m K. This may be due to effects of tube–tube contacts, or also of the incomplete graphitization in their samples. Moreover, Kim et al. also measured the thermal conductivity individual MWNTs, they claimed that $K(T)$ increases as T^2 up to around 100 K, and maximal peak occurs near the 300 K, above this the temperature decreases [190]. This leads the quadratic temperature dependence for the large-diameter nanotubes which essentially act as two-dimensional sheets. Thus, the room-temperature value of $K(T)$ was over 3000 W/m K.

Optical Properties

As CNTs (SWCNTs and MWCNTs) have diameter-dependent direct band gaps, they are promising for the next-generation optoelectronic components [191]. Since a SWNT is a rolled tubule of single-layer graphene with end-caps and a tube diameter of typically ~1 nm and a tube length of ~1 μm; therefore, it can be considered as a long one-dimensional material. It depends on single layer folded and its characteristics can be described from their chiral vector $\vec{C}_h(n, m)$.

Due to the folded structure of the CNTs, the additional quantization arises from electron confinement around the CNT circumference. Therefore, their periodic conditions should be satisfied $\vec{C}_h \cdot k = 2\pi q$, where q is the integer [191]. As consequence, the electronic band structure of a specific CNT make a superposition on the cuts of the electronic energy bands along the corresponding allowed κ lines (cutting lines), as illustrated in Fig. 3.15(a, b). CNTs can be classified as metallic when one of these cuts contains the Dirac (K) point, otherwise it falls in the semiconducting category, a schematic is represents in Fig. 3.16(a, b).

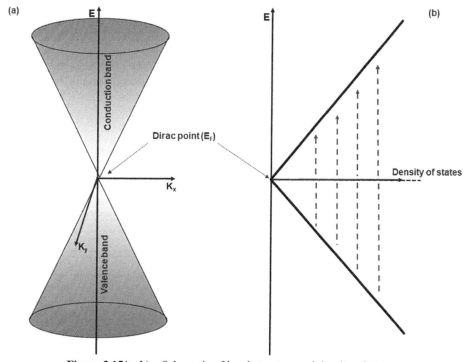

Figure 3.15(a, b) Schematic of band structure and density of state.

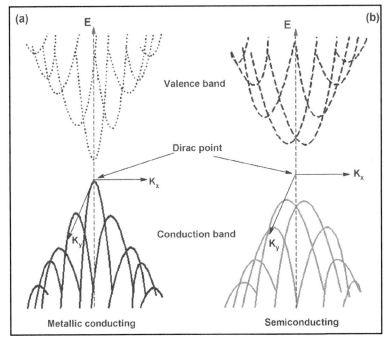

Figure 3.16(a, b)   Schematic for the band structures.

Thus the corresponding electron Valence Band (VB) to the Conduction Band (CB) transition via absorption of one photon of energy $\hbar\omega$ can be described from the band structure and density of the states based on Van-Hove singularities. According to Van-Hove singularities the valance and conductions bands can be expressed as $v_1, v_2, v_3........v_j$ and c_1, c_2, $c_3.....c_j$, at which sharp peaks may occur [192]. The corresponding cuts that creates the additional level of quantization (or sub-bands) which are known as Van Hove singularities depending their polarization (or cross-polarization), as illustrated in Fig. 3.17 [192]. The absorption energy transitions, conduction bands, valence bands and Fermi level are represented here. Specifically the semiconducting and metallic tubes labels are described from S_{ij}, M_{ij} and parallel polarization absorption transitions S_{11}, S_{22}, S_{33}, S_{44}, S_{55} for $p = 1, 2, 4, 5, 7$ and M_{11}, M_{22} correspond to $p = 3, 6$.

Selection Rules with the Density of States

The optical transition between the two states is allowed in the case of photon polarized excitation [193]. If the incident light polarization is parallel to the nanotube main axis, that is a set parallel to z-axis. Then the transition matrix is considered only non-zero for the absorption between VB and CB sub-bands with the same index, while, for the cross-polarization the energy transitions are allowed only between valence and conduction sub-bands with the unity index difference (± 1).

Depolarization Field (or Antenna Effect)

Absorption spectra of a nanotube under Van–Hove singularities both parallel and cross-polarization configurations, specifically, the perpendicular optical absorption is strongly suppressed by the depolarization effect, this is called the antenna effect. The depolarization effect was first described theoretically by combining a tight-binding model and the electrostatic argument [194]. Later Islam et al. observed this phenomenon via absorption measurements on assemblies of aligned nanotubes [195]. They demonstrated that the screened polarizability $\alpha_\perp(\omega)$ can be expressed as a function of the unscreened polarizability $\alpha_{0,\perp}(\omega)$ as:

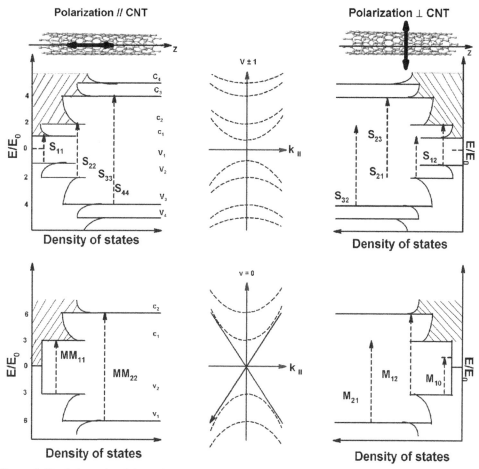

Figure 3.17   Schematic of the carbon nanotubes absorption energy transitions for the parallel ($//$) and perpendicular ($\perp$) light polarization.

$$\alpha_\perp = \left(\frac{1}{\alpha_{0\perp}} + \frac{1}{2Ld^2} \right)^{-1} \tag{3.22}$$

Here L is the nanotube length. The polarizability at frequency ω can be directly related to the absorption cross-section Cabs = k Im(α). Further, Rayleigh experimented at the single nanotube level confirming the strong antenna effect in CNTs [196]. On the contrary, in some photoluminescence experiments weak absorption peaks for perpendicular excitation were detected [197, 198]. Moreover, few groups reported the observation of weaker antenna effect when nanotubes were deposited on various substrates [199–202].

Absorption Transitions in Term of Nanotube Diameter

The total absorption energy transitions yield of ith can be obtained with help of the following relationship:

$$E_{ii} = \frac{p a_0 \gamma_0}{\sqrt{3D}} \tag{3.23}$$

$$P = 1, 2, 3 \ \ldots\ldots\ldots\ldots$$

In this order, the general picture of the absorption in carbon nanotubes is usually described with help of the plain Kataura plot, as depicted in Fig. 3.18. According to this description following conclusions can be drawn:

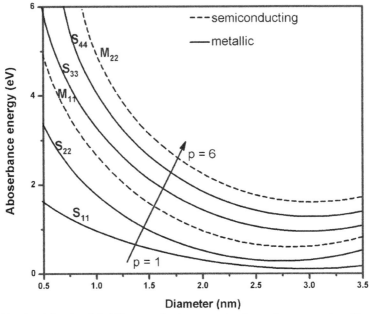

Figure 3.18   Sketch of the simplified Kataura plot between the absorption energy transitions and diameter of the carbon nanotubes. This model has addressed the diameter dependence absorption energy transitions in carbon nanotubes. Considering the linear dispersion of the graphene bands in the K and K valleys, whereas $p = 1, 2, 4, 5$ corresponds to the semi-conducting nanotube $S_{11}, S_{22}, S_{33}, S_{44}, S_{55}$; and $p = 3, 6$ for metallic tube M_{11}, M_{22}.

1. The metallic and semiconducting energy transitions may appear in the order of $S_{11}, S_{22}, M_{11}, S_{33}, S_{44}, M_{22}, S_{55}, \ldots$, that could be conserved.

2. The band gap of the semiconducting CNTs (S_{11}) is in inverse order of nanotubes diameter.. This behavior could be obtained in absorption transitions in both semiconducting and metallic tubes as depicted in the curves in Fig. 3.18. This property may useful to develop an absorption-based characterization tool to discriminate different diameters of the nanotubes. However, the challenge remains in the elaboration of efficient characterization methods to overcome the lack of control in the fabrication process. A large number of techniques have been developed to resolve this problem, notably, the atomic force microscopy, transmission electron microscopy [36], electron diffraction, Raman spectroscopy [203], photoluminescence (PL) [204], absorption/photoluminescence excitation (PLE) [205] and Rayleigh scattering [196].

3. It also led to the conclusion that E_{ii} come closer to each other for the larger diameter nanotubes, therefore, the absorption spectra can weak for the larger diameter nanotubes.

Exciton Properties

Charge carriers confined in one-dimensional systems present strong interactions. Therefore, nanotubes optical properties require taking into account many-body effects. The exciton properties in CNTs are discussed here. To contribute to this Bachilo et al. demonstrated the experiment based on the CNTs chirality [204]. They showed that the absorption and emission properties of SWNTs remain isolated in aqueous surfactant solutions by means of photoluminescence. They indicated that the semiconducting nanotubes are excited in their S_{ii} energy transitions and after non-radiation relaxation processes photo-emission takes place between the first conduction c_1 and valence v_1 sub-bands. They related each peak of the nanotube chirality (n, m), using an empirical Kataura plot that could be obtained [204, 206].

Binding Energy and Band gap (S_{11})

The exciton binding energy band gap was first measured experimentally by two groups in 2005 considering the two-photon absorption technique [207, 208]. According to this concept, the two excitation photons that can form an even state to create an exciton in the $|2p \gg$ state, as a consequence it can relax non-radiatively to the $|1u \gg$ state in PL signal. Thus, the energy difference between them could affect and it can be useful to detect photon energies $E^{2g} - E^{1u}$ which may provide a lower limit for the exciton binding energy E^{exc}. In this order Wang et al. had measured $E_{exc} \approx 420$ meV for the diameter 1.8 nm nanotubes, this corresponds to a third of the band gap (PL signal with energy 1.3 eV) [208]. Maultzch et al. reported slightly smaller values of 325 meV and 240 meV for nanotubes (6,4) ($d = 0.68$ nm) and (7,5) ($d = 0.82$ nm). These significant values are important to compare the typical binding energy recorded for other bulk semiconducting materials (≤ 100 meV) [209]. Although, a single-particle model demonstrated a band gap around 0.9 eV in this diameter range and their self energy E_{e-e} around 700–800 meV was assessed.

Later several theoretical studies tried to describe the optical properties of carbon nanotubes in respect to their structure, including many-body effects [210–213]. Perebeinos et al. proposed the expression for the binding energy:

$$E_b = A_b d^{\alpha-2} m_{eff}^{\alpha-1} \varepsilon_m^{-\alpha} \tag{3.24}$$

Here A_b and α are empirical parameters, ε_m is the effective dielectric constant of the nanotube's vicinity, and m_{eff} the effective mass which depends on the tube chirality [210]. Later the expression $m_{eff} = 2\pi\hbar/2\|C\|v_F$ was modified [214]. Considering that Fermi velocity is identical for all nanotubes, m_{eff} is inversely proportional to the diameter, while, Rayleigh scattering experimental evidence demonstrated that v_F depends on the diameter as well as energy transition S_{ii} [215]. Moreover, including both diameter and chiral angle Capaz et al. provided a detailed description on the exciton properties in chiral nanotubes for transitions E_{11} and E_{22} [213], according to them binding energy yields can be calculated from the following relationship:

$$E_b = \frac{A}{d} + \frac{B + (-1)^\nu \cos 3\theta_c}{d^2} + D \frac{\cos^2 3\theta_c}{d^3} \tag{3.25}$$

Here the first term describes the diameter dependence, while second and third terms can be related to the capture and chirality dependence. In fact the exciton binding energy for higher transitions S_{ii} ($i \geq 2$) is more difficult to access experimentally, however, scanning tunneling spectroscopy yield binding energies of 400 meV and 700 meV for 1.4 nm tubes [216].

Excitons in Transitions (S_{33} and S_{44})

In the beginning the Raman and Rayleigh experiments demonstrated that the third and fourth optical transitions in semiconducting nanotubes can have different behaviors from the S_{11} and S_{22} [217–219]. According to them these transitions are non-excitonic in nature but involve the creation of electron-hole pairs or they have a weak exciton binding energy. Considering this concept, Kane et al. compared the E_{ii} transition to 1 using the following relations [220, 206]:

$$E_{ii} + 0.67 \frac{2p}{3d} \ln\left(\frac{6d}{2p}\right) \tag{3.26}$$

In the experimental evidence it was demonstrated that the first and second transitions in semiconducting CNTs are well defined in contrast to S_{33} and S_{44} transition with the diameter between 1.5 and 2.5 nm. However, many experiments have been conducted to verify the higher order excitons, and photophysics of higher-order optical transitions is debatable due to their importance in the application for nanotubes in the $d \sim$ 1.5–3 nm.

Thus the excitonic nature of higher order transitions in semiconducting nanotubes occurred when bigger exciton binding energy in comparison to lower energy transitions (S_{11}, S_{22}). However, the broadening of the absorption resonances can be related to smaller exciton lifetime.

Metallic Nanotubes Excitons

Since in bulk metallic materials, the free carrier screening of the Coulomb potential prevent the formation of a pair electron-hole, however, it is not the same in carbon nanotubes due to their one-dimensional geometry and the formation of excitons [221, 224]. Wang et al. first measured the exciton in metallic nanotubes considering the absorption of a (21, 21) armchair nanotube with the help of spatial modulation spectroscopy [223]. They found that M_{22} transition (due to the armchair nature, no splitting can exist) optimally fitted including a constant background, a Lorentzian, and an approximated continuum profile.

Hence, the difference between the exciton peak central position and the onset of the continuum gives an estimate of the binding energy, and the measured value of $\sim$50 meV to confirm the weakening of Coulomb interactions in metallic tubes (free carrier screening is orders of magnitude weaker than in bulk).

Exciton Size and Oscillator Strength

Since size (l_χ) directly affects the exciton oscillator strength (f_C) and the radiative lifetime. In one-dimensional materials f_C is inversely proportional to l_χ, and $l_\chi \propto d$ might be the first order approximation with the additional correction accounting in the chirality [66, 78, 96], and their lifetime can be directly related to the inverse of f_C. Capaz et al. calculated the exciton sizes between $\sim$1 and 2 nm for diameters range 0.5–1.0 nm [78], Their estimation provided a good estimate ($l_\chi \sim$ 1 nm) of the value l_χ = 2.0 $\pm$ 0.7 nm [68]. Moreover, Nugraha et al. proposed an advanced model for the study of exciton transitions including transition-dependent exciton size and environmental dielectric screening [212]. In this regard, Choi et al. derived an explicit formula for f_C applicable to relatively large diameter semi-conducting nanotubes. Therefore, the oscillator strength per atom of the $|1u\gg$ for the singlet exciton then it can be correlated as:

$$\frac{f_C}{N_a} = \frac{G(1 + J\xi \cos(3\theta_c)p/d)}{(p+Q)d} \tag{3.27}$$

and

$$l_\chi \sim d\left(\frac{1+Q}{p}\right) \tag{3.28}$$

Here N_a is Avogadro's number, G = 0.29 nm, J = 0.047 nm, Q = 7.5, and $\xi = (-1)^{\mathrm{mod}(n-m,3)+2p/3+\mathrm{mod}(p,3)/3}$. The FWHM of the exciton wave function in real space scales (d/p) can be seen in Fig. 3.19. They demonstrated that the Coulomb potential may be stronger for both high-order energy transitions and larger tubes. According to the first approximation (not considering chirality), f_C increases for both small d and lower transitions p, it indicates that oscillator strength can be damped as the electron-hole Coulomb interaction is enhanced [96, 212]. While chirality can be correlated to the second term of Eq. (3.27), therefore, at the end only the scaling relation $f_C \propto d^{-1}p^{-1}$ may not sufficient to describe the evolution of f_C in SWNTs.

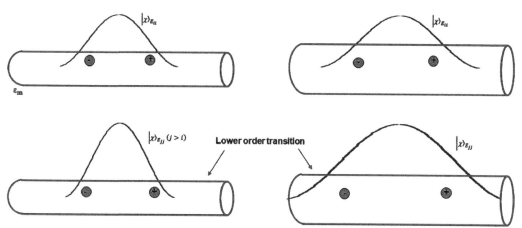

Figure 3.19 Sketch for the exciton size evolution with diameter and energy transition. The exciton wavefunction $|\chi>|$ represents in real space for different resonance S_{ii}. The exciton size can be measured from the FWHM of $|\chi>|$.

Electrical Properties

Electrical Transport in SWCTs

The adequate band structure of the SWCNTs allows to define the energy dependence Drude conductivity σ_{2D} in terms of their surface constitutes, i.e. $\left(\dfrac{2e^2}{h}\right)\dfrac{E}{\hbar v_F}l_e$. In which the elastic scattering length (l_e) of the carriers is proportional to the electron-phonon scattering and it usually increases with decreasing temperature (i.e. $l_e \propto T^{-P}$, for $P > 1$).Usually such characterizations of the electrical conductivity follow two schemes:

(i) *Low Temperatures* ($k_BT < E_F$)
In this, the electrical conductivity of the energy (E) is replaced from the Fermi energy (E_F). Such conductivity can be a category of metallic conductivity. At the finite zero-temperature the magnitude of this can have static disorder [225].

(ii) *High Temperatures* ($k_BT > E_F$)
In this condition system electrical conductivity Energy (E) is replaced from the term k_BT. The electrical conductivity and the charge density become directly proportional to temperature (T). Therefore, it is not significant to measure the intrinsic resistance of a SWNT, because contact addition of the two ends of the carbon nanotubes may be destroyed during the performance [226]. According to theoretical interpretations, for a one dimensional SWCNTs the Landauer formula has predicted an intrinsic resistance (R_{int}^o) is independent of the length that equals to $\dfrac{h}{e^2}\dfrac{1}{T(E_F)}$, with the corresponding resistance $25.8\,\Omega$, in case of perfect transmission through ideal Ohmic contact. This kind of contact resistance occurs only through an intrinsic mismatch between the external contacts to the wire (depending on dimensionality) and the one-dimensional nanotube system. Therefore, both the two-fold spin and band degeneracy of a nanotube with intrinsic resistance (R_{int}) becomes $\dfrac{h}{4e^2}\dfrac{1}{T(E_F)}$ in an individual case. When one incorporates Transmission (T) through the contacts into the one-dimensional channel and then to the next contact, in this case $T = \dfrac{l_e}{l_e + L}$, where l_e is the mean free path length of the scattering and L one-dimensional conductor length [226]. Thus, the resistance is equal to:

$$\frac{h}{4e^2}\frac{l_e+L}{l_e} \equiv \frac{h}{4e^2}\left(1+\frac{l_e}{L}\right) \tag{3.29}$$

Here first term represents the R_{int} and the second term can be related to Ohmic resistance (R_{Ohint}) associated with scattering. In the presence of dynamically scattering impurities, the acoustical optical photons are inevitably present at any temperature above 0 K, therefore, the Ohmic resistance could be considered. In the case of large and small mean paths $l_e \to \infty$ and $l_e \to 0$ a ballistic situation can occur and the Ohmic resistance almost vanishes. However, the material resistance of the contact also has an additional term (R_c), therefore, the total resistance in the external circuit is equal to $= R_{int} + R_{Ohmic} + R_c$. This leads maximum resistance in a MWCNT that can be equal to $\frac{h}{4e^2}$(~ 6.5 kΩ), in the single channel conductivity. But due to the fact that, in practice imperfect contacts (which lead to $T < 1$) and the presence of impurities always lead to a larger resistance values being achieved, while deviations in dimensionality or multiple channels of conduction (MWCNT) may lead to smaller values of the resistance.

Transport Characteristics in MWCNTs

Electrical conductivity of MWNTs has been described considering the tube diameter (d_1) is smaller than the elastic mean free path (l_e), therefore, one-dimensional ballistic transport predominant. While, when d_1 is larger than l_e, the current flow could be diffused to a two-dimensional transport. The additional important quantity is the phase coherence length (l_ϕ), which can be determined from the effective experiment by exploiting the Aharonov-Bohm effect for the MWCNTs tube diameter ~ 250 nm or more [227]. Though value of the (l_ϕ) can be obtained from the direct I–V measurement for the tube diameter ~ 20 nm, but a discrepancy may appear in terms of poor Ohmic contacts. Another factor which may produce the difference, i.e. the quality of the MWNTs; such as the higher temperature synthesized (arc-growth) MWCNTs that exhibit metallic temperature dependence, their resistance linearly decreases with temperature. Therefore, the common low temperature SWCNTs Coulomb Blockade effects are not particularly relevant for MWNTs.

Quantized conductance in a single MWCNT multiples [$G_0(2e^2/h)$] and ballistic transport has been observed at room temperature, by a Scanning Probe Microscope (SPM) [228]. However, the 2G conduction could be measured in the absence of a magnetic field and spin degeneracy may be resolved through electron lattice structure coupling. Another experiment of the MWNT was grown *in situ* on a tungsten contact to get better contact resistance and probed with a W tip, a conductance of up to 490 G_0 [229]. A characteristic of multi-channel quasi-ballistic transport was observed. Usually in measurements nanotubes placed on/below metal electrode contacts on substrates suffer from non-reliable Ohmic contacts, which is a recurring theme in electrical characterization. While, in the STM lithography measurement at low temperature (~ 20 mK), the electron interference typically effects disordered conductors that are present in the transport characteristics [230]. A logarithmic decrease of the conductance with temperature up to saturation was observed, this could be correlated to two-dimensional weak localization effects [231]. Also the evidence of a localization phenomenon in a negative Magneto-Resistance (MR) at a low value (<20 nm) for the l_ϕ was also noticed [232]. Additional, in the doped MWCNTs, the Fermi energy (E_F) may shift by changing the total number of participating conduction channels. These localization effects have also been confirmed in MWCNTs [233].

Magneto-Resistance

Carbon nanotubes Magneto-Resistance (MR) can be described from electron-electron interactions measurements. In the typical MWNT MR measurement magnetic field could be perpendicular [234]. The outcome could be related with their fits of the lines and one-dimensional weak localization theory to calculate the conductance change ($\Delta G \approx (2e^2/h)\, l_\phi/L$ of the phase coherence length (l_ϕ). In the strong temperature dependence the conductance could be G proportional to T^δ, where $\delta < 0.3$. According to Onsager reciprocity relations theoretical concept the two-terminal resistance could be symmetric in a magnetic field while in the four terminal resistances it may be asymmetric. In the MR measurements the Universal Conductance Fluctuations (UCF) have also been recognized with the magnitude e^2/h and quantum interference effects [235]. The disorder in the nanotube could also be the cause of the UCF, this could be likely due to chaotic scattering of the electrons, in a nanotube cavity. Moreover, in contrast to parallel magnetic field, in the perpendicular field the weak localization effects monotonically could disappear ($\Delta B = (h/2e)1/A$.

Superconductivity

Superconductivity in one-dimensional SWNT is due to the reversal of the normally repulsive interactions of the Luttinger liquid. This may happen when the curvature of CNTs leads to the creation of new electron-phonon scattering channels and consequent agreeable electron phonon interactions [236, 237]. It was also proposed that the superconducting transition temperature (T_c) could enhance from chemical doping of nanotubes, such as at higher transition temperatures the alkali doped fullerenes. It was also recognized that superconductivity could be induced in a metallic nanotube bundle in close proximity with a superconducting electrode (Re/Ta on Au), at a characteristic length scale, bounded by both the phase coherence length (l_ϕ) and the thermal diffusion length (l_T) [238]. Such induced superconductivity inferred the existence of the Josephson supercurrents, theoretically it could be the magnitude of $\pi\Delta/eR_N$, where Δ is the band gap and R_N is the resistance of the junction. Although the magnitude of the superconducting state has not been strictly observed, according to the theoretically defined values, it depends on the system behavior.

CONCLUSIONS

In this chapter the historical facts and developments in the area of carbon nanotubes as well as their key physical properties have been described. Specifically, this work emphasized on the interpretation of various forms of the carbon materials such as zero-, one-, two- and three-dimensional, owing to fact that carbon nanotubes applicability also depend on their particle sizes and dimensions. Therefore, a comprehensive description was provided on the basic concepts of carbon nanotubes, including the introduction of carbon and their hybridizations as well as an accurate note on graphite and different forms of carbon nanostructures with an extensive detail on fullerene. More specifically, the most extensively demonstrated significant multi-walled carbon nanotubes (MWCNTs) and single-walled carbon nanotubes (SWCNTs) critical physical properties interpretation was also described. Additionally, a detailed description on electronic properties, structural parameters, electronic structure, curvature effects and bundles of nanotubes has been provided. In this work a complete description on various kinds of defects in a distinct form of the carbon nanotubes has also been provided. This includes a description of the different types of defects in carbon nanotubes and their impact on several defective carbon nanotubes physical properties. As every kind of applicable carbon nanotubes-based system are almost all not free from the defects, a vast number of applications depends on the creation of distinct types of defects, which are useful for the specific application. Considering the emergence of defective carbon nanotubes, the changes in the physical properties have also been discussed. Therefore, in this work detailed descriptions on the mechanical properties including elastic properties of CNT (theoretical aspect, experimental interpretation, structural instability of CNT and strength of CNTs), thermal properties (specific heat of CNTs, thermal conductivity), optical properties (selection rules and density of states, antenna effect, absorption transitions with varying nanotube diameter, exciton properties, binding energy and band gap, excitons in metallic nanotubes, etc.) and electrical properties (electrical transport in SWCTs, transport characteristics in MWCNTs, magneto-resistance, superconductivity) have been incorporated. Hence, defect free and defects containing carbon nanotubes physical properties including types of the defects within in such materials have been discussed in detail.

■ References

1. Foa Torres, L.E.F., Roche, S., Charlier, J.C. (2014) Introduction to Graphene-based Nanomaterials: From Electronic Structure to Quantum Transport. Cambridge University Press, United Kingdom.

2. Hemstreet, Louis A. Jr., Fong, C.Y., Cohen, M.L. (1970) Calculation of the band structure and optical constants of diamond using the nonlocal-pseudopotential method. Phys. Rev. B, 2, 2054–2063.

3. Heimann, R., Evsyukov, S., Kavan, L. (1999) Carbyne and Carbynoid Structures. Kluwer Academic, Dordrecht, The Netherlands.

4. Cataldo, F., ed. (2005), Polyynes: Synthesis, Properties, and Applications. Taylor & Francis, London.

5. Kavan, L., Kastner, J. (1994) Carbyne forms of carbon: Continuation of the story. Carbon, 32, 1533–1536.

6. Lagow, R.J., Kampa, J.J., Wei, H.C., Battle, S.L., Genge, J.W., Laude, D.A., et al. (1995) Synthesis of linear acetylenic carbon: The "sp" carbon allotrope. Science, 267, 362–367.

7. Ravagnan, L., Siviero, F., Lenardi, C., Piseri, P., Barborini, E., Milani, P. (2002) Cluster-beam deposition and *in situ* characterization of carbyne-rich carbon films. Phys. Rev. Lett. 89, 285506.

8. Ravagnan, L., Piseri, P., Bruzzi, M., Miglio, S., Bongiorno, G., Basergaet, A., et al. (2007) Influence of cumulenic chains on the vibrational and electronic properties of sp-sp^2 amorphous carbon. Phys. Rev. Lett., 98, 216103.

9. Meyer, J.C., Girit, C.O., Crommie, M.F., Zettl, A. (2008) Imaging and dynamics of light atoms and molecules on graphene. Nature, 454, 319–322.

10. Jin, C., Lan, H., Peng, L., Suenaga, K., Iijima, S. (2009) Deriving carbon atomic chains from graphene. Phys. Rev. Lett., 102, 205501.

11. Chuvilin, A., Meyer, J.C., Algara-Siller, G., Kaiser, U. (2009) From graphene constrictions to single carbon chains. New Journal of Physics, 11, 083019.

12. Wallace, P.R. (1947) The band theory of graphite. Phys. Rev., 71, 622–634.

13. Painter, G.S., Ellis, D.E. (1970) Electronic band structure and optical properties of graphite from a variational approach. Phys. Rev. B, 1, 4747–4752.

14. Saito, R., Dresselhaus, G., Dresselhaus, M.S. (2000) Trigonal warping effect of carbon nanotubes. Phys. Rev. B, 61, 2981–2990.

15. Li, H., He, X., Liu, Y., Huang, H., Lian, S., Lee, S.-T., Kang, Z. (2011) One-step ultrasonic synthesis of water-soluble carbon nanoparticles with excellent photoluminescent properties. Carbon, 49, 605–609.

16. Cao, Q., Rogers, J.A. (2009) Ultrathin films of single-walled carbon nanotubes for electronics and sensors: A review of fundamental and applied aspects. Adv. Mater., 21, 29–53.

17. Choi, W., Lahiri, I., Seelaboyina, R., Kang, Y.S. (2010) Synthesis of graphene and its applications: A review. Crit. Rev. Solid State Mater. Sci., 35, 52–71.

18. Wu, Y., Lin, X., Zhang, M. (2013) Carbon nanotubes for thin film transistor: Fabrication, properties, and applications. J. Nanomater., 2013, 1–16.

19. Hecht, D.S., Hu, L., Irvin, G. (2011) Emerging transparent electrodes based on thin films of carbon nanotubes, graphene, and metallic nanostructures. Adv. Mater., 23, 1482–1513.

20. Wang, X., Zhi, L., Müllen, K. (2008) Transparent, conductive graphene electrodes for dye-sensitized solar cells. Nano Lett., 8, 323–327.

21. Seo, D.H., Han, Z.J., Kumar, S., Ostrikov, K. (2013) Structure-controlled, vertical graphene-based, binder-free electrodes from plasma-reformed butter enhance supercapacitor performance. Adv. Energy Mater., 3, 1316–1323.

22. Seo, D.H., Pineda, S., Yick, S., Bell, J., Han, Z.J., Ostrikov, K.K. (2015) Plasma-enabled sustainable elemental lifecycles: Honeycomb-derived graphenes for next-generation biosensors and supercapacitors. Green Chem., 17, 2164–2171.

23. Stancu, E.C., Stanciuc, A.-M., Vizireanu, S., Luculescu, C., Moldovan, L., Achour, A., Dinescu, G. (2014) Plasma functionalization of carbon nanowalls and its effect on attachment of fibroblast-like cells. J. Phys. D Appl. Phys., 47, 1–10.

24. Goenka, S., Sant, V., Sant, S. (2014) Graphene-based nanomaterials for drug delivery and tissue engineering. J. Control. Release, 173, 75–88.

25. Yang, K., Zhang, S., Zhang, G., Sun, X., Lee, S.-T., Liu, Z. (2010) Graphene in mice: Ultrahigh in vivo tumor uptake and efficient photothermal therapy. Nano Lett., 10, 3318–3323.

26. Chen, Y., Star, A., Vidal, S. (2013) Sweet carbon nanostructures: Carbohydrate conjugates with carbon nanotubes and graphene and their applications. Chem. Soc. Rev., 42, 4532–4542.

27. Dizaj, S.M., Mennati, A., Jafari, S., Khezri, K., Adibkia, K. (2015) Antimicrobial activity of carbon-based nanoparticles. Adv. Pharm. Bull., 5, 19–23.

28. Kroto, H.W., Heath, J.R., O'Brien, S.C., Curl, R.F., Smalley, R.E. (1985) C_{60}: Buckminsterfullerene. Nature, 318, 162–163.

29. Dresselhaus, M., Dresselhaus, G., Eklund, P. (1996) Science of Fullerenes and Carbon Nanotubes: Their Properties and Applications. Academic Press, New York.

30. Harris, P. (1999) Carbon Nanotubes and Related Structures: New Materials for the XXI Century. Cambridge University Press, Cambridge.

31. Kroto, H.W., McKay, K. (1988) The formation of quasi-icosahedral spiral shell carbon particles. Nature, 331, 328–331.

32. Ugarte, D. (1992), Curling and closure of graphitic networks under electron-beam irradiation. Nature, 359, 707–709.

33. Chuvilin, A., Kaiser, U., Bichoutskaia, E., Besley, N.A., Khlobystov, A.N. (2010) Direct transformation of graphene to fullerene. Nature Chemistry, 2, 450–453.

34. Oberlin, A., Endo, M., Koyama, T. (1976) Filamentous growth of carbon through benzene decomposition. Journal of Crystal Growth, 32, 335–349.

35. Iijima, S. (1991) Helical microtubules of graphitic carbon. Nature, 354, 56–58.

36. Iijima, S., Ichihashi, T. (1993) Single-shell carbon nanotubes of 1 nm diameter. Nature, 363, 603–605.

37. Bethune, D.S., Klang, C.H., de Vries, M.S., Gorman, G., Savoy, R., Vazquez, J., et al. (1993) Cobalt-catalysed growth of carbon nanotubes with single-atomic-layer walls. Nature, 363, 605–607.

38. Zheng, L.X., O'Connell, M.J., Doorn, S.K., Liao, X.Z., Zhao, Y.H., Akhadov, E.A., et al. (2004) Ultralong single-wall carbon nanotubes. Nature Materials, 3, 673–676.

39. Saito, R., Dresselhaus, G., Dresselhaus, M. (1998) Physical Properties of Carbon Nanotubes. Imperial College Press, London.

40. Thess, A., Lee, R., Nikolaev, P., Dai, H., Petit, P., Robert, J., et al. (1996) Crystalline ropes of metallic carbon nanotubes. Science, 273, 483–487.

41. Dresselhaus, M.S., Dresselhaus, G., Avouris, P., Hugdahl, J., Lynum, S., Ebbesen, T.W., (eds) (2001) Carbon Nanotubes: Synthesis, Structure, Properties, and Applications. Topics in Applied Physics Vol. 80, Springer, Berlin.

42. Krishnan, A., Dujardin, E., Treacy, M.M.J., Hugdahl, J., Lynum, S., Ebbesen, T.W. (1997) Graphitic cones and the nucleation of curved carbon surfaces. Nature, 388, 451–454.

43. Smith, B.W., Monthioux, M., Luzzi, D.E. (1998) Encapsulated C60 in carbon nanotubes. Nature, 396, 323–324.

44. Iijima, S., Yudasaka, M., Yamada, R., Bandow, S., Suenaga, K., Kokai, F., et al. (1999) Nano-aggregates of single-walled graphitic carbon nano-horns. Chemical Physics Letters, 309, 165–170.

45. Liu, J., Dai, H., Hafner, J.H., Colbert, D.T., Smalley, R.E., Tans, S.J., et al. (1997) Fullerene "crop circles". Nature, 385, 780–781.

46. Georgakilas, V., Perman, J.A., Tucek, J., Zboril, R. (2015) Broad family of carbon nanoallotropes: Classification, chemistry, and applications of fullerenes, carbon dots, nanotubes, graphene, nanodiamonds, and combined superstructures. Chemical Reviews, 115(11), 4744–4822.

47. Nasir, S., Hussein, M.Z., Zainal, Z., Yusof, N.A., Zobir, S.A.M., Alibe, I.M. (2019) Potential valorization of by-product materials from oil palm: A review of alternative and sustainable carbon sources for carbon-based nanomaterials synthesis. BioResources, 14, 1–37.

48. Abdul Khalil, H.P.S., Fizree, H.M., Jawaid, M., Alattas, O.S. (2011) Preparation and characterization of nano-structured materials from oil palm ash: A bio-agricultural waste from oil palm mill. BioResources, 6, 4537–4546.

49. Fowler, P.W., Manolopoulos, D.E. (2006) An Atlas of Fullerenes. Dover Publications, New York.

50. Radushkevich, L.V., Lukyanovich, V.M. (1952) The structure of carbon forming in thermal decomposition of carbon monoxide on an iron catalyst. Zurn. Fisic. Chim., 26, 88–95.

51. Monthioux, M., Kuznetsov, V.L. (2006) Who should be given the credit for the discovery of carbon nanotubes? Carbon, 44, 1621–1623.

52. Moor, K.J., Osuji, C.O., Kim, J.-H. (2016) Dual-functionality fullerene and silver nanoparticle antimicrobial composites via block copolymer templates. ACS Appl. Mater. Interfaces, 8, 33583–33591.

53. Skariyachan, S., Parveen, A., Garka, S. (2016) Nanoparticle fullerene (C_{60}) demonstrated stable binding with antibacterial potential towards probable targets of drug resistant Salmonella typhi–A computational perspective and in vitro investigation. J. Biomol. Struct. Dyn., 34, 1–20.

54. Dostalova, S., Moulick, A., Milosavljevic, V., Guran, R., Kominkova, M., Cihalova, K., et al. (2016) Antiviral activity of fullerene C_{60} nanocrystals modified with derivatives of anionic antimicrobial peptide maximin H_5. Mon. Chem. Chem. Mon., 147, 905–918

55. Chen, Q., Ma, Z., Liu, G., Wei, H., Xie, X. (2016) Antibacterial activity of cationic cyclen-functionalized fullerene derivatives: Membrane stress. Dig. J. Nanomater. Biostruct., (DJNB) 11, 753–761.

56. Moor, K.J., Osuji, C.O., Kim, J.-H. (2015) Antimicrobial photodynamic therapy with fulleropyrrolidine: Photoinactivation mechanism of Staphylococcus aureus, in vitro and in vivo studies. Appl. Microbiol. Biotechnol., 99, 4031–4043.

57. Kleandrova, V., Luan, F., Speck-Planche, A., Cordeiro, M. (2015) Review of structures containing fullerene-C_{60} for delivery of antibacterial agents. Multitasking model for computational assessment of safety profiles. Curr. Bioinform., 10, 565–578.

58. Sharma, S.K., Chiang, L.Y., Hamblin, M.R. (2011) Photodynamic therapy with fullerenes in vivo: Reality or a dream? Nanomedicine, 6, 1813–1825.

59. Naddeo, J., Ratti, M., O'Malley, S.M., Griepenburg, J.C., Bubb, D.M., Klein, E.A. (2015) Antibacterial properties of nanoparticles: A comparative review of chemically synthesized and laser-generated particles. Adv. Sci. Eng. Med., 7, 1044–1057.

60. Boonstra, J., Post, J.A. (2004) Molecular events associated with reactive oxygen species and cell cycle progression in mammalian cells. Gene, 337, 1–13.

61. Ishaq, M., Bazaka, K., Ostrikov, K. (2015) Pro-apoptotic NOXA is implicated in atmospheric-pressure plasma-induced melanoma cell death. J. Phys. D Appl. Phys., 48, 464002

62. Zhou, R., Zhou, R., Zhang, X., Zhuang, J., Yang, S., Bazaka, K., Ostrikov, K. (2016) Effects of atmospheric-pressure N_2, He, Air, and O_2 microplasmas on mung bean seed germination and seedling growth. Sci. Rep., 6, 32603.

63. Ishaq, M., Bazaka, K., Ostrikov, K. (2015) Intracellular effects of atmospheric-pressure plasmas on melanoma cancer cells. Phys. Plasmas, 22, 122003.

64. Prasad, K., Lekshmi, G., Ostrikov, K., Lussini, V., Blinco, J., Mohandas, M., Vasilev, K., Bottle, S., Bazaka, K., Ostrikov, K. (2017) Synergic bactericidal effects of reduced graphene oxide and silver nanoparticles against Gram-positive and Gram-negative bacteria. Sci. Rep., 7, 1–11.

65. Navarro, E., Baun, A., Behra, R., Hartmann, N.B., Filser, J., Miao, A.-J., Quigg, A., Santschi, P.H., Sigg, L. (2008) Environmental behavior and ecotoxicity of engineered nanoparticles to algae, plants, and fungi. Ecotoxicology, 17, 372–386.

66. Markovic, Z., Trajkovic, V. (2008) Biomedical potential of the reactive oxygen species generation and quenching by fullerenes (C_{60}). Biomaterials, 29, 3561–3573.

67. Fang, J., Lyon, D.Y., Wiesner, M.R., Dong, J. Alvarez. (2007) Effect of a fullerene water suspension on bacterial phospholipids andmembrane phase behavior. Environ. Sci. Technol., 41, 2636–2642.

68. Tang, Y.J., Ashcroft, J.M., Chen, D., Min, G., Kim, C.-H., Murkhejee, B., Larabell, C., Keasling, J.D., Chen, F.F. (2007) Charge-associated effects of fullerene derivatives on microbial structural integrity and central metabolism. Nano Lett., 7, 754–760.

69. Mashino, T., Nishikawa, D., Takahashi, K., Usui, N., Yamori, T., Seki, M., Endo, T., Mochizuki, M. (2003) Antibacterial and antiproliferative activity of cationic fullerene derivatives. Bioorg. Med. Chem. Lett., 13, 4395–4397.

70. Mashino, T., Usui, N., Okuda, K., Hirota, T., Mochizuki, M. (2003) Respiratory chain inhibition by fullerene derivatives: Hydrogen peroxide production caused by fullerene derivatives and a respiratory chain system. Bioorg. Med. Chem., 11, 1433–1438.

71. Hsu, P.-C., Jefferies, D., Khalid, S. (2016) Molecular dynamics simulations predict the pathways via which pristine fullerenes penetrate bacterial membranes. J. Phys. Chem. B, 120, 11170–11179.

72. Krishnamoorthy, K., Veerapandian, M., Zhang, L.-H., Yun, K., Kim, S.J. (2012) Antibacterial efficiency of graphene nanosheets against pathogenic bacteria via lipid peroxidation. J. Phys. Chem. C, 116, 17280–17287.

73. Yang, X., Ebrahimi, A., Li, J., Cui, Q. (2014) Fullerene-biomolecule conjugates and their biomedicinal applications. Int. J. Nanomed., 9, 77–92.

74. Huang, L., Wang, M., Dai, T., Sperandio, F.F., Huang, Y.-Y., Xuan, Y., Chiang, L.Y., Hamblin, M.R. (2014) Antimicrobial photodynamic therapy with decacationic monoadducts and bisadducts of [70] fullerene: In vitro and in vivo studies. Nanomedicine, 9, 253–266.

75. Zhang, Y., Dai, T., Wang, M., Vecchio, D., Chiang, L.Y., Hamblin, M.R. (2015) Potentiation of antimicrobial photodynamic inactivation mediated by a cationic fullerene by added iodide: In vitro and in vivo studies. Nanomedicine, 10, 603–614.

76. McCluskey, D.M., Smith, T.N., Madasu, P.K., Coumbe, C.E., Mackey, M.A., Fulmer, P.A., Wynne, J.H., Stevenson, S., Phillips, J.P. (2009) Evidence for Singlet-Oxygen Generation and Biocidal Activity in Photoresponsive Metallic Nitride Fullerene—Polymer Adhesive Films, ACS Appl. Mater.Interfaces 1, 882–887.

77. Kausar, A. (2017) Advances in polymer/fullerene nanocomposite: A review on essential features and applications. Polym. Plast. Technol. Eng., 56, 594–605.

78. Zhou, G., Li, Y., Xiao, W., Zhang, L., Zuo, Y., Xue, J., Jansen, J.A. (2008) Synthesis, characterization, and antibacterial activities of a novel nanohydroxyapatite/zinc oxide complex. J. Biomed. Mater. Res. Part A, 85, 929–937.

79. Huh, A.J., Kwon, Y.J. (2011) Nanoantibiotics: A new paradigm for treating infectious diseases using nanomaterials in the antibiotics resistant era. J. Control. Release, 156, 128–145.

80. Lyon, D.Y., Alvarez, P.J. (2008) Fullerene water suspension (nC_{60}) exerts antibacterial effects via ROS-independent protein oxidation. Environ. Sci. Technol., 42, 8127–8132.

81. Lyon, D.Y., Brunet, L., Hinkal, G.W., Wiesner, M.R., Alvarez, P.J. (2008) Antibacterial activity of fullerene water suspensions (nC_{60}) is not due to ROS-mediated damage. Nano Lett., 8, 1539–1543.

82. Kai, Y., Komazawa, Y., Miyajima, A., Miyata, N., Yamakoshi, Y. (2003) [60]Fullerene as a novel photoinduced antibiotic. Fullerenes, Nanotubes and Carbon Nanostructures, 11, 79–87.

83. Moor, K.J., Snow, S.D., Kim, J.-H. (2015) Differential photoactivity of aqueous [C60] and [C70] fullerene aggregates. Environ. Sci. Technol., 49, 5990–5998.

84. Tegos, G.P., Demidova, T.N., Arcila-Lopez, D., Lee, H., Wharton, T., Gali, H., Hamblin, M.R. (2005) Cationic fullerenes are effective and selective antimicrobial photosensitizers. Chem. Biol., 12, 1127–1135.

85. Badireddy, A.R., Hotze, E.M., Chellam, S., Alvarez, P., Wiesner, M.R. (2007) Inactivation of bacteriophages via photosensitization of fullerol nanoparticles. Environ. Sci. Technol., 41, 6627–6632.

86. Yin, R., Agrawal, T., Khan, U., Gupta, G.K., Rai, V., Huang, Y.-Y., Hamblin, M.R. (2015) Antimicrobial photodynamic inactivation in nanomedicine: Small light strides against bad bugs. Nanomedicine, 10, 2379–2404.

87. Al-Jumaili, A., Alancherry, S., Bazaka, K., Jacob, M.V. (2017) Review on the antimicrobial properties of carbon nanostructures. Materials, 10, 2–26.

88. Lyon, D.Y., Adams, L.K., Falkner, J.C., Alvarez, P.J. (2006) Antibacterial activity of fullerene water suspensions: Effects of preparation method and particle size. Environ. Sci. Technol., 40, 4360–4366.

89. Kroto, H.W., Heath, J.R., O'Brien, S.C., Curl, R.F., Smalley, R.E. (2018) C_{60}: Buckminsterfullerene. Nature, 318, 162–163.

90. Bacon, R. (1960) Growth, structure, and properties of graphite whiskers. J. Appl. Phys., 31, 283.

91. Sinnott, S.B., Andrews, R. (2001) Carbon nanotubes: Synthesis, properties, and applications. Crit. Rev. Solid State Mater. Sci., 26, 145–249.

92. Rahmandoust, M., Ayatollahi, M.R. (2015) Characterization of Carbon Nanotube Based Composites under Consideration of Defects. Springer, Switzerland.

93. Miranda, A., Barekar, N., McKay, B.J. (2019) MWCNTs and their use in Al-MMCs for ultra-high thermal conductivity applications: A review. J. Alloys and Comp., 774, 820–840.

94. Radushkevich, L.V., Lukyanovich, V.M. (1952) The structure of carbon forming in thermal decomposition of carbon monoxide on an iron catalyst. Zurn. Fisic. Chim., 26, 88–95.

95. Monthioux, M., Kuznetsov, V.L. (2006) Who should be given the credit for the discovery of carbon nanotubes? Carbon, 44, 1621–1623.

96. Khan, I., Huang, S., Wu, C. (2017) Multi-walled carbon nanotube structural instability with/without metal nanoparticles under electron beam irradiation. New J. Phys., 19, 123016–15.

97. Collins, P.G. (2010) Defects and disorder in carbon nanotubes (Chapter 2). *In*: Narlikar, A., Fu, Y., (eds), Oxford Handbook of Nanoscience and Technology: Volume 2: Materials: Structures, Properties and Characterization Techniques. Oxford University Press, Oxford.

98. Sinnott, S.B., Andrews, R. (2001) Carbon nanotubes: Synthesis, properties, and applications. Crit. Rev. Solid State Mater. Sci., 26, 145–249.

99. Lehman, J.H., Terrones, M., Mansfield, E., Hurst, K.E., Meunier, V. (2011) Evaluating the characteristics of multiwall carbon nanotubes. Carbon, 49, 2581–2602.

100. Dai, H. (2002) Carbon nanotubes: Synthesis, integration, and properties. Acc. Chem. Res., 35, 1035–1044.

101. Yetgin, S.H. (2019) Effect of multi walled carbon nanotube on mechanical, thermal and rheological properties of polypropylene. J. Mater. Res. Tech., 8, 4725–4735.

102. Bom, D., Andrews, R., Jacques, D., Anthony, J., Chen, B., Meier, M.S., Selegue, J.P. (2002) Thermogravimetric analysis of the oxidation of multiwalled carbon nanotubes: Evidence for the role of defect sites in carbon nanotube chemistry. Nano Lett., 2, 615–619.

103. DiLeo, R.A., Landi, B.J., Raffaelle, R.P. (2007) Purity assessment of multiwalled carbon nanotubes by Raman spectroscopy. J. Appl. Phys., 101, 64307.

104. Jia, N., Wang, L., Liu, L., Zhou, Q., Jiang, Z. (2005) Bamboo-like CN_x nanotubes for the immobilization of hemoglobin and its bioelectrochemistry. Electrochem. Commun., 7, 349–354.

105. Chopra, N.G., Benedict, L.X., Crespi, V.H., Cohen, M.L., Louie, S.G., Zettl, A. (1995) Fully collapsed carbon nanotubes. Nature, 377, 135.

106. Elliott, J.A., Sandler, J.K.W., Windle, A.H., Young, R.J., Shaffer, M.S.P. (2004) Collapse of single-wall carbon nanotubes is diameter dependent. Phys. Rev. Lett., 92, 095501.

107. Peng, B., Yao, Y., Zhang, J. (2010) Effect of the reynolds and richardson numbers on the growth of well-aligned ultralong single-walled carbon nanotubes. J. Phys. Chem. C, 114, 12960–12965.

108. Wang, X., Li, Q., Xie, J., Jin, Z., Wang, J., Li, Y., et al. (2009) Fabrication of ultralong and electrically uniform single-walled carbon nanotubes on clean substrates. Nano Lett., 9, 3137–3141.

109. Hamada, N., Sawada, S.I., Oshiyama, A. (1992) New one-dimensional conductors: Graphitic microtubules. Phys. Rev. Lett., 68, 1579–1581.

110. White, C.T., Mintmire, J.W. (1998) Density of states reflects diameter in nanotubes. Nature, 394, 29–30.

111. Kane, C.L., Mele, E.J. (1997) Size, shape, and low energy electronic structure of carbon nanotubes. Phys. Rev. Lett., 78, 1932–1935.

112. Ouyang, M., Huang, J.-L., Cheung, C.L., Lieber, C.M. (2001) Energy gaps in "metallic single-walled carbon nanotubes. Science, 292, 702–705.

113. Mintmire, J.W., Dunlap, B.I., White, C.T. (1992) Are fullerene tubules metallic? Phys. Rev. Lett., 68, 631–634.

114. Dunlap, B.I. (1994) Relating carbon tubules. Phys. Rev. B, 49, 5643–5651.

115. Lambin, P., Fonseca, A., Vigneron, J., Nagy, J., Lucas, A. (1995) Structural and electronic properties of bent carbon nanotubes. Chemical Physics Letters, 245, 85–89.

116. Saito, R., Dresselhaus, G., Dresselhaus, M.S. (1996) Tunneling conductance of connected carbon nanotubes. Phys. Rev. B, 53, 2044–2050.

117. Charlier, J.C., Ebbesen, T.W., Lambin, P. (1996) Structural and electronic properties of pentagon–heptagon pair defects in carbon nanotubes. Phys. Rev. B, 53, 11108–11113.

118. Chico, L., Crespi, V.H., Benedict, L.X., Louie, S.G., Cohen, M.L. (1996) Pure carbon nanoscale devices: Nanotube heterojunctions. Phys. Rev. Lett., 76, 971–974.

119. Terrones, M., Banhart, F., Grobert, N., Charlier, J.C., Terrones, H., Ajayan, P.M. (2002) Molecular junctions by joining single-walled carbon nanotubes. Phys. Rev. Lett., 89, 075505.

120. Biro, L.P., Mark, G.I., Koos, A.A., Nagy, B.J., Lambin, P. (2002) Coiled carbon nanotube structures with supraunitary nonhexagonal to hexagonal ring ratio. Phys. Rev. B, 66, 165405.

121. Harris, B., Bunsell, A.R. (1977) Structure and Properties of Engineering Materials. Longman, London, New York.

122. Kelly, B.T. (1981) Physics of Graphite. Applied Science Publishers, London.

123. Ruoff, R.S., Lorents, D.C. (1995) Mechanical and thermal properties of carbon nanotubes. Carbon, 33, 925–930.

124. Tibbetts, G.G. (1984) Why are carbon filaments tubular. Journal of Crystal Growth, 66, 632–638.

125. Robertson, D.H., Brenner, D.W., Mintmire, J.W. (1992) Energetics of nanoscale graphitic tubules. Phys. Rev. B, 45, 12 592–12595.

126. Yakobson, B.I., Brabec, C.J., Bernholc, J. (1996) Nanomechanics of carbon tubes: Instabilities beyond linear response. Phys. Rev. Lett., 76, 2511–2514.

127. Cornwell, C.F., Wille, L.T. (1997) Elastic properties of single-walled carbon nanotubes in compression. Solid State Commun., 101, 555–558.

128. Hernandez, E., Goze, C., Bernier, P., Rubio, A. (1998) Elastic properties of C and $B_xC_yN_z$ composite nanotubes. Phys. Rev. Lett., 80, 4502–4505.

129. Lu, J.P. (1997) Elastic properties of carbon nanotubes and nanoropes. Phys. Rev. Lett., 79, 12971300.

130. Yao, N., Lordi, V. (1998) Young's modulus of single-walled carbon nanotubes. J. App. Phys., 84, 1939.

131. Salvetat, J.P., Bonard, J.M., Thomson, N.H., Kulik, A.J., Forro, L., Benoit, W., Zuppiroli, L. (1999) Mechanical properties of carbon nanotubes. Appl. Phys. A, 69, 255–260.

132. Stallard, J.C., Tan, W., Smail, F.R., Gspann, T.S., Boies, A.M., Fleck, N.A. (2018) The mechanical and electrical properties of direct-spun carbon nanotube mats. Extreme Mechanics Lett., 21, 65–75.

133. Lourie, O., Wagner, H.D. (1998) Evaluation of Young's modulus of carbon nanotubes by micro-Raman spectroscopy. J. Mater. Res., 13, 2418–2422.

134. Yu, M.F., Files, B.S., Arepalli, S., Ruoff, R.S. (2000) Tensile loading of ropes of single wall carbon nanotubes and their mechanical properties. Phys. Rev. Lett., 84, 5552–5555.

135. Yu, M.F., Lourie, O., Dyer, M.J., Moloni, K., Kelly, T.F., Ruoff, R.S. (2000) Strength and breaking mechanism of multiwalled carbon nanotubes under tensile load. Science, 287, 637–640.

136. Wong, E.W., Sheehan, P.E., Lieber, C.M. (1997) Nanobeam mechanics: Elasticity, strength, and toughness of nanorods and nanotubes. Science, 277, 1971–1975.

137. Salvetat, J.P., Kulik, A.J., Bonard, J.M., Briggs, G.A.D., Stockli, T., Metenier, K., Bonnamy, S., Beguin, F., Burnham, N.A., Forro, L. (1999) Elastic modulus of ordered and disordered multiwalled carbon nanotubes. Adv. Mater., 11, 161–165.

138. Salvetat, J.P., Briggs, G.A.D., Bonard, J.M., Bacsa, R.R., Kulik, A.J., Stockli, T., Burnham, N.A., Forro, L. (1999) Elastic and shear moduli of single-walled carbon nanotube ropes. Phys. Rev. Lett., 82, 944–947.

139. Treacy, M.M.J., Ebbesen, T.W., Gibson, J.M. (1996) Exceptionally high Young's modulus observed for individual carbon nanotubes. Nature, 381, 678–680.

140. Krishnan, A., Dujardin, E., Ebbesen, T.W., Yianilos, P.N., Treacy, M.M.J. (1998) Young's modulus of single-walled nanotubes. Phys. Rev. B, 58, 14013–14019.

141. Poncharal, P., Wang, Z.L., Ugarte, D., de Heer, W.A. (1999) Electrostatic deflections and electro-mechanical resonances of carbon nanotubes. Science, 283, 1513–1516.

142. Yu, M.F., Dyer, M.J., Chen, J., Bray, K. (2001) Multiprobe nanomanipulation and functional assembly of nanomaterials inside a scanning electron microscope. *In*: International Conference IEEE-NANO2001, Maui.

143. Dikin, D.A., Chen, X., Ding, W., Wagner, G.J., Ruoff, R.S. (2003) Resonance vibration of amorphous SiO_2 nanowires driven by mechanical or electrical field excitation. J. Appl. Phys., 93, 226.

144. Yakobson, B.I., Brabec, C.J., Bernholc, J. (1996) Nanomechanics of carbon tubes: Instabilities beyond linear response. Phys. Rev. Lett., 76, 2511–2514.

145. Yakobson, B.I., Smalley, R.E. (1997) Fullerene nanotubes: C-1000000 and beyond. Am. Sci., 85, 324–337.

146. Yakobson B.I., Avouris P. (2001) Mechanical properties of carbon nanotubes. *In*: Dresselhaus, M.S., Dresselhaus G., Avouris P. (eds), Carbon Nanotubes. Topics in Applied Physics, Vol 80. Springer, Berlin, Heidelberg, pp. 287–327.

147. Bernholc, J., Brabec, C., Nardelli, M.B., Maiti, A., Roland, C., Yakobson, B.I. (1998) Theory of growth and mechanical properties of nanotubes. Appl. Phys. A, 67, 39–46.

148. Qian, D., Liu, W.K., Ruoff, R.S. (2002) Bent and kinked multi-shell Carbon nanotubes-treating the interlayer potential more realistically. *In*: 43rd AIAA/ASME/ASCE/AHS Structures, Structural Dynamics, and Materials Conferences, Denver, CO.

149. Qian, D., Liu, W.K., Subramoney, S., Ruoff, R.S. (2003) Effect of interlayer interaction on the mechanical deformation of multiwalled carbon nanotube. J. Nanosci. Nanotechnol., 3, 185–191.

150. Despres, J.F., Daguerre, E., Lafdi, K. (1995) Flexibility of graphene layers in carbon nanotubes. Carbon, 33, 87–89.

151. Iijima, S., Brabec, C., Maiti, A., Bernholc, J. (1996) Structural flexibility of carbon nanotubes. J. Chem. Phys., 104, 2089–2092.

152. Ruoff, R.S., Lorents, D.C., Laduca, R., Awadalla, S., Weathersby, S., Parvin, K., Subramoney, S. (1995) Proc. Electrochem. Soc., 95–10, 557–562.

153. Subramoney, S., Ruoff, R.S., Laduca, R., Awadalla, S., Parvin, K. (1995) Proc. Electrochem. Soc., 95–10, 563–569.

154. Falvo, M.R., Clary, G.J., Taylor, R.M., Chi, V., Brooks, F.P., Washburn, S., Superfine, R. (1997) Bending and buckling of carbon nanotubes under large strain. Nature, 389, 582–584.

155. Hertel, T., Martel, R., Avouris, P. (1998) Manipulation of individual carbon nanotubes and their interaction with surfaces. J. Phys. Chem. B, 102, 910–915.

156. Lourie, O., Cox, D.M., Wagner, H.D. (1998) Buckling and collapse of embedded carbon nanotubes. Phys. Rev. Lett., 81, 1638–1641.

157. Ruoff, R.S., Tersoff, J., Lorents, D.C., Subramoney, S., Chan, B. (1993) Radial deformation of carbon nanotubes by Van-Der-Waals forces. Nature, 364, 514–516.

158. Tersoff, J., Ruoff, R.S.(1994) Structural-properties of a carbon-nanotube crystal. Phys. Rev. Lett., 73, 676–679.

159. Lopez, M.J., Rubio, A., Alonso, J.A., Qin, L.C; Iijima, S.(2001) Novel polygonized single-wall carbon nanotube bundles. Phys. Rev. Lett., 86, 3056–3059.

160. Chopra, N.G., Benedict, L.X., Crespi, V.H., Cohen, M.L., Louie, S.G., Zettl, A. (1995) Fully collapsed carbon nanotubes. Nature, 377, 135–138.

161. Benedict, L.X., Chopra, N.G., Cohen, M.L., Zettl, A., Louie, S.G., Crespi, V.H. (1998) Microscopic determination of the interlayer binding energy in graphite. Chem. Phys. Lett., 286, 490–496.

162. Hertel, T., Walkup, R.E., Avouris, P. (1998) Deformation of carbon nanotubes by surface van der Waals forces. Phys. Rev. B, 58, 13870–13873.

163. Avouris, P., Hertel, T., Martel, R., Schmidt, T., Shea, H.R., Walkup, R.E. (1999) Carbon nanotubes: nanomechanics, manipulation, and electronic devices. Appl. Surface Sci., 141, 201–209.

164. Yu, M.F., Dyer, M.J., Ruoff, R.S. (2001) Structure and mechanical flexibility of carbon nanotube ribbons: An atomic-force microscopy study. J. Appl. Phys., 89, 4554–4557.

165. Yu, M.F., Kowalewski, T., Ruoff, R.S. (2001) Structural analysis of collapsed, and twisted and collapsed, multiwalled carbon nanotubes by atomic force microscopy. Phys. Rev. Lett., 86, 87–90.

166. Lordi, V., Yao, N. (1998) Radial compression and controlled cutting of carbon nanotubes. J. Chem. Phys., 109, 2509–2512.

167. Gao, G.H., Cagin, T., Goddard, W.A. (1998) Energetics, structure, mechanical and vibrational properties of single-walled carbon nanotubes. Nanotechnology, 9, 184–191.

168. Shen, W.D., Jiang, B., Han, B.S., Xie, S.S. (2000) Investigation of the radial compression of carbon nanotubes with a scanning probe microscope. Phys. Rev. Lett., 84, 3634–3637.

169. Yu, M.F., Kowalewski, T., Ruoff, R.S. (2000) Investigation of the radial deformability of individual carbon nanotubes under controlled indentation force. Phys. Rev. Lett., 85, 1456–1459.

170. Chesnokov, S.A., Nalimova, V.A., Rinzler, A.G., Smalley, R.E., Fischer, J.E. (1999) Mechanical energy storage in carbon nanotube springs. Phys. Rev. Lett., 82, 343–346.

171. Kelly, B.T. (1981) Physics of Graphite. Applied Science, London.

172. Yu, M.F., Dyer, M.J., Chen, J., Qian, D., Liu, W.K., Ruoff, R.S. (2001) Locked twist in multi-walled carbon nanotube ribbons. Phys. Rev. B, 64, 241403R.

173. Dresselhaus, M.S., Dresselhaus, G., Sugihara, K., Spain, I.L., Goldberg, H.A. (1988) Graphite Fibers and Filaments. Springer, Berlin, Heidelberg.

174. Nardelli, M.B., Yakobson, B.I., Bernholc, J. (1998) Mechanism of strain release in carbon nanotubes. Phys. Rev. B, 57, R4277–R4280.

175. Lauginie, P., Conard, J. (1997) New growing modes for carbon: Modelization of lattice defects, structure of tubules and onions. J. Phys. Chem. Solids, 58, 1949–1963.

176. Yakobson, B.I. (1998) Mechanical relaxation and "intramolecular plasticity" in carbon nanotubes. Appl. Phys. Lett., 72, 918.

177. Nardelli, M.B., Yakobson, B.I., Bernholc, J. (1998) Brittle and ductile behavior in carbon nanotubes. Phys. Rev. Lett., 81, 4656-4659.

178. Ashcroft, N.W., Mermin, N.D. (1976) Solid State Physics. Harcourt Brace, New York.

179. Saito, R., Takeya, T., Kimura, T., Dresselhaus, G., Dresselhaus, M.S. (1998) Raman intensity of single-wall carbon nanotubes. Phys. Rev. B, 57, 4145–4153.

180. Sanchez-Portal, D., Artacho, E., Solar, J.M., Rubio, A., Ordejon, P. (1999) Ab initio structural, elastic, and vibrational properties of carbon nanotubes. Phys. Rev., B, 59, 12678–12688.

181. Benedict, L.X., Louie, S.G., Cohen, M.L. (1996) Heat capacity of carbon nanotubes. Solid State Commun., 100, 177–180.

182. Hone, J., Llaguno, M.C., Biercuk, M.J., Johnson, A.T., Batlogg, B., Benes, Z., Fischer, J.E. (2002) Thermal properties of carbon nanotubes and nanotube-based materials. Appl. Phys. A, 74, 339–343.

183. Rego, L.G.C., Kirczenow, G. (1998) Quantized thermal conductance of dielectric quantum wires. Phys. Rev. Lett., 81, 232–235.

184. Schwab, K., Henriksen, E.A., Worlock, J.M., Roukes, M.L. (2000) Measurement of the quantum of thermal conductance. Nature, 404, 974–977.

185. Heremans, J., Beetz, C.P. (1985) Thermal-conductivity and thermopower of vapor-grown graphite fibers. Phys. Rev. B, 32, 1981–1986.

186. Peierls, R.E. (1955) Quantum Theory of Solids. Oxford University Press, London.

187. Hone, J., Llaguno, M.C., Nemes, N.M., Johnson, A.T., Fischer, J.E., Walters, D.A., Casavant, M.J., Schmidt, J., Smalley, R.E. (2000) Electrical and thermal transport properties of magnetically aligned single wall carbon nanotube films. Appl. Phys. Lett., 77, 666–668.

188. Hone, J., Whitney, M., Piskoti, C., Zettl, A. Thermal conductivity of single-walled carbon nanotubes. Phys. Rev. B, 1999, 59, R2514–R2516.

189. Yi, W., Lu, L., Zhang, D.L., Pan, Z.W., Xie, S.S. (1999) Linear specific heat of carbon nanotubes. Phys. Rev. B, 59, R9015–R9018.

190. Kim, P., Shi, L., Majumdar, A., McEuen, P.L. (2001) Thermal transport measurements of individual multiwalled nanotubes. Phys. Rev. Lett., 8721, art. no. 215502.

191. Yamashita, S. (2019) Nonlinear optics in carbon nanotube, graphene, and related 2D materials. APL Photon., 4, 034301–24.

192. Blancon, J.C. (2013) Optical absorption and electronic properties of individual carbon nanotubes. Physics [physics]. Université Claude Bernard - Lyon I.

193. Barros, E.B., Jorio, A., Samsonidze, G.G., Capaz, R.B., Souza Filho, A.G., Mendes Filho, J.M., Dresselhaus, G., Dresselhaus, M.S. (2006) Review on the symmetryrelated properties of carbon nanotubes. Phys. Rep., 431, 261–302.

194. Benedict, L.X., Louie, S.G., Cohen, M.L. (1995) Static polarizabilities of single-wall carbon nanotubes. Phys. Rev. B, 52, 8541–8549.

195. Islam, M.F., Milkie, D.E., Kane, C.L., Yodh, A.G., Kikkawa, J.M. (2004) Direct measurement of the polarized optical absorption cross section of single-wall carbon nanotubes. Phys. Rev. Lett., 93, 037404.

196. Sfeir, M.Y., Wang, F., Huang, L., Chuang, C.C., Hone, J., O'brien, S.P., Heinz, T.F., Brus, L.E. (2004) Probing electronic transitions in individual carbon nanotubes by rayleigh scattering. Science, 306(5701), 1540–1543.

197. Lefebvre, J., Finnie, P. (2007) Polarized photoluminescence excitation spectroscopy of single-walled carbon nanotubes. Phys. Rev. Lett., 98, 167406.

198. Miyauchi, Y., Ajiki, H., Maruyama, S. (2010) Electron-hole asymmetry in single-walled carbon nanotubes probed by direct observation of transverse quasidark excitons. Phys. Rev. B, 81, 121415.

199. Christofilos, D., Blancon, J.C., Arvanitidis, J., San Miguel, A., Ayari, A., Del Fatti, N., Vallee, F. (2012) Optical imaging and absolute absorption cross section measurement of individual nano-objects on opaque substrates: Single-wall carbon nanotubes on silicon. J. Phys. Chem. Lett., 3, 1176–1181.

200. Tsen, A.W., Donev, L.A.K., Kurt, H., Herman, L.H., Park, J. (2009) Imaging the electrical conductance of individual carbon nanotubes with photothermal current microscopy. Nat. Nano., 4, 108–113.

201. Barkelid, M., Steele, G.A., Zwiller, V. (2012) Probing optical transitions in individual carbon nanotubes using polarized photocurrent spectroscopy. Nano Lett., 12, 5649–5653.

202. Uryu, S., Ando, T. (2012) Environment effect on cross-polarized excitons in carbon nanotubes. Phys. Rev. B, 86, 125412.

203. Rao, A.M., Richter, E., Bandow, S., Chase, B., Eklund, P.C., Williams, K.A., Fang, S., Subbaswamy, K.R., Menon, M., Thess, A., Smalley, R.E., Dresselhaus, G., Dresselhaus, M.S. (1997) Diameter-selective raman scattering from vibrational modes in carbon nanotubes. Science, 275, 187–191.

204. Bachilo, S.M., Strano, M.S., Kittrell, C., Hauge, R.H., Smalley, R.E., Weisman, R.B. (2002) Structure-assigned optical spectra of single-walled carbon nanotubes. Science, 298, 2361–2366.

205. O'Connell, M.J., Bachilo, S.M., Huffman, C.B., Moore, V.C., Strano, M.S., Haroz, E.H., Rialon, K.L., Boul, P.J., Noon, W.H., Kittrell, C., Ma, J., Hauge, R.H., Weisman, R.B., Smalley, R.E. (2002) Band gap fluorescence from individual single-walled carbon nanotubes. Science, 297, 593–596.

206. Weisman, R.B., Bachilo, S.M. (2003) Dependence of optical transition energies on structure for single-walled carbon nanotubes in aqueous suspension: An empirical kataura plot. Nano Lett., 3, 1235–1238.

207. Maultzsch, J., Pomraenke, R., Reich, S., Chang, E., Prezzi, D., Ruini, A., Molinari, E., Strano, M.S., Thomsen, C., Lienau, C. (2005) Exciton binding energies in carbon nanotubes from two-photon photoluminescence. Phys. Rev. B, 72, 241402.

208. Wang, F., Dukovic, G., Brus, L.E, Heinz, T.F. (2005) The optical resonances in carbon nanotubes arise from excitons. Science, 308, 838–841.

209. Klingshirn, C.F. (2005) Semiconductor Optics. Springer-Verlag, Berlin, Heidelberg.

210. Perebeinos, V., Tersoff, J., Avouris, P. (2004) Scaling of excitons in carbon nanotubes. Phys. Rev. Lett., 92, 257402.

211. Ando, T. (1997) Excitons in carbon nanotubes. Jpn. J. Appl. Phys., 66, 1066–1073.

212. Nugraha, A.R.T., Saito, R., Sato, K., Araujo, P.T., Jorio, A., Dresselhaus, M.S. (2010) Dielectric constant model for environmental effects on the exciton energies of single wall carbon nanotubes. Appl. Phys. Lett., 97, 91905.

213. Capaz, R.B., Spataru, C.D., Ismail-Beigi, S., Louie, S.G. (2006) Diameter and chirality dependence of exciton properties in carbon nanotubes. Phys. Rev. B, 74, 121401.

214. Charlier, J.C., Blase, X., Roche, S. (2007) Electronic and transport properties of nanotubes. Rev. Mod. Phys., 79, 677–732.

215. Liu, K., Deslippe, J., Xiao, F., Capaz, R.B., Hong, X., Aloni, S., et al. (2012) An atlas of carbon nanotube optical transitions. Nat. Nano., 7, 325–329.

216. Lin, H., Lagoute, J., Repain, V., Chacon, C., Girard, Y., Lauret, J.S., Ducastelle, F., Loiseau, A., Rousset, S. (2010) Many-body effects in electronic bandgaps of carbon nanotubes measured by scanning tunnelling spectroscopy. Nat. Mater., 9, 235–238.

217. Sfeir, M.Y., Beetz, T., Wang, F., Huang, L., Henry Huang, X.M., Huang, M., Hone, J., O'Brien, S., Misewich, J.A., Heinz, T.F., Wu, L., Zhu, Y., Brus, L.E. (2006) Optical spectroscopy of individual single-walled carbon nanotubes of defined chiral structure. Science, 312, 554–556.

218. Araujo, P.T., Doorn, S.K., Kilina, S., Tretiak, S., Einarsson, E., Maruyama, S., Chacham, H., Pimenta, M.A., Jorio, A. (2007) Third and fourth optical transitions in semiconducting carbon nanotubes. Phys. Rev. Lett., 98, 067401.

219. Michel, T., Paillet, M., Meyer, J.C., Popov, V.N., Henrard, L., Sauvajol, J.L. (2007) E33 and E44 optical transitions in semiconducting single-walled carbon nanotubes: Electron diffraction and raman experiments. Phys. Rev. B, 75, 155432.

220. Popov, V.N. (2004) Curvature effects on the structural, electronic and optical properties of isolated single-walled carbon nanotubes within a symmetry-adapted nonorthogonal tight-binding model. New J. Phys., 6, 17.

221. Spataru, C.D., Ismail-Beigi, S., Benedict, L.X., Louie, S.G. (2004) Excitonic effects and optical spectra of single-walled carbon nanotubes. Phys. Rev. Lett., 92, 077402.

222. Malic, E., Maultzsch, J., Reich, S., Knorr, A. (2010) Excitonic rayleigh scattering spectra of metallic single-walled carbon nanotubes. Phys. Rev. B, 82, 115439.

223. Wang, F., Cho, D.J., Kessler, B., Deslippe, J., Schuck, P.J., Louie, S.G., Zettl, A., Heinz, T.F., Shen, Y.R. (2007) Observation of excitons in one-dimensional metallic single-walled carbon nanotubes. Phys. Rev. Lett., 99, 227401.

224. Choi, S.K., Deslippe, J., Capaz, R.B., Louie, S.G. (2013) An explicit formula for optical oscillator strength of excitons in semiconducting single-walled carbon nanotubes: Family behaviour. Nano Lett., 13, 54–58.

225. Bandaru, P.R. (2007) Electrical properties and applications of carbon nanotube structures. J. Nanosci. Nanotechnol., 7, 1–29.

226. Datta, S. (1995) Electronic Transport in Mesoscopic Systems. Cambridge University Press, New York.

227. Bachtold, A., Strunk, C., Salvetat, J.P., Bonard, J.M., Forro, L., Nussbaumer, T., Schonenberger, C. (1999) Aharonov–Bohm oscillations in carbon nanotubes. Nature, 397, 673–675.

228. Frank, S., Poncharal, P., Wang, Z.L., de Heer, W.A. (1998) Carbon nanotube quantum resistors. Science, 280, 1744–1746.

229. Li, H.J., Lu, W.G., Li, J.J., Bai, X.D., Gu, C.Z. (2005) Multichannel ballistic transport in multiwall carbon nanotubes. Phys. Rev. Lett., 95, 086601.

230. Langer, L., Bayot, V., Grivei, E., Issi, J.P., Heremans, J.P., Olk, C.H., Stockman, L., Haesendonck, C.V., Bruynseraede, Y. (1996) Quantum transport in a multiwalled carbon nanotube. Phys. Rev. Lett., 76, 479–482.

231. Kelly, M.J. (1995) Low-Dimensional Semiconductors. Oxford University Press, New York.

232. Song, S.N., Wang, X.K., Chang, R.P.H., Ketterson, J.B. (1994) Electronic properties of graphite nanotubules from galvanomagnetic effects. Phys. Rev. Lett., 72, 697–700.

233. Liu, K., Avouris, P., Martel, R., Hsu, W.K. (2001) Electrical transport in doped multiwalled carbon nanotubes. Phys. Rev. B, 63, 161404.

234. Schonenberger, C., Bachtold, A., Strunk, C., Salvetat, J.P., Forro, L. (1999) Interference and interaction in multi-wall carbon nanotubes. Appl. Phys. A, 69, 283–295 .

235. Lee, P.A., Stone, A.D., Fukuyama, H. (1987) Universal conductance fluctuations in metals: Effects of finite temperature, interactions, and magnetic field. Phys. Rev. B, 35, 1039–1070.

236. Benedict, L.X., Crespi, V.H., Louie, S.G., Cohen, M.L. (1995) Static conductivity and superconductivity of carbon nanotubes: Relations between tubes and sheets. Phys. Rev. B, 52, 14935–14940.

237. Tinkham, M. (2004) Introduction to Superconductivity. Dover Publications Inc., Mineola, New York.

238. Kasumov, A.Y., Deblock, R., Kociak, M., Reulet, B., Bouchiat, H., Khodos, I.I., Gorbatov, Y.B., Volkov, V.T., Journet, C., Burghard, M. (1999) Supercurrents through single-walled carbon nanotubes. Science, 284, 1508–1511.

Chapter 4

Introduction of Graphene

INTRODUCTION

The carbon has been taken from the name from the latin word carbo, this literal meaning is charcoal. This element has unique electronic structure that allows for hybridization to build up sp^3, sp^2, and sp networks. This kind hybridization ability allows to form more known stable allotropes than any other element. The carbon most common allotropic form is graphite which is an abundant natural mineral and together with diamond, it has known since ancient times. Graphite has sp^2 hybridized carbon atomic layers from stacked structure with weak van der Waals forces. In which the single layers of carbon atoms tightly packed into a two-dimensional (2D) honeycomb crystal lattice, that is called graphene. This name was given to Boehm, Setton, and Stumpp in 1994. Graphite has a remarkable anisotropic behavior in term of their thermal and electrical conductivity. Graphite is highly conductive in the direction parallel to the graphene layers due to in-plane metallic character, while, it has a poor conductivity in the direction perpendicular to the layers due to occurrence of the weak van der Waals interactions between them. In the graphene layer the carbon atoms form three σ bonds with neighboring carbon atoms by overlapping of sp^2 orbitals, whereas, the remaining p_z orbitals overlapping form a band of filled π orbitals, i.e. the valence band and a band of empty π^* orbitals as well as the conduction band. The filled, unfilled valance and conduction bands including empty band are playing significant role in the high in-plane conductivity. Since the interplanar spacing of the graphite (0.34 nm) is not large enough to host molecules/ions or other inorganic species. To resolve this problem various theoretical groups have contributed actively and speculated the inter-calculations approaches/or methods discussed. The aim is how to enlarge the interlayer galleries of graphite from 0.34 nm to higher values, it can reach more than 1nm in some cases, depending on the size of the guest species. The first intercalation of potassium in graphite have been demonstrated with a plethora of chemical species that tested to construct the known graphite intercalation compounds. This outcome has been provided a path to targeted species can be stabilized between the graphene layers through the channel of the ionic or polar interactions without influencing original graphene structure. This kind of compound formation is only one element it can be more such lithium, potassium, sodium, and other alkali metals, but also with anions such as nitrate, bisulfate, or halogens.

According to IUPAC, the term Graphene means "A single carbon layer of the graphite structure, describing its nature by analogy to a polycyclic aromatic hydrocarbon of quasi infinite size". In the current scientific world, it is a one atom thick hexagonal carbon array that known for its intriguing properties and exciting applications. Therefore, graphene is the elite allotrope in the carbon fraternity which has received immense attention of the scientific community owing to its intriguing structure and properties. In the more precise words, graphene is a flat monolayer allotrope of carbon with sp^2 hybridized atoms fashioned in a honeycomb like lattice. Their honeycomb like lattice can be moulded into other carbon allotropic structures and it may consider as the mother of all graphitic forms. Graphene can also be wrapped to form 0D Fullerenes, rolled to 1D carbon nanotubes, stacked to form 3D graphite. Normally graphene is a dense material to such an extent that even the smallest helium atom cannot pass through it. Graphene is known to be thinnest material known till date; it also has a strongest entity. Its electrical conductivity comparable to copper as well as outshines all other materials in thermal conductivity. In the pristine form graphene can be considered as a semimetal or zero gap semiconductor with unique electronic, mechanical properties and with unexpectedly high opacity with an astonishingly low absorption ratio of white light. Usually, graphene synthesis depending on the number of layers, purity and point of application. The isolation of this single graphene sheets successfully synthesized from Geim and coworkers and the explosion of research interests followed therein gave impetus to the synthesis of high quality, cost effective graphene sheets. Predominately, top-down approach involves separating individual graphene layers in Graphite by overcoming Van der Waals force of attraction between the layers. The key hurdles from this method is to separate the layers efficiently without damage and also preventing the restacking of graphene sheets after exfoliation. Another key approach is bottom down employs direct synthesis of graphene using carbon precursors as building blocks to form graphene layers. The bottom up approach demands high quality graphitization which can be achieved only at high temperatures. Although procedures of bottom up methods are simpler than top up methods, their yield graphene with greater defects.

Considering various significant properties of the graphene including synthesis, structure and extraordinary physical properties, this study indented to introductory study on graphene with their different form such bilayer, graphone, graphyne, graphdiyne, graphene including kind of defects in graphene structure by describing topological defects and their forms stone wales defects, vacancies, ripples and Ad-atoms defects. Moreover, the knowledge of line defects in graphitic materials are valuable for their technical performances. Therefore, the line defects in graphene has also addressed. Since defects in graphene depends on their number layers stacking, therefore, separately defects in the bilayer graphene is also discussed. In this order the cracks formation in graphene is also accommodated in this study. In terms of the physical properties, graphene the thermal, optical, mechanical electronic, chemical and semiconducting properties of the graphene are also extensively discussed in this study. Hence, with this chapter work a detailed view on graphene history, structure, different layer structures, defects and cracks formations within graphene including their different physical properties are presented that would be useful to understand the motive of this whole book i.e. composite materials under a glassy regime, (as we discussed in chapter 5 to 7).

GRAPHENE

Boehm was the first to use graphene, which he described as an allotrope of carbon for the single planar sheet of sp^2-bonded carbon atoms, which were densely packed in a honeycomb [1, 2]. According to him, the structural relationship of graphite can be correlated with various developed forms in which the honeycomb structure is the basic building block of other important allotropes of carbon, such as:

(i)   Graphite three-dimensional stacked honeycomb structure
(ii) In the two-dimensional arrangement it is known as graphene

(iii)   The rolled-up honeycomb structural form is designated as one-dimensional carbon nanotube.

(iv)   The wrapped honeycomb structures are defined as zero-dimensional fullerenes.

Normally, a single layer graphene sheet contains in-plane σ-bonds and out-of-plane π-bonds, as schematic shown in Fig. 4.1. Contribution of π-bonds for the electron conduction in graphene provides a weak interaction between graphene layers, while the covalent σ-bonds form a rigid backbone of the hexagonal structure through the c-axis plane. This means the π-bonds are responsible for controlling the communication between different graphene layers. Usually, a single layer graphene exhibits the three σ-bonds/atoms in one plane, whereas, the π-orbitals are perpendicular to the plane σ-bonds/atom.

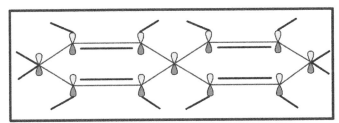

Figure 4.1   Schematic for single-layer graphene.

Historically, graphene is known as graphite that is composed of many layers of graphene stacked together. Such stacking makes a three-dimensional structure that is graphite, while graphene is a two-dimensional, one-atom-thick material. Evidence of the use of graphite has been recognized in pottery decorations from graphite about 6,000 years ago. The modern concept of graphite is also almost 500 years old, the first graphite ore was found and mined in England in the 16th century.

In the beginning, people used it to mark sheep with it and it was believed that the ore of the lead, known as 'plumbago'. In 1779 Scheele demonstrated that plumbago is actually carbon, not lead. Later in 1789, a German scientist named it graphite (a Greek word for 'writing'). It has been used as a writing material in pencils since the 18th century (Fig. 4.2). Due to its layered morphology and weak dispersion forces between adjacent sheets, it has been used as a solid lubricant.

Figure 4.2   A lead pencil tip made of graphite.

The term 'graphene' was first introduced in 1987 to describe single sheets of graphite [3]. The IUPAC (International Union of Pure and Applied Chemistry) replaced the term 'graphite layers' from the 'graphene'. According to the modern definition "graphene is a two-dimensional monolayer of carbon atoms that consists basic building block of graphitic materials (like, fullerene, carbon nanotubes, graphite)". More precisely, it has been recognized that, graphene is a two-dimensional material which has a single layer of carbon atoms arrangement in a honeycomb-like structure, as shown in Fig. 4.3(a, b). The carbon-carbon bond length in graphene has been found to be 0.142 nm, while their layer height just above 0.33 nm. It is the thinnest and strongest material known so far. Graphene is almost entirely transparent and it has a dense structure in which even the smallest atom helium cannot pass through it. It can conduct electricity as efficiently as copper as well as acts like a heat conductor.

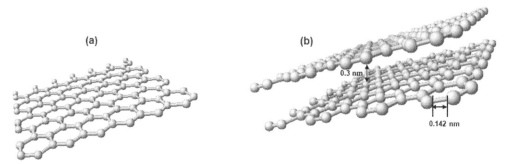

Figure 4.3(a, b) Schematic of the single layer and two layers of graphene.

In 1859 a British chemist, Benjamin Bordie demonstrated that a highly lamellar structure from thermally reduced graphite oxide reacted to graphite with potassium chlorate and fuming nitric acid, therefore, forming a suspension of graphene oxide crystallite. The earliest study on the properties of graphene oxide paper was conducted by Kohlschutter and Haenni in 1919. However, earlier it was believed it could not be grown due to their thermodynamics limit, which prevented the formation of two-dimensional crystal in a free state [4]. Though Wallace had tried to study the electronic properties of three-dimensional graphite and demonstrated the band theory of graphite [5]. According to him:

> The structure of the electronic energy bands and Brillouin zones for graphite is developed using the 'tight binding' approximation. Graphite is found to be a semi-conductor with zero activation energy, i.e., there are no free electrons at zero temperature, but they are created at higher temperatures by excitation to a band contiguous to the highest one which is normally filled. The electrical conductivity is treated with assumptions about the mean free path. It is found to be about 100 times as greater as across crystal planes. A large and anisotropic diamagnetic susceptibility is predicted for the conduction electrons; this is greatest for fields across the layers. The volume optical absorption is accounted for.

Moreover, the next milestone work on graphene was reported in the first TEM image of a few layers of graphene by Ruess and Vogt in 1948 [6]. In 1961 Hanns-Peter Boehm and coworkers identified the isolated single graphene sheets using the TEM and XRD. This work was published in 1962. Boehm was also a member of the IUPAC (International Union of Pure and Applied Chemistry) and formally defined the term graphene in 1994. However, it is surprising that many reviews and papers have mentioned that graphene was only discovered in 2004. The TEM taken by Boehm et al. remained the best observation in the last 40 years.

Hence in the last 40 years (between 1960 and 2000) research on graphene has grown slowly in multifarious directions, including synthesis. The aim to observe superior electrical properties from thin graphite or graphene layers, while deducing graphene can be considered to be a formidable task in both theoretical and experimental aspects. Such as in the graphite intercalation systems, large molecules were inserted between atomic planes, to generate isolated graphene layers in a three-dimensional matrix, and the subsequent removal of the larger molecules to produce a mixture of stacked (or scrolled) graphene layers without disturbing the structure. In this period of research, the main concern was high conductivity of graphite intercalation compounds and future applications.

There were attempts to grow graphene using the same approach which was used for the growth of carbon nanotubes, but allowed the formation of thicker than ≈100 layers graphite films [7]. In this order, Hess and Ban used for the first time a Chemical-Vapor-Deposition (CVD) technique, in which carbon atoms were supplied from a gas phase for the formation of monolayer graphite or graphene [8]. However, Land et al., made efforts for the epitaxially growth of few-layer graphene from the chemical vapor deposition of hydrocarbons on metal substrates as well as on top of other materials, they successfully deposited the SiC by the thermal decomposition method [9, 10].

The epitaxial graphene consists a single-atom-thick hexagonal lattice of sp^2 bonded carbon atoms, like a free-standing graphene. There is significant charge transfer from the substrate to the epitaxial graphene, whereas, in some cases hybridization between the d orbitals of the substrate atoms and π orbitals of graphene can significantly affect the electronic structure of the epitaxial graphene. This experimental demonstration verified the fact that electric current would be effectively carried by mass less charge carriers in graphene as pointed out theoretically by Semenoff et al. [11].

For use as a dry lubricant, the layered morphology and weak dispersion forces between adjacent sheets have made graphite an ideal material, compared to similar structured and more expensive compounds like hexagonal boronnitride and molybdenum disulfide. The high electrical ($104 \ \Omega^{-1} \ cm^{-1}$) property along the plane and its thermal conductivity (3000 W/mK) enable graphite to be used as electrodes for the heating elements in industrial blast furnaces [12].

At the beginning of the 21st century, many important discoveries were reported related to graphene. Such as in 2003, Enoki explained the anisotropy of graphitic material [13]. Furthermore, the bulk graphite was first intercalated in 2002 [14], therefore, graphene planes can be separated by layers of intervening atoms or molecules. This has provided new three-dimensional materials. Large molecules could be inserted between atomic planes, which can lead to greater separation in certain cases. These kinds of compounds could be considered as isolated graphene layers embedded in a three-dimensional matrix. Among these and other remarkable innovations and discoveries in the area of graphene, the first Noble Prize was awarded in 2004 to Andre Geim and Kostya Novoselov for their novel research work "Nano-scaled graphene plates". Later several milestones were achieved in the field of graphene science and technology and this journey continues to investigate different forms of the graphene and their possible technical use for future requirements.

BILAYER GRAPHENE

Usually two layers graphene is called bilayer graphene. This can be formed either in twisted configurations such as two layers rotate relative to each other or graphitic Bernal stacked configurations from half atoms in one layer lie on top of half the atoms to other. The bilayer and few-layer (more than two layers) graphene has been classified as pseudo two-dimensional sp^2 hybridized carbon structures. The bilayer and few-layer graphene properties are different from monolayer graphene and graphite.

Carbon atoms are stacked in bilayer graphene uniquely, in hexagonal or AA stacking, or AB stacking, as shown in Fig. 4.4(a, b). It is evident that at the end of each hexagonal structure of the graphene layer (mono or bilayer) contains hydrogen atoms to satisfy its four atoms valiancy. This explains that in a graphene layer each hexagonal structure at its interface contains three hydrogen atoms.

Moreover, in this kind of graphene the band gap can be created by applying an external electrical field, and their magnitude can be controlled by the magnitude of the electric field. In the bilayer graphene band gap can be changed from zero to infrared energies. Their band gap can also be precisely tuned in this range (zero to infrared energies). This offers a great flexibility in the design and optimization of semiconducting devices, the corresponding schematic is illustrated in Fig. 4.4(c).

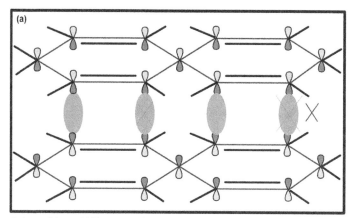

Figure 4.4(a) AA-stacking of bilayer graphene; out of plane the π-bond of the top layer of graphene overlaps with the π-bond of the lower layer, formation a band separation at a zero distance.

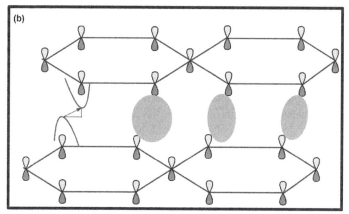

Figure 4.4(b) AB stacking of bilayer graphene; zero band gap indirect type band formation, in which carbon lower layer does not lie exactly below the carbon of the top layer.

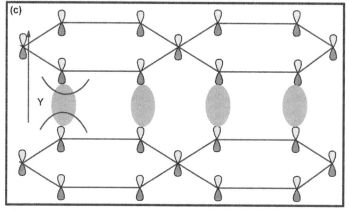

Figure 4.4(c) Bilayer graphene band formation under an applied electric field; a slight increase in the band gap (along the Y-axis).

Since in monolayer graphene there is no place for the p_z orbital (π bond) to interact with the carbon atom underneath, while, in the case of bilayer graphene such a possibility may exist. Therefore, bilayer graphene can form a band gap with zero energy. If an electric field is applied (*c*-axis), there is a possibility that their band gap can shift from zero to a maximum of 250 meV. Such a shift in band gap can be maneuvered with the magnitude of the electric field. The condition is predominantly favored when the bilayer has AA type stacking. In the case of AB type stacking, the p_z of both carbon atoms cannot exactly fall below each other. In this situation, it would be preferred that the interaction leads to the formation of indirect band gap and their magnitude may also change from zero to 250 meV. Therefore, it is required to establish the behavior with AB type graphene. Additionally, it is also necessary to define whether the shift in the band gap is due to the change in the distance 0.3354 nm or is due to the shift in the magnitude of wavelength of π-electrons.

GRAPHONE

The half hydrogenated graphene that consists of a carbon sheet with a twist is called graphone. Particularly bilayer graphone has non-magnetic and metallic characteristics due to their delocalized orbital above the carbon chains. This can occur due to half carbon atoms hydrogenated, thereby the π-bonding network disrupted, which leads to the localized and unpaired electrons in non-hydrogenated carbon atoms. Therefore, it allows the magnetic ordering (ferromagnetic, antiferromagnetic) as well as the band gap values that can be controlled by the hydrogenation patterns. These features make them promising materials for spintronic applications [15]. However, spintronics are also known as magneto-electronics, in which the property of spin of electrons is associated with magnetic moment and charges are used to develop the solid-state devices, that can store and transfer information in digital forms.

It is also recognized that in triangular graphone, all electrons can be localized in hydrogen atoms. No electrons can exist in the space between two hydrogen atoms. In between the two hydrogen atoms, a very weak interaction makes the triangular graphone metastable. While the rectangular graphone electron density between any two hydrogen atoms can be large, this allows a strong interaction between hydrogen atoms. According to electronic band structure calculation, the rectangular graphone can be related to anti-ferromagnetic, having an indirect band gap of 2.5 eV as compared to ferromagnetic triangular graphone, that possesses a very small band gap or nearly zero [16].

The bond lengths of C–C and C–H in triangular graphone were calculated to be 1.495 Å and 1.157 Å by Feng and Zhang [16]. The distance between two carbon planes is 0.322 Å. While the bond length of C–H for the rectangular graphone were estimated to be 1.109 Å, that is smaller than the C–H bond length for the triangular graphone. Moreover, if the C–H bonds are not perpendicular to carbon planes then all H atoms could be in a similar plane. And the distance between H atoms are 2.030 Å, 2.524 Å and 2.319 Å, respectively. While the distance between two carbon planes is 0.399 Å. Thus, the calculated distance is a little bit larger than the triangular graphone.

Graphyne

Graphyne is also an allotrope of carbon and its crystal structure is sp and sp^2 bonded carbon atoms arrangement in the lattice for one atom thick planar sheet. More than two-dimensional planar carbons besides graphene can also exist, which is defined as graphyne, its existence was also predicted [17–20].

But the fact remains that as yet only a few molecular fragments have been synthesized [21, 22]. It is believed that a vast number graphynes may be possible, in which each of them can have double and triple bonds with slightly different arrangements, as shown in Fig. 4.5. Theoretically the existence of graphynes has been known since the 1980s.

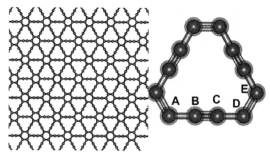

Figure 4.5 Geometry of γ-graphyne; the left side represents a γ-graphyne monolayer plane and right hand side shows a γ-graphyne molecule.
(Reproduced from Peng, Q., Wei, J., De, S. 2012. Mechanical properties of graphyne monolayer: A first-principles study. Physical Chemistry Chemical Physics, 14, 13385–13391, with permission, copyright PCCP Owner Societies)

Baughman et al. first proposed graphyne in 1987, as part of a larger investigation of the properties of carbon novel forms that were sporadically reported, but were not systematical investigations [18]. Specifically, graphyne received a great deal of attention due to their electronic structure that was quite different from the existing materials, including carbon-based materials like diamond and graphite. The most common form of graphyne is γ-graphyne, but it is not limited to this and may exist in several others forms [17].

Graphdiyne

The discovery of graphene and the prediction of graphyne have provided a path for additional two-dimensional materials which has received significant research attention. Along with these, graphdiyne has also been recognized as a variant of graphyne that contains two acetylenic linkages in each unit cell rather than the one linkage of graphyne, as illustrated in Fig 4.6(a, b) [23]. In graphdiyne the two acetylenic linkages can have double the length of carbon chains for those connected with hexagonal rings [24].

This could be connected to non-sharing mechanical properties behavior of graphynes. Therefore, graphdiyne is a softer material than graphyne or graphene having in plane stiffness of 120 N/m, that is equal to Young's modulus of 375 GPa when their thickness is at 0.320 nm [24].

(a)   **(b)**

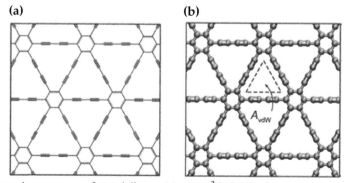

Figure 4.6 Schematic structure of graphdiyne. **(a)** sp–sp^2-hybridized carbon atoms two-dimensional chemical structure of graphdiyne, with the characteristic hexagons of graphene by diacetylenic (single- and triple-bond) linking that forms a repeating and regular nanomesh. **(b)** An atomistic model for the graphdiyne including atomistic triangular pores, with van der Waals openings (Avdw).
(Reproduced from Cranford, S.W., Buehler, M.J. (2012) Selective hydrogen purification through graphdiyne under ambient temperature and pressure. Nanoscale, 2012, 4, 4587–4593, with permission of The Royal Society of Chemistry. Copyright © 2012)

In 1997 Haley et al. first speculated about graphdiyne [25]. At the beginning, research work focused on synthesis of the material having organic molecules that were alike. Similar materials computational models were used to estimate the properties of graphdiyne [25]. Although graphdiyne belongs to graphyne family; but due to its interesting properties, it is considered separately [24, 26].

The ab initio simulation calculations predicted the in plane stiffness and Poisson's ratio 0.453. Under the strain free state, the band gap has been calculated between 0.47 eV to 1.12 eV by using different methods. According to the generalized gradient approximation the PerdewBurke-Ernzerh exchange correlation functions underestimated the band gap, while the Heyd-Scuseria-Ernzerh exchange correlation functions overestimated the band gap. However, these results have bound the range of possible band gaps, and limited the non-zero values. This can have a distinct advantage over graphene for transistors with high on–off ratios.

Additionally, the theoretical demonstration had also shown that the band gap is proportional (with a positive proportionality constant) to externally applied strain [26]. Their band gap can vary from 0.28 eV to 0.71 eV for the $\varepsilon = -0.05$, $\varepsilon = 0.06$, with the strain free value at 0.47 eV. This indicates that the band gap of graphdiyne can be readily tuneable for specific applications. Thus, the near mass less behavior of electrons, as predicted by the Dirac cone shape of the electron band gap, can be verified, making them an attractive semi conductive material.

Moreover, Cui et al. conducted an investigation augmenting the wave method [27]. They showed that the equal biaxial strain increases with the band gap, which has been confirmed by previous results, while, demonstrating that the uniaxial strain decreases the band gap with changing the electron densities around the carbon atoms in the diacetylenic linkages. Such coupling of mechanical loading in electronic properties is significant and useful for engineering applications.

Due to the similarity in graphyne and graphdiyne, a common thickness (3.20 Å) can be used for their qualitative comparison. But when thickness and stiffness of graphdiyne has been achieved approximately by 30% of the graphyne then their strength is noticeably reduced [23, 25]. The reductions can be directly correlated to extra acetylenic linkage that effectively reduced the atomic coordination number, as a consequence weakening the bonding. This also impacted on the mechanical behavior of graphdiyne that appeared isotropic [25]. This demonstrates that graphdiyne is not suitable for nanocomposites synthesis.

Besides such mechanical properties their electronic properties were predicted to be exceptional. Though several graphdiyne have been synthesized, most of the properties have been predicted with the help of computational studies rather than experimental outcomes. The special interest is its low effective electron mass that can give the Dirac cone structure of the electron bands. The effective masses of the order of 0.08 are an actual electron and their band gap in a semi conduction range [25]. To determine the effects of finite width on the properties of the material, graphdiyne nanoribbons have been extensively explored. Their band gap shows increases with decreasing ribbon width, under a similar armchair and zigzag orientations trend [28]. Subsequently, with increased width infinite sheets of graphdiyne do not have directional dependence. An identical study also demonstrated in terms of the response of band gap to electric fields. The corresponding electric field perpendicular to nanoribbon has shown a decreased band gap up to a closing field strength 0.11 V/Å

Graphane

Graphane is also similar to graphone, which is a hydrogenated sheet of graphene. Their primitive cell contains two carbon atoms and two hydrogen atoms [29]. The significant difference is that it is a 50% hydrogenated sheet compared to 100% hydrogenated graphene sheet, while their structural parameters are identical. However, they have roughly 5° difference in the C–C–C bond angle orientation and less than 0.1 Å difference in the C–C bond lengths. Additionally, with the notable difference, each carbon in graphane has been bound to a single hydrogenated atom under complete sp^3 hybridization, while, graphones are mixtures of sp^2 and sp^3 hybridized bonds and graphenes sp^2 hybrid structure. Moreover, at finite temperatures graphane does not has intrinsic thermal ripples that are usually present in graphene, the typical geometry of graphene is represented in Fig. 4.7 [30].

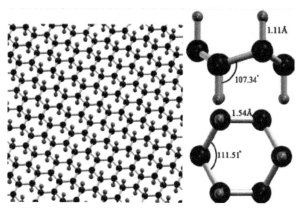

Figure 4.7 Schematic for the geometry of graphene.
(Reproduced from Peng, Q., et al., (2013) A theoretical analysis of the effect of the hydrogenation of graphene to graphene on its mechanical properties. Physical Chemistry Chemical Physics, 15, 2003–2011, with permission of The Royal Society of Chemistry. Copyright © 2013)

Unlike graphone, graphane can be synthesized easily and their electronic, optical, mechanical and thermal properties have been investigated extensively [31–41]. Graphane is attracting much attention due to the fact that hydrogenation of graphene into graphane can be reversed through annealing at high temperatures; as a consequence it can restore the original properties of graphene [42]. Graphane can be applicable in the areas of hydrogen storage, bio-sensing, transistors and spintronic devices [43–47].

DEFECTS IN GRAPHENE STRUCTURE

Infinite number of distinct structural defects may occur in synthesized graphene that may be introduced by energetic particles irradiation or by chemical treatment [48, 49]. The structural defects in graphene are dependent on dimensionalities, as shown in Fig. 4.8.

According to the concept of zero-dimensional point defects, typically includes vacancies, adatoms and substitutions. Each individual vacancy can have various possible configurations without undercoordinated atoms due to sp^2-hybridized carbons may arrange themselves into a variety of different polygons and not restricted only in hexagons [50]. Beside the reconstructed defect has no dangling bonds, it may locally increase the reactivity of the structure that can work as an adsorption site for atoms, molecules and clusters of graphene [51]. Meanwhile, the foreign atoms can also be introduced into the hexagonal carbon lattice by substitution [52, 53]. Additional, the Stone Wales (SW) defects (explained later) reveals from CC bond rotation without atoms gained or lost could also considered as zero-dimensional defects. Since all such point defects can migrate in the plane that can aggregate to form complex higher dimensional structural defects. Therefore, one-dimensional line defects emerge in different situations from those in bulk crystals owing to reduced dimensionality of graphene. This is not only a dislocation but also grain boundaries of one-dimensional lines, in which atoms can be arranged anomalously. However, two-dimensional stacking faults may also exist in few-layer graphene similar to graphite [54]. It is well established that all those defects can be directly observed by Transmission Electron Microscopy (TEM), specifically, in the development of aberration corrected electron optics [55]. More significantly, imaging electrons can transfer some energy to C atoms when they pass through the graphene sheets, that may stimulate structural changes for further investigation on formation and evolution of structural defects [56, 57]. Therefore, TEM is the most adopted experimental tool to characterize the defective atomic structure of graphene.

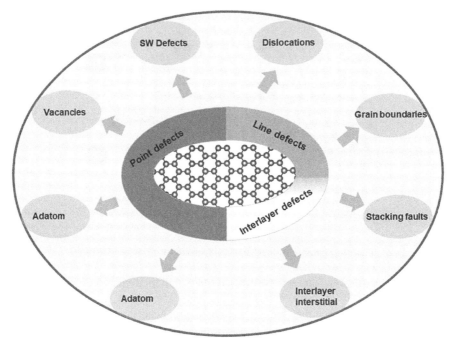

Figure 4.8 Schematic representation of structural defects in graphene depending on different dimensions such as zero-dimensional point defects (Stone Wales defects, vacancies, adatoms and substitutions). One-dimensional line defects (dislocations and grain boundaries). Interstitial defects between the layers can bridge the adjacent layers that may form higher-dimensional structures with another stacking fault defect in few-layer graphene.

Though investigators have predicted that structurally perfect graphene possesses extraordinary properties, such as extreme mechanical strength, excellent electronic and thermal conductivities. The properties of graphene can be significantly degraded owing to structural defects as shown in Table 4.1. It has been demonstrated that both vacancies and dopant atoms can change the scattering of electrons and phonons; therefore, they can greatly affect the electrical and thermal conductivity with the increasing concentration of defects. The weaker bonds around defects may also reduce the mechanical strength. According to Molecular Dynamics (MD) simulation prediction, Young's modulus gently decreases with increasing defect concentration, while the thermal conductivity decrease quickly up to 25% of the perfect graphene, as a consequence, concentration of single vacancy may increase up to 0.5% (or SW defects increase to 1%) [58]. Additionally, the physicochemical properties of graphene not only depend on the presence of defects but also on the defect types and the distribution of defects.

Table 4.1 Structurally perfect graphene and defective graphene performance parameter difference [59].

	Perfect graphene	Defective graphene
Young modulus (TPa)	~1	0.15–0.95
Fracture strength (GPa)	90–130	47–117
Electrical conductivity (S/m)	10^8	$1–10^5$
Electron mobility (cm^2/V/s)	$10^5–10^6$	$1–10^4$
Thermal conductivity (W/m/K)	$10^3–10^4$	$10–10^3$

Usually defects in graphene have been categorized into two different groups: the first one is intrinsic defects that are composed of non-sp^2 orbital hybrid carbon atoms in graphene. These kinds of defects are usually due to the existence of non-hexagonal rings surrounded by hexagonal rings; while the second kind of defects are extrinsic defects. Hence, the crystalline order is perturbed with non-carbon atoms in graphene [60].

Topological Defects

Stone–Wales (SW) Defects

In graphene, Stone–Wales (SW) defects are created due to rotation of a single pair of carbon atoms, this gives pentagonal and heptagonal rings in the adjacent pairs. These kinds of defect formations is not due to the introduction or removal of carbon atoms or dangling bonds. The defect formation energy can be around 5 eV [61, 62]. Such kinds of defects can be introduced using the electron radiation or rapid cooling under the high-temperature environments, as shown in Fig. 4.9 [63, 64]. The reason for the formation of these defects may be due to electron impact.

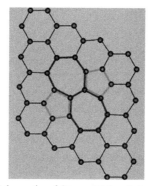

Figure 4.9   Schematic of Stone–Wales defects in graphene.
(Reproduced from Banhart, F. et al., (2011) Structural defects in graphene, ACS Nano, 5, 1, 26–41, with permission of American Chemical Society, Copyright © 2011)

SW defect in graphene can modify its vibrational mode which may be identified from the Raman spectrum. Therefore, in such defects C–C bond rotation can be shorter and stiffer than the unrotated C–C bonds; as a consequence, they release stress that leads to the elongation of the bonds with less stiffness in the area around the defect. Hence, in the graphene Raman characterization of the D band appears due to defects hardening, while the G band cause the plane vibration of sp^2-bonded carbon atoms to soften. Due to this, a shift appears in the frequency of the G and D bands that can be directly related to the characterization of the SW defects [65].

The incorporation of SW defects in graphene can also affect its mechanical properties. According to mathematical simulations it was also predicted that the presence of SW defects significantly reduces the failure strain and intrinsic strength of monolayer graphene sheet [66]. Moreover, the initial failure state and strength may be sensitive to the defect orientation and tensile direction [67]. When the rotated bond is perpendicular or parallel to loading directions, the tensile strength of the armchair direction can be much lower than the zigzag direction. However, the strength in the armchair direction can be much larger, therefore, the breaking strength decreases with the increase in the tilting angle between tensile direction and aligned axis of the SWs [67]. Thus, the elastic modulus may decrease gradually with the increase of defects; therefore, the concentration can have a mild impact on elastic modulus compared to thermal conductivity [66].

Creation of SW defects in graphene can also lift the degeneracy of the band at the Fermi level that opens up a non-zero band gap [65], this makes them favorable to be used in transistors. At the center of SW defect, the p_z orbitals of carbon atoms can predominantly contribute to the defect band.

The width of such band depends on the orientation and concentration of defects of the graphene sheet [65]. Similarly, the presence of SW defects can also make thermal conductivity anisotropic due to symmetry break and quench of the thermal conductivity by reducing phonon mean free path at low temperatures. Hence the phonon-phonon umklapp scattering becomes dominant at room temperature and above [68]. Moreover, SW defect does not induce magnetization in the defective system, as all C atoms in the system bonded with three other C atoms, do not leave the possibility for the dangling bond.

The existence of SW defect in graphene sheet has not only modified its physical properties but also changed its chemical properties. The SW defect can also increase the local chemical reactivity that may work as an adsorption site for other atoms and molecules. Their DFT calculations has predicted that, the presence of defects can strongly influence the hydrogen chemical reactivity [69, 70]. As an instance, due to occurrence this kind defects; the hydrogenation energy can shift highly unfavorably in perfect graphene virtual energy to neutral over a SW defect [70]. Therefore, graphene materials with SW defects can have an important application in hydrogen storage. Furthermore, the presence of SW defects can also enhance the adsorption of lithium (Li), sodium (Na) and calcium (Ca) due to their increase charge transfer between adatoms and fundamental defective sheets [71, 72].

Vacancies

Vacancy is the point type of defect in which one or more atoms can be missing from the lattice sites, which may occur naturally in all kinds of crystalline materials including graphene. Vacancy defects in graphene can be divided into single vacancy, double vacancy and multiple vacancies. When a single carbon atom is missed in a carbon hexagon ring, it is called single vacancy in graphene, as shown in Fig. 4.9. Using experimental evidence of the single vacancy defect, their atomic structure can be calculated [73, 74]. Incorporation of this defect graphene undergo a Jahn-Teller distortion for the minimal total energy, in which two of the three dangling bonds can be connected to each other toward the missing atom. However, one dangling bond remains owing to geometrical reasons. The formation of this type of vacancy defect with such a dangling bond requires higher energy than the Stone-Wales defect. The estimated energy calculations for this type of vacancy formation is approximately $E_f \approx 7.5$ eV [75, 76]. Under contraction of this type of defect, the reconstruction can transmit to neighbors of the atom with the dangling bond forcing them together, as a consequence they push the dangling bond atom out of the plane in order to preserve its bond lengths [77]. These out-of-plane displacements can play a significant role in the energy lowering effect.

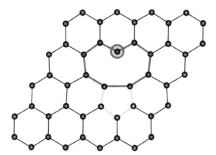

Figure 4.10   Schematic of double bond vacancy defect in graphene.
(Reproduced from Banhart, F. et al., (2011) Structural defects in graphene, ACS Nano, 5, 1, 26–41, with permission of American Chemical Society, Copyright © 2011)

The double bond defects occur due to the coalescence of two single bonds or from the removal of two neighboring atoms, as illustrated in Fig. 4.10. The double bond defect is more stable than single bond defects owing to the absence of dangling bonds in the reconstruction and more complex from several other possible configurations. The most common 5–8–5 configuration of double bond defects in which two opposing pentagons and one shared octagon are formed instead of four hexagons in perfect graphene. This type of configuration is planar with minor perturbations in the bond lengths around the defect. Since the formation energy of a 5–8–5 defect is 8.7 eV, which is much lower than two separated single bond defects [77]. This reveals the 5–8–5 defect is thermodynamically favored over a single bond defect. This type of defect bonds migration energy can also be as high as 7 eV, giving their immobility at a very high temperature [77]. Another frequently observed defect configuration is 555–777, this type of defect configuration is composed of three pentagons and three heptagons. The formation energy of 555–777 configuration defects is 1 eV lower than 5–8–5 defect [78], this could be correlated to their high order stability. Another defect configuration 5555–6–7777 consists of four pentagons and four heptagons in the alternate arrangement around one hexagon. This can also be transformed from the 555–777 configuration through the bond rotation and their defect formation energy is within the range [79]. Additionally, during the dynamical process of migration and transformation, some other configurations can also be observed (such as a 5–7–7–5 defect) due to the intermediate structure with two pentagons and two heptagons [80].

The more complex defect configurations can be due to the removal of more than two atoms or by the migration and aggregation of single bond and double bond defects. Like a tetravacancy can generate from the removal of four carbon atoms or through the coalescence of two double bond defects. Such a tetravacancy defect can also form different configurations owing to their abundant permutations of missing atoms and bond rotation paths. The two most frequently demonstrated tetravacancy are extended linear defect configurations [81], due to the removal of two adjacent dimmers along the same armchair axis, and the other is due to the coalescence of two neighbored 5–8–5 defects by three bond rotations.

Usually a multi-bond defect is defined with the even number of missing atoms, in which all bonds can be saturated after reconstruction that is energetically favored over structures with the odd number of missing atoms following the condition that at least one open bond remains. When a larger number of atoms are removed from a small area, then a hole with unsaturated bonds around its circumference may be created [81]. This can have either an armchair orientation, or zigzag orientation or any other direction between these two may appear at the edge. Under these circumstances, the zigzag edge may be the most stable case as the others might be transformed to zigzag edge under electron irradiation [82]. The additional other edge configurations could appear through the removal and reconstruction of carbon atoms, such as, extended pentagon–heptagon reconstruction [83]. The hydrogen atoms and other chemical groups could saturate dangling bonds at the edge under the suitable conditions, resulting in a dramatic increase in the number of possible edge defects. It could also extend the dislocation line, which may be from a linear arrangement of vacancies that are close to the saturation of the dangling bonds over the line [84]. In a situation of larger number density of vacancies creation in graphene, the formation of amorphous two-dimensional carbon glass or crystalline Haeckelite structure may be more favorable [85, 86].

The experimental evidence on the impact of vacancies demonstrated that it can decrease Young's modulus and tensile strength [87]. The experimental and theoretical studies had also confirmed that vacancies can increase the detrimental effect on the mechanical property of graphene [66, 88]. The mathematical modeling study had demonstrated that the vacancies can be more detrimental to Young's modulus and tensile strength as compared to Stone–Wales defects [66]. However, Young's modulus may gradually reduce with the increasing concentration in various kinds of defects, which can be more detrimental at low concentration (i.e., 2%) under the influence of a great number of under coordinated atoms [66]. Hence, the reconstruction of vacancies could minimize their detrimental effects. As defined by the mathematical modeling simulations, a considerable decrease on the tensile strength and failure strain from the increasing defect concentration in the range of 0.25 to 2% was not noticed in the case of double bond defects, but was considerable in the case of single bond defects [66].

Defect induced graphene reconstruction also has an impact on the distribution of vacancies and elastic modulus significantly. Therefore, it is not a problem to conclude that vacancies aligned in the direction perpendicular to the loading direction can have more of an impact on Young's modulus of defective graphene. Their positions of vacancies can have a significant impact on the shear modulus and Poisson's ratio. In the case of small separation distances of vacancies, the stress concentration may superimpose, resulting in a greater reduction in strength. Moreover, if the adjacent vacancies coalesce form holes locally, a precipitous drop in elastic stiffness and strength may be possible [88].

Such defects creation in graphene not only influences Young's modulus, but also impacts on the thermal conductivity which is more sensitive under the presence of vacancies. As an example only 0.25% concentration of single and double bonds defects could reduce the thermal conductivity up to 50% or even more [58, 66]. In particular thermal conductivity reduction is at a very rapid rate at low defect concentration (0.3%) and it becomes gradual at higher concentration of defects [58]. Moreover, this reduction is rapid in single bond defects compared to double bond and Stone–Wales defects. Since at the same defect density, the defect centers (each defect is treated as an individual scattering center) in single bond defective graphene are about twice as much as compared to the double bond and Stone–Wales defective graphene. This leads to more phonon-defect scattering in single bond graphene, therefore, undergoing more severe deterioration in thermal conductivity than double bond defects. Thus, the reduction in thermal conductivity is dominated by the defect density rather than the defect types with increasing defect density if the defect concentration is large enough.

The vacancy defects can also act as scattering centers for the phonon as well as electron waves. This results in a drop of conductance, however, different configurations of vacancy of defects might affect the electronic transport in different ways. It is also expected that vacancy defects can strongly affect the electronic structure. Therefore, theoretically the Dirac equation has to be modified after the introduction of vacancies that obviously influence the electronic structure. As a consequence, the overlap p_z orbitals can be altered in the vicinity of vacancies; local rehybridization of σ and π orbitals due to reconstruction changes of the electronic structure. DFT study has shown that open band gap due to vacancies may reach up to 0.3 eV [89]. Although, a small amount of experimental evidence is available as compared to theoretical data on the investigation of the electronic properties of vacancy-defective graphene. This is due to the challenge in preparing well-defined defect structures [50].

The defective vacancy graphene can also influence magnetization. Owing to localized unpaired electrons or dangling bonds that may essentially cause magnetization. Since the Stone–Wales defect does not induce magnetization owing to the lack of dangling bonds while the vacancy defect may be magnetic, as an extra π electron can contribute to an unpaired spin. Therefore, vacancy with local magnetic moments can increase the flat bands as well as develop the magnetic ordering. According to DFT calculations, it has been estimated that the single bond defect magnetic moment is in the order of 1.1 μB per vacancy which is mainly contributed from the under coordinated atom. This kind of magnetic moment may disappear when coalesced into (without dangling bonds) double bond defective states [78]. Hence it could be concluded that double bond defective graphene do not have magnetization, but its single bond counterpart can possess magnetization.

Thus, due to the defects associated with dangling bonds, it can enhance the reactivity of graphene. Several functional groups such as, hydroxyl, carboxyl, etc., can easily attach to vacancy defects by under coordinated atoms. The reconstructed vacancies without dangling bonds can increase local reactivity due to the change of local π-electrons density [90, 91]. Hence defective graphene can be useful for various applications such as high-efficiency material for hydrogen storage, high-capacity anode material for ion batteries, adsorption material for environmental protection, etc. [69, 71, 72, 90].

Ripples

Ripples are characterized as intrinsic features of graphene sheets that can strongly influence electronic properties. By inducing the effective magnetic fields as well as change in local potential number of flexural modes per unit of area at a certain temperature T, the corresponding temperature thermal wavelength of flexural modes can be defined from, L_T as follows:

$$N_{ph} = \frac{2\pi}{L_T^2} \ln\left(\frac{L}{L_T}\right) \tag{4.1}$$

$$L_T = \frac{2\pi}{\sqrt{K_B T}} \left(\frac{K}{\sigma}\right)^{\frac{1}{4}} \tag{4.2}$$

At the room temperature ($T = 300$ K), $L \sim 1$Å, this indicates that free-floating graphene should always crumble at room temperature due to the thermal fluctuations associated with flexural phonons [92].

Adatom Defects

In the graphene sheet, the interstitial atoms do not exist owing to the occupation of a C atom to any plane position that is prohibited due to a high energy requirement. Therefore, the additional C atoms property should belong to the third dimension rather than straining the local structure in two dimensions. Spontaneously, the additional adatoms belongs to three important high-symmetry positions, as shown in Fig. 4.11, in which B is the bridge site just above the bond center between two adjacent atoms, T is directly on top of an atom, H site is above the center of a hexagonal ring [93]. According to such a defect description, a C atom can interact with a perfect graphene sheet, therefore, some degree of sp³-hybridization can appear locally; therefore, the adatom could be bonded to the underlying atoms in the graphene. The high symmetry sites formation energy can be largely dependent on the number of formed bonds [94]. The bridge site B can be at an energetically favored position for lower formation energy (less than 7 eV), in which two new

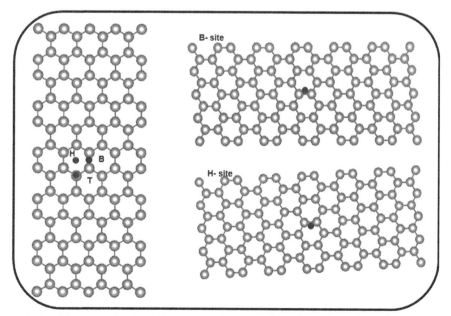

Figure 4.11   Adatoms and substitutions on graphene; configurations of a carbon adatom at various symmetry sites.

covalent bonds can be formed between the adatom and the two adjacent atoms. It has also been demonstrated that the top position T (0.5–1 eV) is less stable than site B, when the adatom bonds directly attached to only one layer atom. The position of H can be the most unstable for C adatoms with the formation energy more than 8 eV, due to the character that the adatom does not make bonds from the layer atoms [94]. Therefore, a small energy barrier (less than 0.5 eV) can allow adatoms to migrate easily over the graphene surface at room temperature.

The migrated adatoms when meet with vacancies; they formed vacancy adatom pair that may be unstable against recombination even below room temperature. The complete or partial recombination depends on the number of adatoms matches with the requirement for complete recombination. In the case of meeting of the two migrated adatoms, they can form a dimer, such a dimer can be incorporated into the network of carbon atoms at the expense of local curvature [95]. However, the out-of-plane defect can also have pairs of pentagons and two heptagons; such defective states are also known as inverse Stone–Wales defect. It should be noted that the arrangement of non-hexagonal rings for the inverse Stone Wales is different from Stone–Wales defect, in which two heptagons are separated by a pair of pentagons. The formation energy of the inverse Stone–Wales defects can be little higher than SW defects. This demonstrates that concentration of inverse Stone–Wales defects could be negligible in the otherwise flat carbon nanostructures. Furthermore, for inverse Stone–Wales defects, they may align in a particular way, therefore, can locally change the curvature of graphene sheets and form corrugate ridges of nearly arbitrary contours or even blisters [95].

Despite C adatoms, other foreign atoms can also be chemisorbed or physically adsorbed on the graphene surface, depending on the bonding between the foreign atoms and graphene. Under strong interactions, they may form covalent bonding between the foreign atom and the nearest C atoms in graphene through the chemisorptions process. While, below the van der Waals interaction the weak bonding occurs through the physical adsorption. These two processes are very similar to C adatoms, high-symmetry positions such as the bridge position B, the position T on top of a carbon atom, or the position H on top of the center of a hexagon. This makes them a favorable site for adsorption. In order to confirm this view, the DFT calculations also predicted that the H site energetically favored the position for most of the metal and transition metal elements. Those possessed a partially filled d shell, whereas the B site can be energetically favorable for most of the non metallic elements and transition metal elements that have a filled or almost filled d shell [96]. Moreover, the T site can be the most stable adsorption site for hydrogen atoms and haloid atoms owing to its electrostatic attraction between the oppositely charged adatom and graphene surface.

The foreign adatoms on perfect graphene can also be pinned through structural defects. This normally acts as a potential doping site due to local increase in reactivity of the π-electron system. Since both Stone Wales defects and reconstructed vacancies can trap foreign atoms even when all the covalent bonds are saturated. In this condition, the strain field around reconstructed defects can result in an interesting interaction over a scale of 1–2 nm in between the defect and an adatom migration on the surface [97].

Defective graphene is also receiving much attention due to its tuneable electrical and magnetic properties by doping [90]. All kinds of defects can lead to scattering of the electron waves, this can change the electron trajectories, and eventually leads to a drop in conductance. However, foreign species on substitutional sites are considered to be unfavorable due to strong scattering of the conduction electrons at such sites that may deteriorate the electronic properties of graphene [50]. Because the dopants can lead to the resonant scattering effects which strongly depend on the distribution of the dopants as well as geometry of the graphene [98]. Therefore, adatoms and substitutions disrupt the sp^2 hybridization of the carbon atoms and local curvatures. The adatom complexes can also have an influence on the rehybridization due to the significant changes in electronic properties of the graphene.

According to theoretical calculations and the demonstrated experimental evidence, the magnetic behavior in graphene can be interpreted in terms of defects with under coordinated carbon atoms (e.g., vacancies, and carbon adatoms) [89]. These kinds of defects have local magnetic moments that could raise the flat bands and consequently develop magnetic ordering. Therefore, magnetism could also appear from impure atoms. The DFT calculations predicted that foreign adatom single bond defect complexes can be magnetic, such as, vanadium (V), chromium (Cr), manganese (Mn), cobalt (Co), gold (Au), copper (Cu), while foreign adatom double bond defects complexes are magnetic for all kinds of transitions metals [88]. The overall magnetic moment range 1–3 µB which mainly come from the metal atom, specifically from Cu and Au. Since Cu and Au have filled d shells, according to an estimate more than 30% of the atom magnetization comes from the s and p states, and about half of the total magnetization is due to neighboring C atoms [88]. It is interesting that even foreign atoms that have a non-magnetic behavior, could under a specific chemical environment have local magnetic moments. As an example, the high-level N-doping (29.82%) could have a significant increase in the magnetization of graphene (up to 0.3 emu/g). The generation of near room temperature ferromagnetism is at a high Curie temperature of 250 K [52].

Line Defects

Dislocations

According to well defined concepts only edge dislocations are possible in two-dimensional graphene sheet, due to fact that Burgers vector reflects the magnitude and direction of the crystalline lattice distortion constrained that lies in the graphene plane. Therefore, this kind of dislocation can be described as a semi-infinite strip with width b (here b corresponds to the magnitude of the Burgers vector) of the layer [99]. The dislocation in the defective graphene could be formed due to the reconstruction of a vacancy chain along either the armchair or zigzag direction [84], as illustrated in Fig. 4.12.

Usually such dislocations have two opposite cores composed along with 5–7 pairs without dangling bonds and appearing at the end of the strip. The simplest case of this kind of dislocation has been demonstrated with b = 52.46 Å inserts in a semi-infinite strip of atoms along the armchair direction, whereas, the Burgers vector has been oriented towards the zigzag direction. Similarly, there are several ways of the configuration construction of the dislocation in a graphene sheet have been identified in the recent years, by using a different orientation even for the longer Burgers vectors [99]. Therefore, it is worth noting that the introduction of dislocations into graphene sheets can extensively ripple the graphene to accommodate the strain in the system.

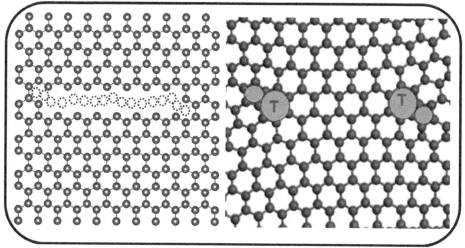

Figure 4.12   Dislocation defect in graphene that forms under reconstruction of a vacancy chain along zigzag direction.

Grain Boundaries

Grain boundaries are the the main line of defects, their boundaries are separated with two grains with lattices tilted relative to each other due to the misorientation angle θ, and the tilt axis is perpendicular to the graphene plane, which is frequently observed in CVD synthesized graphene. The grain boundaries appear when two grains have different orientations coalesce. A defect can be considered as a line of reconstructed point defects. Since the grain boundaries are symmetric and form a mirror symmetry plane between the two crystalline regions. Therefore, it can be inferred as a periodic array of parallel edge dislocations that may result in a symmetric grain boundary with small θ. It has been defined that with increasing θ, the spacing between neighboring dislocations decrease which is equivalent to the increase in dislocation density.

Theoretical description of the atomistic simulations has predicted that only some specific orientations can match periodic boundary conditions [51]. Hence, an asymmetric grain boundary periodicity can only occur when certain proportionate conditions are fulfilled, and the resulted period is typically much longer. Sometimes, out-of-plane corrugation can also occur to accommodate the strain that is generated by the grain boundaries. This simply reduces their formation energies. However, the large angle ($\theta = 20°–30°$) suspended graphene may become flat, while stable configurations of small θ defects ($< 20°$) can be strongly corrugated [99]. Experimentally the corrugation fields can be recognized from TEM [100]. The theoretical studies also focused on grain boundaries configurations within the short period and their experimental evidence can be visualized using the TEM or AC-TEM [101].

This kind of defect consists of an alternating line of pentagon pairs separated by octagons [101]. Additionally, grain boundaries can also form a sharp interface by meandering chains of alternating pentagons and heptagons. In this situation, the grain boundaries loops may be equivalent to point defects in the crystal lattices. The stable loop structure can have highly symmetric flower-shaped defect. This type of the grain boundary may be formed from the rotation of a core of the hexagons. The detailed formation mechanisms have not fully been understood, however, it is worth noting that they may be from the structure due to the relaxation of defective graphene.

Grain boundary configuration also changes under the electron irradiation through bond rotations. The energetic grain boundaries migration from the macroscopic concepts can be interpreted in terms of time-averaged translation to occur only in the presence of significant boundary curvature [102]. However, fully enclosed grain boundaries loops can shrink to the point of disappearance, leading to the restoration of pristine lattice [102].

The physical properties of polycrystalline materials depend on the size and distribution of grains as well as atomic structure of the grain boundaries. Graphene grain boundaries are more significant due to their ability to govern the electronic properties. The introduction of a point defect can lead to the injection of charge into the whole electronic system. Therefore, grain boundaries as lines of reconstructed point defects may have the same impact on a larger scale. Experimentally, it has been confirmed that grain boundaries composed of a pentagon and octagons that may act as conducting metallic wires [101]. Such line defects may raise the localized electronic states in a transverse direction; it may also extend along the line [101]. It can also enhance the conductivity along the line and open up the possibility of the fabrication of all-carbon electronic devices [101]. According to the theoretical description, grain boundaries can impede electronic transport; therefore, all recurrence of the grain boundaries may be divided into two distinct classes [99]. For example, one may have very high transmission probabilities (~ 0.8) of low-energy charge carriers across the grain boundary, while, the other may completely come from the reflection of charge carriers in a rather broad energy range (up to ~1 eV) [99]. The transport characterizations of the isolated individual grain boundary have confirmed their higher electrical resistance. However, the increase of resistance may vary across different grain boundaries [103].

Thus, grain boundaries can naturally alter the mechanical properties of graphene to a large extent. The CVD grown suspended graphene nanoindentation has demonstrated that elastic stiffness of polycrystalline graphene with different grain sizes (1–5 μm and 50–200 μm) can be statistically identical to the single-crystalline graphene [104]. Theoretical simulations of the single grain boundary model revealed the large angle grain boundaries may be stronger as pristine graphene, while small angle defect configurations could be significantly weaker [105]. This concept had been verified from nanoindentation outcomes [106]. The grain boundaries experimental evidence also indicated that the large angle fracture strength is comparable to their single-crystalline counterparts, while the lower angle value (θ = 20 –30%) can result in smaller fracture strength than those of pristine graphene [104, 106]. Thus, the described description indicates that cracks could not only propagate along grain boundaries but also inside the grains, in which junctions may act as nucleation centres [104, 107].

Defects in Bilayer Graphene

The layered graphene construction forms a graphite-like structure. If graphene itself is free of defects, in this case no carbon atoms bond exists chemically in between the layers. Besides the existing intrinsic defects such as holes, dangling bonds or carbon atoms in the migrating state in the graphene. This situation may occur even with only two layers of graphene. Therefore, the defective graphene sheet layers can form new chemical bonds with adjacent carbon atoms [108].

Like graphite, bilayer graphene also has preferable AB stacking sequence along with a weak van der Waals interaction. Though such defects can exist in every individual layer independently, generally, they have a tendency to form covalent interlayer bonds when adatoms are located in between the layers [109]. In the layered defective graphene, the interlayer C atoms are usually less stable in AB stacking, this may allow the adatom vacancy pairs to recombine into perfect graphene under unrestricted barrier circumstances [110]. Moreover, the interlayer adatoms could be stable when basal layer shears with to each other. Such shearing could allow the adatom to forms bond to more atoms with the adjacent layers [110]. This kind of possible configuration is called 'ylid interstitial' in which two bonds can be formed to connect one layer and one to the opposite layer within a vertical plane, as shown in Fig. 4.13(a). Besides the ylid interstitial may be placed around a vacancy, the adatom prefers to link two basal sheets rather than to combine with the vacancy, this preference may form a metastable configuration sometimes and referred to as 'Wigner defects' [110]. Another widely studied configuration is 'spiro interstitial', in such

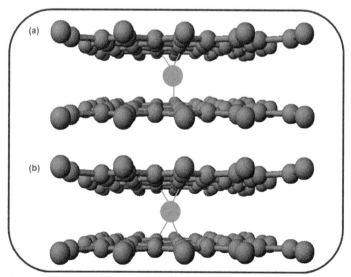

Figure 4.13   Defects in bilayer graphene, **(a)** 'ylid interstitial' connects with two bonds to one layer and one to the opposite. **(b)** 'spiro interstitial' forms four bonds their two bonds connects to each layer.

a construction the bridge atom can form four bonds with two to each layer, as illustrated in Fig. 4.13(b). This kind of defect may be important to control the morphology of graphene layers under irradiation and heat treatment conditions. Moreover, the rotational stacking fault is also a significant defect in bilayer graphene. In which the Moire´ pattern may occur when two basal layers rotated relative to each other, this can be more complex in few layered graphene [54, 111].

Hence as the stacking process involves more layers of graphene, the structural defects become more complex. The complex defects may ultimately affect the macrostructure of the building material. Additionally, it can also affect the physical and chemical properties of the material. Since the monolithic graphene and graphene nanosheets are not infinitely large in space scale, therefore, graphene in different stacking regions should involve concurrent domain processes in the construction of the graphite structure. In this case, if the domain processes are not good, it could result in a lack of long-range order in the material. This could also cause materials defects [112].

Cracks in Graphene

Cracks in graphene are also a significant parameter specifically when electronic devices made from graphene are produced. Usually the molecular dynamics simulation has been used to investigate cracks and their crack-paths in single-layer graphene. It has been demonstrated that, the crack-path strongly depends on the orientation of the initial crack [113]. This depends on the use of the cut-off function in the specified potential region [114]. The additional key factor that governs the fracture of single-layer graphene is the competition between bond breaking and bond rotation at a crack tip. With these two factors, the symmetric cleavage fracture can also be induced due to the breakage of the first bond at the crack tip as well as asymmetric cleavage fracture by breaking the bond adjacent to the first bond at the crack tip [115]. The coupled quantum/continuum mechanics approach has also been used to investigate/ or describe the crack propagation in graphene. This approach is useful to find out the crack propagation in armchair and zigzag graphene sheets. The outcome of this approach in the examined systems reveals that initially the cracks are perpendicular zigzag, whereas, the armchair edges, which eventually grow as irregular in armchair sheets but continue as self-similar crack growth in zigzag sheets.

Cracks in graphene are also a key factor that alter or affect their uses for high mechanical strength consideration. To establish the velocity and instability of crack motion in the hexagonal lattice of graphene under pure opening loads requires extensive investigation, although it had also been done using the atomistic molecular dynamics simulations. This result had indicated that the brittle crack in zigzag direction could propagate supersonically at 8.82 km/s under uniform normal loading of edge displacements. However, the crack moving in a straight line with low speed can produce atomically smooth edges, whereas, kinking may occur beyond a critical speed around 8.20 km/s, which is equivalent to 65% of Rayleigh wave speed in graphene [115].

PROPERTIES OF GRAPHENE

Human beings have always been curious to know more about the behaviour of different mechanisms at the sub-atomic level. More specifically, a deeper understanding of many of the newly discovered materials even at a much smaller level. Sub-atomic level innovations attracted much attention due to innovated instruments to look into the world of atoms, molecules and the change in the properties of materials when they combine. This kind of congregation of the molecular arrangement gave birth to one of the materials with the most potential of the century; known as a graphene. As the specific properties of graphene stem from the unique world of their chemical configuration, which could be exploited bythe abundance of commercial and fundamental applications. Graphene has a very complex compositions behavior that depends on what kinds of interactions it makes with other chemicals, such as the hydrogen. Different types of graphene have been investigated in the past few years, such as, unilayer graphene, bilayer graphene, graphene, graphone, etc. They all have novel properties with remarkable differences from graphite. The monolayer graphene was first prepared from graphite and its properties had been studied extensively.

Thermal Properties

Usually graphene has unmatched thermal conductivity at a temperature 27°C, i.e. ≈ 5000 W/mK. The adequate thermal properties of graphene is due to strong in-plane carbon bonds. This kind of in-plane bond arrangement has made them an excellent heat conductor. The two-dimensional graphene has a little or no phonon scattering characteristics. Generally, the low-energy phonons of the system are involved in heat transfer, because graphene offers excellent high thermal conductivity.

Graphene also has an ambipolar electric field behavior, such as the charge carriers can be tuned continuously between electrons and holes for the high concentrations 10^{13} cm^{-2} whereas the mobilities μ is in excess of 15,000 cm^2 V^{-1} s^{-1} under ambient conditions. Though at room temperature, thermal conductivity of graphene has been estimated between $(4.84 \pm 0.44) \times 10^3$ to $(5.30 \pm 0.48) \times 10^3$ W·m^{-1}·K^{-1} [116]. However, the CVD-grown graphene thermal conductivity values obtained were much lower (≈ 2500 W/mK) [117]. The argument on the thermal conductivity has also been provided: graphene can possess a type of structure, such as, AA or AB types and number of layers present in the graphene that can control thermal conductivity. Their high thermal conductivity also makes them useful for the electronic circuit applicability as a heat sink. This allows to measure thermal conductivity values through a non-contact optical technique. It has been reported that the obtained values are larger than those obtained from carbon nanotubes or diamond. Thus, it indicates that thermal conduction is phonon-dominated. In this view the two-dimensional graphene has three acoustic phonon modes. In which two in-plane modes consisting of a linear dispersion relation, while the out-of-plane mode has a quadratic dispersion relation. Due to this, the T^2 dependent thermal conductivity of the linear mode dominated at low temperatures from the factor $T^{1.5}$ contribution to the out-of-plane mode.

Thermal management is one of the key factors for reliable performance of electronic devices at a time when considerable amount of heat is generated during the operation. Since graphene can be a major component in electronic devices in the future due to its high thermal conductivity (up to 5000 W/mK) at room temperature, in the case of single layer defect-free graphene [116]. Their strong CAC covalent bonds and phonon scattering can also contribute in high thermal conductivity performance. It was also reported that the thermal conductivity of pure single-layer graphene is much higher than the past reported thermal conductivity of other carbon allotropes at room temperature, such as carbon nanotubes (3000 W/mK for MWCNT and 3500 W/mK for SWCNT) [118, 119]. The graphene thermal conductivity may also be affected by factors such as defects, edge scattering and isotopic doping [120, 121]. Usually, these factors can harm the conductivity due to phonon scattering at defect and phonons modes localization due to doping.

If we assume a point defect or a single vacancy defect in the graphene, then thermal conductivity can rapidly decrease up to 20% of the former value. With the increasing defect concentration, the thermal conductivity normally drops slowly. This is due to the low defects' concentration of graphene, the defects become the center of heat flow scattering and it weakens the heat dissipation potential of graphene [122].

The influence of extrinsic defects on the thermal conductivity of graphene has also been described using molecular dynamics simulation. As an example, if some of the carbon atoms on the graphene become sp^3 hybridized, carbon atoms connected to the other three carbon atoms and one hydrogen atom, then the foreign hydrogen defects can result in the decrease in thermal conductivity. Such an incorporated defect in graphene is only about 2.5% , but it can reduce the thermal conductivity up to 40%. There is also some evidence that random dispersion of hydrogen atoms can largely influence the thermal conductivity of graphene [123].

Optical Properties

Graphene is also recognized as a transparent material, as a result; it can have applications in many photonic devices for those requiring conducting but transparent thin films. The monolayer graphene can absorb white light 2.3 $\pi\alpha \approx 2.3\%$ with the 97.7% transmittance, here $\alpha \sim 1/37$ is the fine-structure constant. The stacking ordering and their orientation can also influence the optical properties of graphene; therefore, bilayer graphene could have rather interesting optical properties compared to the single layer graphene.

Experimentally graphene can be imaged from the optical image contrast on Si/SiO_2 substrate owing to interference, the image contrast increases with the number of layers. In the case of monolayer graphene, the absorption has been defined flat in between the wave length range of 300 to 2500 nm, the corresponding absorbance peak at ~250 nm in the UV region, can be related to the inter band electronic transition from the unoccupied π-states. Moreover, highly transparent graphene possesses low resistivity (10^{-6} Ωcm), making it suitable also for electrodes in liquid crystal devices.

Therefore, the increasing market demand of flexible transparent conductors such as, touch screens, flexible displays, printable electronics, solid-state lighting, especially organic light emitting diodes and thin film photovoltaics applicability makes them emerging materials. Additionally, touch screens with transparent windows in smart phones, ATM machines and portable entertainment devices with Indium Tin Oxide (ITO) are key potential uses. This is due to the ITO film transparency needs to be in the extent of 90% and up to a wavelength of 550 nm, with sheet resistances of 10–30 Ω/sq.

The difficulty with using ITO films is the expensive scarcity of the Indium cost that is about US$ 1000/kg. Additionally, this element can be finished very quickly. Their preparation methods such as sputtering, evaporation, pulsed laser deposition and electroplating are expensive. By nature, ITO is a brittle, crystalline material that can fracture easily. Therefore, to resolve these ITO problems scientists have looked at alternatives. To overcome the ITO associated problems, graphene can be a potential alternative due to its lightweight, robust, flexible, chemically stable and low cost. In this order, touch screens based on graphene sheets have been already introduced in the market. Specifically, graphene's mechanical strength and flexibility has made them superior compared to indium tin oxide and graphene films can be deposited from a solution process over a large area. The pyrolyzing camphor method synthesized graphene layered thin films, optical and electrical sheet resistance and transmittance were obtained at 860 Ω/sq cm and 91% (at 550 nm wavelength), respectively [124]. Moreover, more than 80% transparency was also achieved in the range of 250 to 1750 nm compared to ITO glass from 250–800 nm.

Mechanical Properties

Graphene sheets are highly flexible, and can stretch like a balloon, and can even can stand pressure differences of several atmospheres. They also remain impermeable to small atoms like helium.

Graphene has a very light weight, being only about 0.77 mg/m^2. Graphene has tensile high Young Modulus (1.1 TPa), high breaking or fracturing strength (130 GPa). Yet, it is the strongest known material (200 times stronger than structural steel). The mechanical properties of single, bilayer and multiple layers of graphene have been investigated by several researchers using different methods, as summarized in Table 4.2. In order to define the utility of this novel material, the production of photovoltaic cells using flexible graphene coating with a layer of nanowires was demonstrated [125]. Considering the superior flexibility of the graphene, it was also predicted that in future transparent and flexible solar cells would be possible on windows and roofs of buildings. Theoretically, Young's modulus of pure graphene reaching as high as 0.7–1 TPa, but different kinds of the defects can affect the modulus. The point defects and single vacancy defects may decrease Young's modulus of graphene, while increasing the density of two defects [58]. A relationship between the density of single vacancy defects and the percentage change of Young's modulus (Young's modulus of defective graphene/non-defective graphene) has also been defined, showing it to be a linear relationship [58]. While, a non-linear Young's modulus relationship between point defects density was interpreted. Moreover, with increasing density of defects, Young's modulus gradually changes and represents a platform. This could be directly correlated to Young's modulus that is not sensitive with the point defect density. Further, the mechanical properties of the graphene under the presence of sp^3 hybrid carbon atoms and vacancy defects have also been described [88]. The obtained evidence revealed that the elastic modulus of the graphene is insensitive to defect density with sp^3 hybrid carbon atoms. In contrast, for the case of the vacancy defects, the result is quite the opposite. The existence of the vacancy defects can produce a significant reduction in the elastic modulus of graphene [88].

Table 4.2 Graphene mechanical properties

Types of material	Mechanical property	Technique	References
Graphene	Strain ~ 1.3% in tension Strain ~ 0.7% in compression	Raman Spectroscopy	Tsoukleri et al. 2009
Monolayer graphene	$E = \pm 0.1$ TPa $\sigma_{int} = 130 \pm 10$ GPa at σ_{int} 0.25	AFM	Lee et al. 2008
Monolayer bilayer trilayer graphene	$E = 1.02$ TPa; $\sigma = 130$ GPa $E = 1.04$ TPa; $\sigma = 126$ GPa $E = 0.98$ TPa; $\sigma = 130$ GPa	AFM	Li et al. 2009

Several investigations on the effect of the extrinsic defects on the mechanical properties of graphene have also been done by different research groups. It was demonstrated that Young's modulus with C–O–C heteroatom defects may be 42.4% lower than the non-defective graphene. However, their tensile strengths are almost unchanged. This behavior could be due to the introduction of oxygen atoms. It may be the reason that foreign oxygen atoms bend the graphene sheet. Therefore, a deformation in graphene increases after applying the load. This description leads to the tensile strength of graphene depending on the strength of the C–C bond. In this case, two carbon atoms could connect to the oxygen atom and still be connected with each other. Due to this, despite the existing C–O–C defects, the change of graphene tensile strength is small [126]. Thus, based on the above discussion, it is not difficult to find that the intrinsic defects of graphene, such as vacancy defects, which greatly affect the tensile strength of graphene, on the other hand the extrinsic defects that can only influence the graphene deformation modulus.

Electronic Properties

One of the novel electronic properties of graphene is that it can sustain huge electrical currents. Preferably π-bonds of the graphene contribute to the electron conduction that provides a weak interaction between graphene layers. The transport of charge carriers in graphene usually could be described from the Dirac rather than the Schrödinger equation. As the two equivalent carbon sub-lattices in the honeycomb lattice, in which the cone-like valance and conduction bands intersects at the Fermi level at the wave vector K and K_0 points of the Brillouin zone, as depicted in 4.14(a,b,c).

The massless Dirac fermions can reveal numerous exceptional properties. According to the description of graphene, it can have a zero-band gap with a two-dimensional semiconducting property under a well-defined ambipolar electric field effect. This means quasi particles with a large mean-free-path. Moreover, the Dirac point energy dispersion in two dimensions also indicates that graphene has gapless band gap with the semiconducting property. Their density of states vanished linearly when approaching the Fermi energy. Graphene can also conduct either electrons or holes with the high concentration 10^{13} cm^{-2}. It also has an extraordinary carrier mobility in the order of $\approx$ 500,000 cm^2 V^{-1} s^{-1}. The high mobility of the graphene is due to the electrons' movement through the perfect honeycomb lattice in smooth sailing.

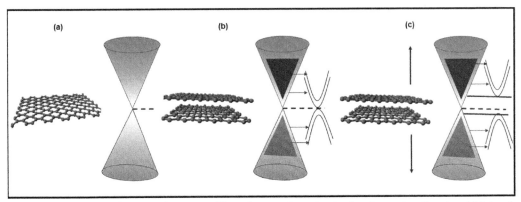

Figure 4.14 **(a)** Monolayer graphene cone-like valance and conduction bands construction; in which bands intersect at the Fermi level without the band gap. **(b)** Bilayer graphene bands construction for a gapless semiconductor. **(c)** Bilayer graphene band gaps construction under applied electric field in the direction perpendicular to σ-bond.

The electronic properties of the graphene widely depend on their thickness. Hence, it is obvious to control the thickness during synthesizing the graphene by controlling the growth parameters. Graphene also has a high electron (or hole) mobility and low Johnson noise (electronic noise created due to thermal stir of the charge carriers). This is one of the key parameters to minimize the dark current in *p-n* junction.

According to the band theory of solids it can be described from the tight binding approximation and many physical properties of graphite have been defined. Due to large (3.37Å) spacing of lattice planes of graphite compared to hexagonal spacing in the layer 1.42Å. It was described that interactions of electrons between the planes can be neglected and conduction by electrons takes place only in layers (i.e., in the π-plane). Moreover, it was demonstrated that electron transport in graphene can be explained by Dirac's (relativistic) equation [127]. According to this concept, the charge carriers in graphene imitate relativistic particles with zero rest mass with an effective speed of light, $c \approx 10^6$ m s^{-1}. The two-dimensional Dirac fermions characteristic has revealed that even if the concentration of charge carrier moves to zero; the conductivity of graphene does not become zero. This is the fundamental difference between the Dirac fermions and Schrödinger equation concepts. According to the earlier assumption, electrons propagating in the honeycomb lattice lose their effective mass; as a consequence the quasi-particles are different.

A drastic variation in the electronic properties of monolayer and bilayer graphene sheet has been recognized. Particularly, in the monolayer graphene, the electron exhibits high mobility and travels with the speed of light. Simply high mobility of the graphene monolayer electrons does not make it a very significant material until and unless it provides the possibility of controlling its band gap, similar to a semiconductor in which one could control and modulate the electron movement to achieve the desired outcome. In more appropriate words, graphene could not be conducted unless energy is supplied to enhance electrons across the gap between the valence band and the conduction band. Hence in the case of no band gap, graphene is like an automobile moving with no control on either speed or turning right or left or stopping at a traffic light.

Defects in graphene can also influence the bond length of the interatomic bonds' valence. They may also change the type of the hybrid trajectories of the partial carbon atoms. The changes in bond length and orbital reveal that the graphene defect domain induced electrical properties.

The most common point defects and single vacancy defects in graphene form an electron scattering center on the surface of the graphene. These newly formed centers affect the electron transfer [128], therefore, resulting in a decrease in the conductivity of graphene. It is a fact that defects free (like point and single vacancy defects) synthesis of graphene is unavoidable in practical. This tells us why the practical concavity of graphene is different from the ideally defined value. Thus, the non-avoidable of defect circumstances indicates a direction for the study including reduction of intrinsic defects in graphene to improve the thermal conductivity.

In contrast to intrinsic defects, the foreign atoms defects effect on electrical properties of graphene can be more complicated and interesting. Essentially the graphene oxide is not a conductive material, and its square resistance can reach more than 10^{12} Ω [129]. Thus one can speculate that the graphene conductivity can decrease due to existence of oxygen atoms and oxygen-containing functional groups. Although, a number of theoretical studies have also pointed out that oxygen atom defects in graphene, like C–O–C defects, can make the graphene support metal conductive under the reasonable position [130].

A large number of reports have also pointed out that in-plane foreign atom defects can be formed from the nitrogen and boron atoms to improve the conductivity of graphene. To explore different forms of oxygen atoms defects in graphene, investigators demonstrated that the nitrogen and boron atoms resonance scattering on graphene affects its the electrical properties. Moreover, scientific reports have also focused on the position of nitrogen and boron atoms, with the description of the two-dimensional width of graphene and its own symmetry that can affect its electrical properties [131].

Graphene electronic properties can be strongly affected from the adsorption of molecules. This property makes graphene an attractive material for gas sensing applications of different areas. Graphene sensors can be extensively used to detect the gas concentration in the specified composition. Additionally, the structure and electronic properties of graphene-molecule adsorption adducts are widely dependent on graphene configuration. Furthermore, researchers have also demonstrated that defective graphene has the highest adsorption energy with CO, NO and NO_2 molecules.

Chemical Properties

Graphene has a high chemical stability due to its honeycomb network in which the strong in-plane sp^2 hybrid bonds exist. The chemically inert behavior of graphene makes it useful to prevent the metal and metal alloys from oxidation. The oxidation resistance behavior was first introduced with help of Cu and Cu/Ni coating with graphene using the CVD technique [132]. Possessing such a chemical stability and inert behavior, predicted that it can improve the durability of potential optoelectronic devices [133].

Graphene oxide has soluble in polar and non-polar solvents characteristics. Due to the presence of each graphene atom on the surface, their atoms are capable of interacting with any molecule of the target gas or vapor species. Therefore, graphene can have a very special property in any chemical reaction. This feature offers the opportunity to adjust the conductivity of graphene by selecting a molecule to adsorb at the surface of the graphene. All these features of graphene could make it a prominent candidate for NEMS applications, such as, pressure sensors and resonators.

Zhou and Bongiorno demonstrated the origin of chemical and kinetic stability in graphene. According to this study, "at the moderate temperatures (# 706°C), thermal reduction of graphene oxide is inefficient and after its synthesis the material enters in a metastable state. They used the first-principles and statistical calculations to investigate both the low-temperature processes leading to the decomposition of graphene oxide and the role of ageing on the structure and stability. Their study has shown that the key factor underlying the stability of graphene oxide is the tendency of the oxygen functionalities to agglomerate and form highly oxidized domains surrounded by areas of pristine graphene. Further, within the agglomeration of functional groups, the primary decomposition reactions can be hindered by both geometrical and energetic factors. The number of reacting sites can also reduce due to occurrence of local order in the oxidized domains and the close packing of the oxygen functionalities, therefore, the decomposition reactions become–on average–endothermic by more than 0.6 eV".

It can be noticed that each atom of graphene could be exposed for a chemical reaction from two sides owing to their two-dimensional structural characteristics. Additionally, a carbon atom at the edge of a graphene sheet should show special chemical reactivity, while the presence of defects within the sheet may also increase the chemical reactivity. The chemical reactivity as well as the electronic properties of graphene depends on the number of layers as well as the relative position of atoms in adjacent layers (stacking order).

A bilayer graphene stacking ordering usually in either AA (in which each atom is on top of another atom) or AB, where a set of atoms in the second layer sits on top of the empty center of a hexagon in the first layer, as shown in Fig. 4.15(d). With the increasing number of layers, the stacking ordering becomes more complicated. However, the carbon atoms of monolayer graphene could completely satisfy with its valence state, meaning that dangling bonds are not created, as shown in Fig. 4.15(a). This could be correlated by the AA-type stacking monolayer structure. On the other hand, the carbon atoms at the surface of the bilayer (of either AA or AB structure) may contain some weak dangling bonds owing to the existence of van der Waal force. This could be because one layer below the surface layer was separated at a distance of 0.32 nm and the force responsible to keep the second layer attached with the top surface layer is van der Waal force, as shown in Fig. 4.15(b, d). The absence of the van der Waal force on the top carbon atoms allows the presence of electrons on the surface of top layers in different environments rather than the carbon of the second layer. Therefore, electron interactions with the top layer act differently to electrons of the bilayer (Fig. 4.15(b)). This behavior is due to the AA-type structure, in which each carbon atom is present just below the top layer carbon atom, while the AB-type structure has dangling bonds in the second layer. This leads to the electron interactive behavior with the bilayer graphene of AA-type structure different to the AB-type structure.

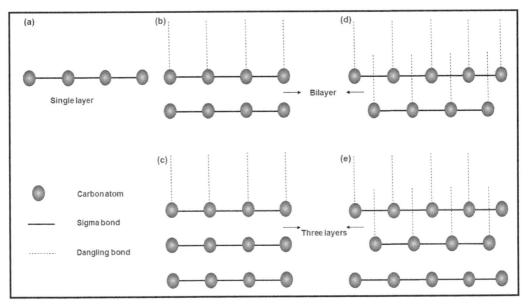

Figure 4.15 **(a)** Monolayer graphene; **(b)** Bilayer graphene of AA-type stacking structure with the dangling band on top layer; **(c)** Schematic of trilayer graphene under AA-type stacking structure with the dangling bond effectively on top layer; **(d)** AB-type stacking structure of bilayer graphene that forms a dangling band under the influence of first and second layer of carbon; **(e)** Trilayer graphene structure with formation of a dangling bond between top two successive layers of the carbon atom.

In the case of the trilayer graphene, the scenario becomes different. In this case, the AA-type structure under the effect of the dangling bond could act more or less like the bilayer graphene, but their magnitude of electron interaction to the top surface of graphene could be different from the AA-type bilayer graphene (see Fig.4.15(c)). The situation becomes more complex with the AB-type three layers structures (see Fig. 4.15(e)). Here the dangling bonds are influenced from the top layer as well as second layer carbon atoms. Their combination of dangling bonds also influences the electron interaction of bilayer graphene in a different way. Hence, the reactivity of graphene depends on the number of layers and type of structures (AA or AB type).

Moreover, the presence of the dangling bonds could promote the temporary chemical reaction, in other words the functionalization of graphene. This could be one of the reasons that functionalization can take place easily with bilayer graphene rather than monolayer graphene.

The study on the relationship between defects and the chemical properties of graphene has mainly focused on the introduction of extrinsic defects in graphene. Due to the fact that without heteroatom defects graphene is chemically inert, even if it contains intrinsic defects. Therefore, it is not very easy to carry out chemical reactions from the pure or intrinsic defects containing graphene.

To raise the chemical application prospects of extrinsic defects in graphene has attracted much attention, specifically incorporation of oxygen atoms in graphene. Mainly because of two key reasons: (i) Oxidized graphene carries some oxygen-containing groups such as hydroxyl and carboxyl. It can make the oxidized graphene hydrophilic; therefore, it can disperse uniformly in water. The oxidized graphene and various salt compounds or hydrophilic polymers could form hydrogen bonds or ionic bond interactions. This makes it easy for substances that can uniformly load on graphene in water. This kind of composite materials can be applied in catalytic, lithium, super capacitor, drug introduction, etc. [134–142]. (ii) Oxidized graphene can have sound self-assembly and film-forming properties. This makes oxidized graphene good in the area of flexible films.

Semiconducting Properties

The interference of p_z electrons in the honeycomb is responsible for gapless semiconductor properties, under a point-like Fermi surface. Although graphene has a zero energy band gap but its band gap can be altered according to changing the size, shape, thickness and fraction of the sp^2 domains [143, 144]. Moreover, the bilayer graphenes electronic band structure can be modified significantly under the influence of the electric field effect. Therefore, their energy band gap can be tuned continuously from zero to ≈ 0.3 eV with the SiO_2 as a dielectric. More recently IBM has provided evidence where the energy band gap was tuned to the order of 0.13 eV using the structure semiconducting graphene [145].

CONCLUSIONS

In this chapter a systematic review on the development of the graphene including their different structural forms such as single and bilayer graphene were presented. The structural difference as well as properties of various kinds of graphene, such as, graphone, graphyne, graphdiyne, graphane have also been described. Defects in a graphene is an essential requirement for their technique application, therefore, a description on the importance of defects in graphene, including topological defects with their forms of Stone–Wales defects, vacancies, ripples, adatom defects, moreover, are also discussed including the line defects and their forms dislocations and grain boundaries. Since a characteristic of graphene defects depend on their shape, size and number stacking layers, therefore, defects in the bilayer graphene has also been mentioned separately. To define the technical use of any graphene based device it is required to know the deformation or cracks key factor, therefore the cracks formation in graphene and their impact on the performance have been described. Furthermore, the general properties of graphene have also been discussed with a brief description of the thermal properties including the pure graphene of thermal conductivity as well as reduction in it under the influence of the defects, optical properties of pure graphene as well as the impact of the defects on it, mechanical properties for both defective as well as non-defective graphene, chemical properties for both defective and non-defective kinds of graphene and semiconducting properties of the graphene including opening their band gaps for the defective states. Hence this study has demonstrated that, although graphene has remarkable advantages with two-dimensional structural properties, however, their implication requires different kinds of defective creation within the structure, which could reduce the performing ability of the devices based on this. This could be a debatable issue for further innovation to improve the performances of defects containing graphene based working systems.

References

1. Boehm, H.P., Clauss, A., Fischer G., Hofmann, U. (1962) *In*: Proceedings of the Fifh Conference on Carbon. Pergamon Press, Oxford.

2. Boehm, H.P., Clauss A., Fischer, G.O., Hofmann, U. (1962) Thin Carbon Leaves. Z. Naturf., 17, 150–153.

3. Mouras, S., Hamm, A., Djurado, D., Cousseins, J.C. (1987) Synthesis of first stage graphite intercalation compounds with fluorides. Revue de Chimie Minerale, 24, 572–582.

4. Landau, L.Z. Phys. (1930) Diamagnetismus der Metalle. Z. Physik, 64, 629–637.

5. Wallace, P.R. (1947) The band theory of graphite. Phys. Rev., 71, 622–634.

6. Ruess, G., Vogt, F. (1948) Höchstlamellarer kohlenstoff aus graphitoxyhydroxyd. Monatshefe für Chemie, 78, 3–4, 222–242.

7. Krishnan, A., Dujardin, E., Treacy, M.M.J., Hugdahl, J., Lynum, S., Ebbesen, T.W. (1997) Graphitic cones and the nucleation of curved carbon surfaces. Nature, 388, 451–454.

8. Hess, W.M., Ban, L.L. (1966) Proc. 6th Int. Congr. on Electron Microscopy (Kyoto), 1, 569.

9. Land, T.A., Michely, T., Behm, R.J., Hemminger, J.C., Consa, G. (1992) STM investigation of single layer graphite structures produced on Pt(111) by hydrocarbon decomposition. Surf. Sci., 264, 261–270.

10. Oshima, C., Nagashima, A. (1997) Ultra-thin epitaxial films of graphite and hexagonal boron nitride on solid surfaces. J. Phys. Condens. Matter, 9, 1–20.

11. Semenoff. G.W. (1984) Condensed-matter simulation of a three-dimensional anomaly. Phys. Rev. Lett., 53, 2449–2452.

12. Bouchard, E., Lavenac, J., Roux, J.C., Langlais, F., Delhaès, P. (2001) Pyrocarbon deposits on a graphite surface observed by STM. Chemical Vapor Deposition. 7, 125–130.

13. Enoki, T., Suzuki, M., Endo, M. (2003) Graphite Intercalation Compounds and Applications. Oxford University Press, U.K.

14. Dresselhaus, M.S., Dresselhaus, G. (2002) Intercalation compounds of graphite. Adv. Phys., 51, 1–186.

15. Ray, S.C., Soin, N., Makgato, T., Chuang, C.H., Pong, W.F., Roy, S.S., Ghosh, S.K., Strydom, A.M., McLaughlin, J.A. (2014) Graphene supported graphone/graphane bilayer nanostructure material for spintronics. Sci. Rep., 4, 3862–7.

16. Feng, L., Zhang, W.X. (2012) The structure and magnetism of graphone. AIP Advances, 2, 042138-1–042138-6.

17. Malko, D., Neiss, C., Viñes, F., Görling, A. (2012) Competition for graphene: Graphynes with direction-dependent dirac cones. Phys. Rev. Lett., 108, 086804.

18. Baughman, R.H., Eckhardt, H., Kertesz, M. (1987) Structure-property predictions for new planar forms of carbon-layered phases containing sp^2 and sp atoms. J. Chem. Phys., 87, 6687.

19. Coluci, V.R., Braga, S.F., Legoas, S.B., Galvão, D.S., Baughman, R.H.(2003) Families of carbon nanotubes: Graphyne-based nanotubes. Phys. Rev. B, 68, 035430.

20. Baughman, R.H., Galvao, D.S., Cui, C.X., Wang, Y., Tomanek, D. (1993) Fullereneynes: A new family of porous fullerenes. Chem. Phys. Lett., 204, 8–14.

21. Diederich, F. (1994) Carbon scaffolding: Building acetylenic all-carbon and carbon-rich compounds. Nature, 369, 199–207.

22. Bunz, U.H., Rubin, Y., Tobe, Y. (1999) Polyethynylated cyclic pi-systems: Scaffoldings for novel two and three-dimensional carbon networks. Chem. Soc. Rev., 28, 107–119.

23. Cranford, S.W., Buehler, M.J. (2012) Selective hydrogen purifcation through graphdiyne under ambient temperature and pressure. Nanoscale, 4, 4587–4593.

24. Ivanovskii, A.L. (2013) Graphynes and graphdiynes. Prog. Solid State Chem., 41, 1–19.

25. Haley, M.M., Brand, S.C., Pak, J.J. (1997) Carbon networks based on dehydrobenzoannulenes: synthesis of graphdiyne substructures. Angew. Chem. Int. Ed. Engl., 36, 836–838.

26. Pei, Y. (2012) Mechanical properties of graphdiyne sheet. Physica B, 407, 4436–4439.

27. Cui, H.J., Sheng, X.L., Yan, Q.B., Zheng, Q.R., Su, G. (2013) Strain-induced dirac cone-like electronic structures and semiconductor-semimetal transition in graphdiyne. Phys. Chem. Chem. Phys., 15, 8179–8185.

28. Kang, J., Wu, F., Li, J. (2012) Modulating the bandgaps of graphdiyne nanoribbons by transverse electric felds. J. Phys. Condens. Matter., 24, 165301.

29. Sahin, H., Ataca, C., Ciraci, S. (2010) Electronic and magnetic properties of graphane nanoribbons. Phys. Rev. B, 81, 205417.

30. Costamagna, S., Neek-Amal, M., Los, J.H., Peeters, F.M. (2012) Thermal rippling behavior of graphane. Phys. Rev. B, 86, 041408.

31. Zou, W., Yu, Z., Zhang, C.X., Zhong, J.X., Sun, L.Z. (2012) Transport properties of hybrid graphene/graphane nanoribbons. Appl. Phys. Lett., 100, 103109.

32. Sahin, H., Ataca, C., Ciraci, S. (2010) Electronic and magnetic properties of graphane nanoribbons. Phys. Rev. B, 81, 205417

33. Cudazzo, P., Tokatly, I.V., Rubio, A. (2011) Dielectric screening in two-dimensional insulators: Implications for excitonic and impurity states in grapheme. Phys. Rev. B, 84, 085406.

34. Yang, Y.E., Yang, Y.R., Yan, X.H. (201) Universal optical properties of graphane nanoribbons: A frst-principles study. Physica E, 44, 1406–1409.

35. Topsakal, M., Cahangirov, S., Ciraci, S. (2010) The response of mechanical and electronic properties of graphane to the elastic strain. Appl. Phys. Lett., 96, 091912.

36. Peng, Q., Liang, C., Ji, W., De, S. (2013) A theoretical analysis of the effect of the hydrogenation of graphene to graphane on its mechanical properties. Phys. Chem. Chem. Phys., 15, 2003–2011.

37. Cadelano, E., Palla, P.L., Giordano, S., Colombo, L. (2010) Elastic properties of hydrogenated grapheme. Phys. Rev. B, 82, 235414.

38. Cadelano, E., Colombo, L.(2012) Effect of hydrogen coverage on the Young's modulus of grapheme. Phys. Rev. B, 85, 245434.

39. Leenaerts, O., Peelaers, H., Hernández-Nieves, A.D., Partoens, B., Peeters, F.M. (2010) First-principles investigation of graphene fluoride and graphane. Phys. Rev. B, 82, 195436.

40. Munoz, E., Singh, A.K., Ribas, M.A., Penev, E.S., Yakobson, B.I. (2010) The ultimate diamond slab: GraphAne versus graphEne. Diamond Relat. Mater., 19, 368–373.

41. Neek-Amal, M., Peeters, F.M. (2011) Lattice thermal properties of graphane: Thermal contraction, roughness, and heat capacity. Phys. Rev. B, 83, 235437.

42. Lu, Y.H., Feng, Y.P. (2009) Band-gap engineering with hybrid graphane-graphene nanoribbons. J. Phys. Chem. C, 113, 20841–20844.

43. Hussain, T., Sarkar, A.D., Ahuja, R. (2012) Strain induced lithium functionalized graphane as a high capacity hydrogen storage material. Appl. Phys. Lett., 101, 103907.

44. Pujari, B.S., Kanhere, D.G. (2009) Density functional investigations of defectinduced mid-gap states in graphane. J. Phys. Chem. C, 113, 21063–21067.

45. Tan, S.M., Sofer, Z., Pumera, M. (2013) Biomarkers detection on hydrogenated graphene surfaces: Towards applications of graphane in biosensing. Electroanalysis, 25, 703–705.

46. Huang, L., Zheng, Z. (2012) Patterning graphene nanostripes in substratesupported functionalized graphene: A promising route to integrated, robust, and superior transistors. Front Phys., 7, 324–327.

47. Da, H., Feng, Y.P., Liang, G. (2011) Transition-metal-atom-embedded graphane and its spintronic device applications. J. Phys. Chem. C, 115, 22701.

48. Xu, T., Yin, K., Xie, X., He, L., Wang, B., Sun, L. (2012) Size-dependent evolution of graphene nanopores under thermal excitation. Small, 8, 3422–6.

49. Bagri, A., Mattevi, C., Acik, M., Chabal, Y.J., Chhowalla, M., Shenoy, V.B. (2010) Structural evolution during the reduction of chemically derived graphene oxide. Nat. Chem., 2, 581–7.

50. Yang, G., Li, L., Lee, W.B., Ng, M.C. (2018) Structure of graphene and its disorders: A review. Sci. Technol. Adv. Mater. 19:1, 613–648.

51. Chen, L., Hu, H., Ouyang, Y., Pan, H.Z., Sun, Y.Y., Liu, F. (2011) Atomic chemisorption on graphene with stone thrower wales defects. Carbon, 49, 3356–61.

52. Liu, Y., Shen, Y., Sun, L., Li, J., Liu, C., Ren, W. et al. (2016) Elemental superdoping of graphene and carbon nanotubes. Nature Communications, 7, 10921.

53. Deng, D., Chen, X., Yu, L., Wu, X., Liu, Q., Liu, Y. et al. (2015) A single iron site confined in a graphene matrix for the catalytic oxidation of benzene at room temperature. Sci. Adv., 1, 1500462.

54. Warner, J.H., Rummeli, M.H., Gemming, T., Buchner, B., Briggs, G.A.D. (2008) Direct imaging of rotational stacking faults in few layer graphene. Nano. Lett., 9, 102–6.

55. Chen, J., Shi, T., Cai, T., Xu, T., Sun, L., Wu, X. et al. (2013) Self healing of defected graphene. Appl. Phys. Lett., 102, 103107–5.

56. Shen, Y., Sun, L. (2015) Setting up a nanolab inside a transmission electron microscope for twodimensional materials research. J. Mater. Res., 30, 3153–76.

57. Sun, L., Banhart, F., Warner, J. (2015) Two-dimensional materials under electron irradiation. MRS Bull., 40, 29–37.

58. Hao, F., Fang, D., Xu, Z. (2011) Mechanical and thermal transport properties of graphene with defects. Appl. Phys. Lett., 99, 041901.

59. Xu, T., Sun, L. (2018) Structural defects in graphene. *In*: Stehr, J., Buyanova, I., Chen., W. (eds), Defects in Advanced Electronic Materials and Novel Low Dimensional Structures: A Volume in Woodhead Publishing Series in Electronic and Optical Materials. Elsevier, UK, pp. 137–160. DOI: https://doi.org/10.1016/B978-0-08-102053-1.00005-3.

60. Henderson, M.A. (1999) Surface perspective on self-diffusion in rutile TiO_2. Surf. Sci., 419, 174–187.

61. Banhart, F., Li, J.X., Krasheninnikov, A.V. (2005) Carbon nanotubes under electron irradiation: Stability of the tubes and their action as pipes for atom transport. Phys. Rev. B, 71, 241408.

62. Ma, J., Alfè, D., Michaelides, A., Wang, E. (2009) Stone–Wales defects in graphene and other planar sp^2-bonded materials. Phys. Rev. B, 80, 033407.

63. Meyer, J.C., Kisielowski, C., Erni, R., Rossell, M.D., Crommie, M.F., Zettl, A. (2008) Direct imaging of lattice atoms and topological defects in graphene membranes. Nano. Lett., 8, 3582–3586.

64. Banhart, F., Kotakoski, J., Krasheninnikov, A.V. (2011) Structural defects in graphene. ACS Nano, 5, 26–41.

65. Shirodkar, S.N., Waghmare, U.V. (2012) Electronic and vibrational signatures of stonewales defects in graphene: First-principles analysis. Phys. Rev. B, 86, 165401.

66. Mortazavi, B., Ahzi, S. (2013) Thermal conductivity and tensile response of defective graphene: A molecular dynamics study. Carbon, 63, 460– 470.

67. He, L., Guo, S., Lei, J., Sha, Z., Liu, Z. (2014) The effect of stone-thrower-wales defects on mechanical properties of graphene sheets—A molecular dynamics study. Carbon, 75, 124–132.

68. Krasavin, S.E., Osipov, V.A. (2015) Effect of stone-wales defects on the thermal conductivity of graphene. J. Phys. Condens. Matter, 27, 425302.

69. Yadav, S., Zhu, Z., Singh, C.V. (2014) Defect engineering of graphene for effective hydrogen storage. Int. J. Hydrogen. Energ., 39, 4981–4995.

70. Duplock, E.J., Scheffler, M., Lindan, P.J.D. (2004) Hallmark of perfect graphene. Phys. Rev. Lett., 92, 225502.

71. Datta, D., Li, J., Koratkar, N., Shenoy, V.B. (2014) Enhanced lithiation in defective graphene. Carbon, 80, 305–310.

72. Datta, D., Li, J., Shenoy, V.B. (2014) Defective graphene as a high-capacity anode material for Na- and Ca-ion batteries. ACS Appl. Mater. Interfaces., 6, 1788–1795.

73. Meyer, J.C., Kisielowski, C., Erni, R., Rossell, M.D., Crommie, M.F., Zettl, A. (2008) Direct imaging of lattice atoms and topological defects in graphene membranes. Nano. Lett., 8, 3582–3586.

74. Banhart, F., Kotakoski, J., Krasheninnikov, A.V. (2011) Structural defects in graphene. ACS Nano, 5, 26–41.

75. Krasheninnikov, A.V., Lehtinen, P.O., Foster, A.S., Nieminen, R.M. (2006) Bending the rules: Contrasting vacancy energetics and migration in graphite and carbon nanotubes. Chem. Phys. Lett., 418, 132–136.

76. El-Barbary, A.A., Telling, R.H., Ewels, C.P., Heggie, M.I., Briddon, P.R. (2003) Structure and energetics of the vacancy in graphite. Phys. Rev. B, 68, 144107.

77. El-Barbary, A., Telling, R., Ewels, C., Heggie, M., Briddon, P. (2003) Structure and energetics of the vacancy in graphite. Phys. Rev. B, 68, 144107.

78. Lee, G.D., Wang, C., Yoon, E., Hwang, N.M., Kim, D.Y., Ho, K. (2005) Diffusion, coalescence, and reconstruction of vacancy defects in graphene layers. Phys. Rev. Lett., 95, 205501.

79. Tian, W., Li, W., Yu, W., Liu, X. (2017) A review on lattice defects in graphene: Types, generation, effects and regulation. Micromachines, 8, 163–15.

80. Girit, C.O., Meyer, J.C., Erni, R., Rossell, M.D., Kisielowski, C., Yang, L. et al. (2009) Graphene at the edge: Stability and dynamics. Science, 323, 1705–8.

81. Xu, T., Xie, X., Yin, K., Sun, J., He, L., Sun, L. (2014) Controllable atomic-scale sculpting and deposition of carbon nanostructures on graphene. Small, 10, 1724–8.

82. Girit, C.O., Meyer, J.C., Erni, R., Rossell, M.D., Kisielowski, C., Yang, L, et al. (2009) Graphene at the edge: Stability and dynamics. Science, 323, 1705–8.

83. Kim, K., Coh, S., Kisielowski, C., Crommie, M.F., Louie, S.G., Cohen, M.L. et al. (2013) Atomically perfect torn graphene edges and their reversible reconstruction. Nature Communications, 4, 2723.

84. Jeong, B.W., Ihm, J., Lee, G.D. (2008) Stability of dislocation defect with two pentagonheptagon pairs in graphene. Phys. Rev. B, 78, 165403.

85. Kotakoski, J., Krasheninnikov, A., Kaiser, U., Meyer, J. (2011) From point defects in graphene to two-dimensional amorphous carbon. Phys. Rev. Lett., 106, 105505.

86. Terrones, H., Terrones, M., Hernandez, E., Grobert, N., Charlier, J.C., Ajayan, P.M. (2000) New metallic allotropes of planar and tubular carbon. Phys. Rev. Lett., 84, 1716–19.

87. Sammalkorpi, M., Krasheninnikov, A., Kuronen, A., Nordlund, K., Kaski, K. (2004) Mechanical properties of carbon nanotubes with vacancies and related defects. Phys. Rev. B, 70, 245416.

88. Zandiatashbar, A., Lee, G.H., An, S.J., Lee, S., Mathew, N., Terrones, M. et al. (2014) Effect of defects on the intrinsic strength and stiffness of graphene. Nat. Commun., 5, 3186–9.

89. Topsakal, M., Akturk, E., Sevincli, H., Ciraci, S. (2008) First-principles approach to monitoring the band gap and magnetic state of a graphene nanoribbon via its vacancies. Phys. Rev. B, 78, 235435.

90. Georgakilas, V., Otyepka, M., Bourlinos, A.B., Chandra, V., Kim, N., Kemp, K.C. et al. (2012) Functionalization of graphene: Covalent and non-covalent approaches, derivatives and applications. Chem. Rev., 112, 6156–214.

91. Jia, Y., Zhang, L., Du, A., Gao, G., Chen, J., Yan, X. et al. (2016) Defect graphene as a trifunctional catalyst for electrochemical reactions. Adv. Mater., 28, 9532–8.

92. Castro Neto, A.H., Guinea, F., Peres, N.M.R., Novoselov, K.S., Geim, A.K. (2009) The electronic properties of graphene. Rev. Mod. Phys., 81, 109–162.

93. Xu, T., Sun, L. (2018) Structural defects in graphene. *In*: Stehr, J., Buyanova, I., Chen., W. (eds), Defects in Advanced Electronic Materials and Novel Low Dimensional Structures: A volume in Woodhead Publishing Series in Electronic and Optical Materials. Elsevier, UK, pp. 137–160. DOI: https://doi.org/10.1016/B978-0-08-102053-1.00005-3.

94. Li, L., Reich, S., Robertson, J. (2005) Defect energies of graphite: Density-functional calculations. Phys. Rev. B, 72, 184109.

95. Lusk, M.T., Carr, L.D. (2008) Nanoengineering defect structures on graphene. Phys. Rev. Lett., 100, 175503.

96. Nakada, K., Ishii, A. (2011) Migration of adatom adsorption on graphene using DFT calculation. Solid State Commun., 151, 13–16.

97. Cretu, O., Krasheninnikov, A.V., Rodrıguez-Manzo, J.A., Sun, L., Nieminen, R.M., Banhart, F. (2010) Migration and localization of metal atoms on strained graphene. Phys. Rev. Lett., 105, 196102.

98. Biel, B., Blasé, X., Triozon, F., Roche, S. (2009) Anomalous doping effects on charge transport in graphene nanoribbons. Phys. Rev. Lett., 102, 096803.

99. Yazyev, O.V., Louie, S.G. (2010) Topological defects in graphene: Dislocations and grain boundaries. Phys. Rev. B, 81, 195420.

100. Lehtinen, O., Kurasch, S., Krasheninnikov, A., Kaiser, U. (2013) Atomic scale study of the life cycle of a dislocation in graphene from birth to annihilation. Nat. Commun., 4, 2098.

101. Lahiri, J., Lin, Y., Bozkurt, P., Oleynik, II., Batzill, M. (2010) An extended defect in graphene as a metallic wire. Nat. Nanotechnol., 5, 326–9.

102. Kurasch, S., Kotakoski, J., Lehtinen, O., Skakalova, V., Smet, J., Krill, C.E. et al. (2012) Atom-byatom observation of grain boundary migration in graphene. Nano. Lett., 12, 3168–73.

103. Yu, Q., Jauregui, L.A., Wu, W., Colby, R., Tian, J., Su, Z. et al. (2011) Control and characterization of individual grains and grain boundaries in graphene grown by chemical vapour deposition. Nat. Mater., 10, 443–9.

104. Lee, G.H., Cooper, R.C., An, S.J., Lee, S., van der Zande, A., Petrone, N. et al. (2013) High-strength chemical-vapor-deposited graphene and grain boundaries. Science, 340, 1073–6.

105. Wei, Y., Wu, J., Yin, H., Shi, X., Yang, R., Dresselhaus, M. (2012) The nature of strength enhancement and weakening by pentagon_heptagon defects in graphene. Nat. Mater., 11, 759–63.

106. Rasool, H.I., Ophus, C., Klug, W.S., Zettl, A., Gimzewski, J.K. (2013) Measurement of the intrinsic strength of crystalline and polycrystalline graphene. Nat. Commun., 4, 2811.

107. Song, Z., Artyukhov, V.I., Yakobson, B.I., Xu, Z. (2013) Pseudo hall-petch strength reduction in polycrystalline graphene. Nano. Lett., 13, 1829–33.

108. Wu, Z.S., Ren,W., Xu, L., Li, F., Cheng, H.M. (2011) Doped graphene sheets as anode materials with superhigh rate and large capacity for lithium ion batteries. ACS Nano, 5, 5463–5471.

109. Telling, R.H., Ewels, C.P., El-Barbary, A.A., Heggie, M.I. (2003) Wigner defects bridge the graphite gap. Nat. Mater., 2, 333.

110. Li, L., Reich, S., Robertson, J. (2005) Defect energies of graphite: Density-functional calculations. Phys. Rev. B, 72, 184109.

111. Hass, J., Varchon, F., Millan-Otoya, J.E., Sprinkle, M., Sharma, N., de Heer, W.A. et al. (2008) Why multilayer graphene on 4H-SiC (0001) behaves like a single sheet of graphene. Phys. Rev. Lett., 100, 125504.

112. Teobaldi, G., Ohnishi, H., Tanimura, K., Tanimura, K., Shluger, A.L. (2010) The effect of van der Waals interactions on the properties of intrinsic defects in graphite. Carbon, 48, 4145–4161.

113. Omeltchenko, A., Yu, J., Kalia, R.K., Vashishta P. (1997) Crack front propagation and fracture in a graphite sheet: A molecular-dynamics study on parallel computers. Phys. Rev. Lett., 78, 2148–2151.

114. Khare, R., Mielke, S.L., Paci, J.T., Zhang, S., Ballarini, R., Schatz, G.C., Belytschko, T. (2007) Coupled quantum mechanical/molecular mechanical modeling of the fracture of defective carbon nanotubes and graphene sheets. Phys. Rev. B, 75, 075412.

115. Terdalkar, S.S., Huang, S., Yuan, H., Rencis, J.J., Zhu, T., Zhang, S. (2010) Nanoscale fracture in graphene. Chem. Phys. Lett., 494, 218–222.

116. Balandin, A.A., Ghosh, S., Bao, W., Calizo, I., Teweldebrhan, D., Miao, F., Lau, N. (2008), Superior thermal conductivity of single-layer graphene. Nano Lett., 8, 902–907.

117. Cai, W., Moore, A.L., Zhu, Y., Li, X., Chen, S., Shi, L., Ruoff, R.S. (2010), Thermal transport in suspended and supported monolayer graphene grown by chemical vapor deposition. Nano. Lett., 10, 1645–1651.

118. Pop, E., Mann, D., Wang, Q., Goodson, K., Dai, H. (2006) Thermal conductance of an individual single-wall carbon nanotube above room temperature. Nano Lett., 6, 96–100.

119. Kim, P., Shi, L., Majumdar, A., McEuen, P.L. (2001) Thermal transport measurements of individual multiwalled nanotubes. Phys. Rev. Lett., 87, 215502.

120. Bai, J.W., Zhong, X., Jiang, S., Huang, Y., Duan, X.F. (2010) Graphene nanomesh. Nat. Nanotechnol., 5, 190–194.

121. Nika, D.L., Pokatilov, E.P., Askerov, A.S., Balandin, A.A. (2009) Phonon thermal conduction in graphene: Role of umklapp and edge roughness scattering. Phys. Rev., 79, 155413–12.

122. Reddy, C.D., Ragendran, S., Liew, K.M. (2006) Equilibrium configuration and continuum elastic properties of finite sized graphene. Nanotechology, 17, 864–870.

123. Chien, S.K., Yang, Y.T., Chen, C.O.K. (2011) Influence of hydrogen functionalization on thermal conductivity of graphene: Nonequilibrium molecular dynamics simulations. Appl. Phys. Lett., 98, 033107.

124. Kalita, G., Masahiro, M., Uchida, H., Wakita, K., Umeno, M. (2010) Few layers of graphene as transparent electrode from botanical derivative camphor. Materials Lett., 64, 2180–2183.

125. Park, H., Chang, S., Smith, M., Gradečak, S., Kong, J. (2013) Interface engineering of graphene for universal applications as both anode and cathode in organic photovoltaics. Sci. Rep., 3, 1581–8.

126. Yazyev, O.V. (2008) Magnetism in disordered graphene and irradiated graphite. Phys. Rev. Lett., 101, 037203.

127. Novoselov, K.S., Geim, A.K., Morozov, S.V., Jiang, D., Katsnelson, M.I., Grigorieva, I.V., Dubonos, S.V., Firsov, A.A. (2005) Two-dimensional gas of massless dirac fermions in graphene. Nature, 438, 197–200.

128. Hernandez, Y., Nicolosi, V., Lotya, M., Blighe, F.M., Sun, Z., De, S., Mcgovern, I.T., Holland, B., Byrne, M., Gun'Ko, Y.K., et al. (2008) High-yield production of graphene by liquid-phase exfoliation of graphite. Nat. Nanotechnol., 3, 563–568.

129. Zhao, J., Pei, S., Ren, W., Gao, L., Cheng, M. (2010) Efficient preparation of large-area graphene oxide sheets for transparent conductive films. ACS Nano, 4, 5245–5252.

130. Xu, Z., Xue, K. (2010) Engineering graphene by oxidation: A first-principles study. Nanotechnology, 21, 045704.

131. Biel, B., Blasé, X., Triozon, F., Roche, S. (2009) Anomalous doping effects on charge transport in graphene nanoribbons. Phys. Rev. Lett., 102, 096803.

132. Chen, S., Brown, L., Levendorf, M., Weiwei, C., Sang-Yong, Ju., Edgeworth, J., Li, X., Magnuson, C.W., Velamakanni, A., Piner, R.D., Kang, J., Park, J., Ruoff, R.S. (2011) Oxidation resistance of graphene-coated Cu and Cu/Ni alloy. ACS Nano, 5, 1321–1327.

133. Blake, P., Brimicombe, P.D., Nair, R.R., Booth, T.J., Da, Jiang., Schedin, F., Ponomarenko, L.A., Morozov, S.V., Gleeson, H.F., Hill, E.W., Geim, A.K., Novoselov, K.S. (2008) Graphene-based liquid crystal device. Nano Lett., 8, 1704–1708.

134. Yeh, C.C., Chen, D.H. (2014) Ni/reduced graphene oxide nanocomposite as a magnetically recoverable catalyst with near infrared photo thermally enhanced activity. Appl. Catal. B. Environ., 150–151, 298–304.

135. Khai, T.V., Na, H.G., Kwak, D.S., Kwon, Y.J., Ham, H., Shim, K.B., Kim, H.W. (2012) Comparison study of structural and optical properties of boron-doped and undoped graphene oxide films. Chem. Eng. J., 211–212, 369–377.

136. Verma, S., Verma, D., Jain, S.L. (2014) Magnetically separable palladium-graphene nanocomposite as heterogeneous catalyst for the synthesis of 2-alkylquinolines via one pot reaction of anilines with alkenyl ethers. Tetrahedron Lett., 55, 2406–2409.

137. Gao, T., Huang, K., Qi, X., Li, H., Yang, L., Zhong, J. (2014) Free-standing SnO_2 nanoparticles graphene hybrid paper for advanced lithium-ion batteries. Ceram. Int., 40, 6891–6897.

138. Li, S., Wang, B., Liu, J. (2014) In situ one-step synthesis of $CoFe_2O_4$/graphene nanocomposites as high-performance anode for lithium-ion batteries. Electrochimica Acta, 129, 33–39.

139. Zhang, M., Yang, X., Kan, X., Wang, X., Ma, L., Jia, M. (2013) Carbon-encapsulated $CoFe_2O_4$/ graphene nanocomposite as high performance anode for lithium ion batteries. Electrochimica Acta, 112, 727–734.

140. Chen, Y., Huang, Z., Zhang, H., Chen, Y., Cheng, Z., Zhong, Y., Ye, Y., Lei, X. (2014) Synthesis of the graphene/nickel oxide composite and its electrochemical performance for supercapacitors. Int. J. Hydrog. Energy, 39, 16171–16178.

141. Zhang, F., Xiao, F., Dong, Z.H., Shi, W. (2013) Synthesis of polypyrrole wrapped graphene hydrogels composites as supercapacitor electrodes. Electrochimica Acta, 114, 125–132.

142. Liu, J., Cui, L., Losic, D. (2013) Graphene and graphene oxide as new nanocarriers for drug delivery applications. Acta Biomater., 9, 9243–9257.

143. Kosynkin, D.V., Higginbotham, A.L., Sinitskii, A., Lomeda, J.R., Dimiev, A., Price, B., Tour, K., James, M. (2009) Longitudinal unzipping of carbon nanotubes to form graphene nanoribbons. Nature, 458, 872–6.

144. Li. X., Zhang, G., Bai, X., Sun, X., Wang, X., Wang, E., Dai, H. (2008) Highly conducting graphene sheets and langmuir-blodgett films. Nat. Nanotechnol., 3, 538–42.

145. Obeng, Y., Srinivasan, P. (2011) Graphene: Is it the future for semiconductors? An overview of the material, devices, and applications. The Electrochemical Society Interface, 20, 1, 47–52. . http://wp.electrochem.org/dl/interface/spr/spr11/spr11_p047–052.pdf

Chapter 5

Chalcogenide Systems— CNTs Nanocomposites

INTRODUCTION

Nanocomposites materials are composed of two or more constituents, usually possessing different physical and chemical properties. They remain separately distinct at the microscopic level, whereas, collectively comprise a single physical material with any phase dimension less than 100 nm [1–3]. Broadly speaking, the basic concept is to make the nanocomposites integrate one or more discontinuous nano-dimensional phases into a single continuous macro-phase to achieve synergistic properties. Their physical/chemical properties are in a combined entity inherently different from the composite individual entities. The formed single entity composites properties may be superior to those of the individual material constituents. Usually in a composite configuration one of the combined material constituents is in much greater concentration forming a continuous matrix surrounding the others, this could serve as a nanofiller (or reinforcement). In forming of nanocomposites, each distinct phase structure and property integrates to fabricate hybrid materials, therefore, possessing multifunctionalities in terms of both structures and material properties. The scientific and societal needs of the 21st century demand increases in higher-performance sustainable and multifunctional nanomaterials becoming more innovative due to nanotechnology and nanocomposites development. Additionally, the innovation of new materials and novel characterization tools in the nanotechnology domain paved the way for the latest design of next-generation nanocomposites that not only are easily controllable, but also have multiple intrinsic engineering functionalities.

Therefore, the artificial composite materials under the controlled anisotropy potential can offer considerable scope for integration between the processes of material specification and component design. This is a key point for composites, as they represent a departure from the conventional engineering practice. In usual engineering design, a component usually takes material properties to be isotropic. However, this is also sometimes inaccurate for conventional materials; such as, a metal sheet normally has different properties in the plane of the sheet from those through thickness direction [4]. Whereas a composite material can have large anisotropies in stiffness and strength, which should be taken into account during design. This is not only a limited variation in strength, but also other parameters such as direction, the effect of any anisotropy in stiffness on the stresses set up in the component under a given external load should also be taken into account (i.e. mechanical properties). It should be kept in mind that the produced composite material with the way of loading its components is accounted for.

There are several ways to make different types of composite materials. As an example, typical microstructures are key parameters in three main classes, they can be grouped according to the nature of the matrix. A vast number of composites for industrial use are made based on polymeric matrices, like thermosets, thermoplastics, etc. Such composite materials are usually reinforced with aligned ceramic fibers, such as glass or carbon. They normally exhibit marked anisotropy due to their much weaker and less stiffer matrix than fibers. Thus to formulate the composite materials, it is important to consider the properties of potential constituents. More precisely, properties like stiffness (Young's modulus), strength and toughness. The constituent density also has a great significance in many situations, due to the fact that the mass of the component may also be of critical importance. Additionally, the thermal properties of the constituents such as, expansivity and conductivity should also be considered. Since the composite materials are subject to temperature changes (during manufacture and/or in service), a mismatch between the thermal expansion of the constituents could lead to internal residual stresses. These factors can have a strong effect on the mechanical behavior of the newly formed composite materials.

As discussed above, the polymeric materials could be one potential candidate to make a nanocomposite with different kinds of the elements. Therefore, putting nano fillers into polymer systems can result in the polymer nanocomposites with the multifunctional, high-performance polymer with their characteristics beyond the traditionally filled polymeric materials. Hence, under the filler control at the nanoscale level, polymer nanocomposites can exhibit maximize property enhancement of the selected polymer systems or can potentially generate the exceptional novel inside physical phenomenon that could meet the requirements of military, aerospace and commercial applications. The reinforcements of the polymer nanocomposites can reveal the special mechanical, optical, electrical and magnetic properties of the composites, in the circumstances that the polymer matrix provides support for the reinforcements and retains their properties of the constituent polymer. But it is not limited only to polymeric materials, composites can be made with different kinds of materials such as crystalline, etc.

CHALCOGENIDE SYSTEMS COMPOSITES

Since chalcogenide glassy materials are a special class of polymer, therefore, these kinds of materials could be one of the natural potential candidates to make composite materials with other types of materials, including organic and inorganic constituents. The existing theories and concepts of polymorphism formation in chalcogenide glasses demonstrate the concept of a continuous random network that could be explained from the view of clusters of structural-independent polyforms with the other key principles of chalcogenide glasses (such as, kinetic, thermodynamic and structural–chemical) [5–12].

Usually, the objective of different concepts reflects the glass structure, but they are distinct either in excess in general, and could not explain a number of experimental data, such as the concept of polymeric structure [13]. Hence, distinct concepts have contradictory data explanations, such as, crystalline concept; modern diffractometric has demonstrated that in well-synthesized glass, there are no crystallites even with the smallest crystals consisting of a small number of elementary cells [10]. Moreover, according to the structure of amorphous solid matter concept composed of different types of micro-crystals based on 'micro-crystallite' model is not in compliance with the commonly accepted opinion of random network terms [14]. It is believed that this is a close view on the relation between the glass-forming ability and the existence of several crystalline polyforms in the same material. Additionally, demonstrating a view of the structural unit in amorphous material forming a continuous random network (SiO_4 tetrahedron, in the SiO_2 case) is similar in respect to energy, corresponding to the existence of crystalline polyforms. While this point of view has a significant value, but it could not provide the complete conceptual resolution of the problem.

The relation between polymorphism and its ability to form non-crystalline matter has been debated by many investigators in the past several decades. Their works were predominantly connected to the glass formation and polymorphism of substance [15]. According to the initial hypothesis of glass structure it was considered that they consist of a very small different size crystal, and have the tendency to polymorphize the substance during glass formation [16]. Thus amorphous glass is the mixture of the substance with the structural modification, as for examples, S, Se, etc. Several investigators contributed their views and characterized the polymorphism in various systems such as BeF_2, $ZnCl_2$, B_2O_3, GeO_2, P_2O_5, As_2O_3, Sb_2O_3, TeO_2, S and Se. They also explored the cause-and-effect relation between polymorphism and concluded that the concept of glass formation, are clusters structurally independent of polyforms which connected to each other by sterically strained interfaces [11]. Hence the universal correlation between non-crystalline state and polymorphism for elements of the periodic table was established [17]. Later it extensively examined the non-crystalline states (including vitreous) of the substances on the basis of that and recognized the easiest formation of the non-crystalline state for sulfur, selenium, phosphorus, boron and arsenic. More specifically for chalcogenide glasses, it indicated the relation between polymorphism and glass formation [18–20]. In addition to explaining the polymerization of such materials concept of the Short Range Ordering (SRO) was introduced [21, 22]. According to modern theory, this concept of SRO is a local arrangement of atoms around a certain atom taken as a reference point. SRO can be characterized by the coordination number and chemical nature of atoms located in the first coordination sphere of the atom taken as a reference point and the geometry of their arrangement: inter-atom distance values and interbond angle values. The general acceptance of this definition has been confirmed [21, 23]. Further it has established that two distinct SRO-I and SRO-II phases exist in these substances [24]. These two new phases create different Intermediate-Range Orderings (IROs) and different Long-Range Orders (LROs). Where the IRO is correlated to the structure of a fragment of crystal structure in which the structural units of SRO-I joined themselves by structural units of SRO-II. Atoms in such fragments may be arranged in compliance with regularities of their arrangement excluding one regularity—translation symmetry, i.e., LRO. This structural fragment is very much like a crystal, but nevertheless is not a crystal—it is a crystalloid [25]. Thus the crystalloid is a bearer of short-range and intermediate-range orders that profoundly could influence structure and properties of one-component glass-forming at the different angle of the substance.

Particularly, the melt of stable hexagonal polymorphous selenium at high temperature, lead to metastable molecules Se_8 being formed with the typical structure of monoclinic selenium and disintegrates to fragments at cooling [22]. Their molecular experimental evidences indicated that the melted selenium cooled to room temperature is a copolymer consisting of structural fragments of cis- and trans-configurations which are typical for monoclinic and hexagonal [13]. Therefore, such a liquid is apparently different from glass in the degree of polymerization of substance.

Moreover, based on a critical analysis of the existing concepts of glass-forming liquid and glass structure, the following key statements have been established [26]:

1. The one-component vitreous substance formation (an element or a chemical compound) is the process of generation, mutual transformation and copolymerization of structural fragments without crystalloids in disordered polymeric polymorphous-crystalloid structure (network, tangle of chains, ribbons, etc.).

2. The crystalloid is a fragment of crystal structure consisting of a group of atoms connected by chemical bonds. The stereometric ordering rules demonstrate that the inherent one of the crystal polymerization without translational symmetry. This means there is no kind of crystalloids, at even the smallest crystalloids.

3. Crystalloid can be directly connected to notions Short Range Ordering and Intermediate Range Ordering (SRO and IRO) that may apply for both non-crystalline and crystalline substances.

4. Every non-crystalline substance has two or more SROs, and two or more IROs without LRO. The number of IRO types depend on the number of polymorphisms taking part in the formation of the non-crystalline substance. While the crystalline substance may have one, two or more SROs with a restriction of only one type of IRO and LRO.

5. The ordered crystalloids of different polymorphisms join together in accordance with rules of one of the polymorphism (except the translational symmetry). Due to inherent statistical alteration, they may form disordered polymeric-crystalloid structure of vitreous substance in which order and disorder can exist.

6. Such substance structure is not absolutely continuous; therefore, separate broken chemical bonds and other structural defects exist.

7. These materials can also be constructed with crystalloids of different polymorphisms that the degree of co-polymerization decreases at an increased temperature. At a specific temperature, the effects include the disintegration of some crystalloids, which leads to the disappearance of IROs for certain polymorphisms and the formation of separate structural fragments that is only in SRO.

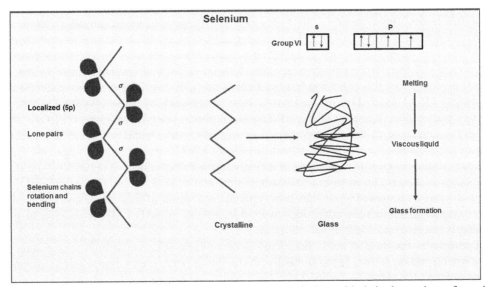

Figure 5.1   Schematic structure of crystalline and amorphous selenium with their electronic configuration.

Additionally, chalcogenide also consists of random mixture of rings and helical chains accompanied by coordination defects, as shown in Fig. 5.1. Usually in a stable phase chalcogenide elements have covalent bonds with the two nearest neighbors following the 8–N rule [27], in which atoms have six electrons in the outermost shell under a configuration s^2p^4. The s state electrons do not participate in bond formation; however, these states have energies below the p states. Therefore, two covalent bonds are formed between p electrons of the chalcogen atoms, however, another electron pair called Lone Pair (LP) which remains unbonded. These unbounded electron pairs could play a key role in interacting with different kinds of inorganic and organic elements, because such structures can typically form organic polymers like polyethylene. Owing to this similarity, amorphous chalcogenide are also known as an inorganic polymer [28]. Hence chalcogenide elements hybrid polymeric behavior with fullerene like ring structures would allow them to form composites with the different kinds of inorganic/organic materials.

Moreover, healing properties of chalcogenides can repair the components and store reinforcement as the self-repairing system of the composite structures. Self-healing polymers and polymer composites can be classified into two categories: intrinsic and extrinsic. The intrinsic self-healing process governs from the surface rearrangement, surface approach, wetting, diffusion and randomization. On the other hand, to fill brittle-walled vessels with a polymerized medium fluid, it should occur at a healing temperature. Therefore, polymerization of the chemicals could damage the area which may play the role of crack elimination. Like bio-inspired self-healing from the hollow fibers embedded within a structure at different length-scales in different materials, such as, bulk polymers and polymer composites. The complete filling of a healing agent into the fibers can be from vacuum assisted capillary action. It could be described in three ways; (i) Single-part adhesive, in which all hollow pipettes contain only one kind of particles, (ii) Two-part adhesive process, in which the curing agent can be used to fill into the neighboring hollow tubes, (iii) Two-part adhesive in this process one component could be incorporated into hollow tubes through other microcapsules.

Chalcogenide composite materials could have remarkable properties, but they have a common shortcoming due to the failure of the matrix-rich interlaminar regions where stress transfer between load-bearing is less efficient and nearly polymer properties dominate. Even though composites often incorporate approximately 60 volume chalcogens, their local matrix regions can persist, this is a critical concern in commercial applications in terms of catastrophic failure damage. Therefore, their hybrid composites that are different from traditional composites through the incorporation of nanofillers into the matrix phase may need multi-scale reinforcement to enhance the stiffness, strength and toughness. The imbued multifunctionality form composites could also enhance the electrical and thermal conductivity or barrier properties depending on the choice of nanoparticle reinforcement. Chalcogenon multi-scale composites can be classified into two types of systems, one system entails independent dispersion of the nanofillers throughout the polymer matrix, due to the mixed inclusion system through the incorporation of various types of nanoparticles, such as rubber particles, elastomeric block copolymers or combinations of the above into the matrix. In another way, the nanofillers can be chemically linked to the chalcogen matrix, such as carbon nanotubes (CNTs) or carbon nanofibers (CNFs) and forms the surface and results in a hybrid composite system or by catalyzing surfaces through the direct attachment of loose CNTs on CNFs to the outer surface of parent matrix. Such regions of CNTs or CNFs of hybrid matrix is known as the forest region owing to their dense packing, preferential alignment and similar lengths. These forest regions can act as an anchor to further strengthening the matrix interface and more efficiently transfer stress. Therefore, to overcome the short comings of traditional chalcogenide systems the composite systems would be suitable.

CARBON NANOTUBES COMPOSITES

Carbon based nanomaterials have had an extraordinary impact over the last three decades to define the reach and applications of nanotechnology. Since the discovery of fullerenes and the carbon nanotubes (CNTs) to graphene and other two-dimensional materials, the scientific and academic world has been flushed with new ideas from inventions with innumerable attempts to find the killer applications for these remarkable nanostructures. More specifically for the CNTs composite materials, the basic question of whether their structures can be tuned as ideal reinforcements for the composite matrices. Whether CNTs mechanical properties really are the right choices to mechanical reinforcement, still remains widely unanswered. Although CNTs are sp^2 allotropes, the structure, morphology and dimensionality of the largely interesting carbon nanostructures are quite different, therefore, it is the nature of their interactions with the adjacent matrix [29]. Thus, the overall composite mechanical behavior could be from the two reinforcement units in a distinct manner. It would really be useful to define the selection rules that choose one over the other in composite applications. Considering this, investigators have made attempts to define a rule for such composite materials, keeping their unique structures, how to make good reinforcements in composites and uncovering the excellent electrical and thermal properties with the mechanical properties [30].

Since CNTs are rolled-up graphenes, they possess in-plane properties that translate into axial properties, this makes them among the stiffest axial fibers ever created. They can also be easily bent, twisted and buckled [30]. The CNTs can also have a nested structure of tubes inside tubes, i.e. single-walled (SWNTs) to multi-walled nanotubes (MWNTs). Their nested structure can produce the notable effect in their physical properties. Generally, the local stiffness is extremely high due to nearly defect-free structure of nanotubes, therefore, this is the major issue to make reinforcements in these structures. Such carbon nanostructures are particulate fillers with their larger dimensions (lengths of nanotubes) reaching several hundred micrometers in most or typical cases (i.e. millimeters). Usually short fibers have poor load carriers in fiber composites, and this effect has been proved experimentally for the composites with CNT dispersions. Moreover, the surfaces of nanotubes can be atomically smooth; therefore, it reflects the lack of dangling bonds or defects (except at the edges of the nanotubes). This means the strong matrix filler bonds are hard to accomplish with poor interfacial load transfer during mechanical deformation [29]. Its high electron and phonon scattering can be associated with the electrical and thermal properties. Hence, the interfacial problem is the major roadblock in carbon fiber composites before industry can figure out the sizing of fibers from chemical modifications. This can be overcome for nanotubes by attempting chemical functionalization of CNTs surfaces that may substantially compromise their intrinsic properties [31]. This can also be related to another key issue in the inhomogeneous dispersion of nanotubes in the matrix. As without proper surface treatments CNTs tend to aggregate easily due to the strong van der Waals interactions between them, therefore, poor dispersed bundles or agglomeration remain in the matrix. This often leads to a poor interfacial connectivity and the formation of mechanical stress concentration or other functionally singular sites, as a final result, it affects the composite performances. To partially overcome the dispersion challenge, non-covalent functionalization methods could be used [32]. However, this method is not appropriate in resolving the interface problem. Therefore, another systematic and carefully engineered approach has been designed for the CNTs composites with their optimal performances, as shown in Fig. 5.2(a). A hypothetical position for the different types of composites modifier was demonstrated in the past [33]. According to this concept, the ideal dispersion of material throughout the composite is application dependent; typically to get better mechanical properties, one should make as high a loading of the filler as possible that should align in the direction of the load. To get good electrical percolation one should ensure a random percolated network with as low a concentration as possible. However, to achieve better multi functionalities, well ordered microstructures are usually required, it is therefore contradictory.

Consequently great efforts were made to develop lightweight and strong composite materials with CNTs as reinforcements. Note that CNTs are considered to be discontinuous short fillers, and they may still possess outstanding mechanical properties. Additionally their composites can have extremely high Young's modulus at nanoscale dimensions, as well as their specific geometries could offer high specific surface area [see Fig.5.2(b)]. This could efficiently tailor the interface properties between the reinforcements and composite matrices. But CNTs nanocomposites may not be as strong or stiff as a continuously reinforced composite such as typical carbon fiber lamination used for primary load-carrying structural applications. Nonetheless, if the extraordinary potential and multi functionality of CNTs are fully functional and their nanocomposites are properly designed, they might become game-changer composite materials. But challenges remains such as dispersion, viscosity control, sizing of the CNTs without compromising intrinsic properties, accurately characterizing the reinforcement's mechanical properties. Among these Young's modulus and strength properties as well as interfacial shear strength at the filler matrix interface, can determine the required critical length of the fillers with the most efficient load-transfer capability. All these factors are very critical for designing short-fiber composite systems, but more attention was paid on improving the practical issues such as dispersion in matrix materials and increasing the loading fractions without suffering viscosity-related processing issues.

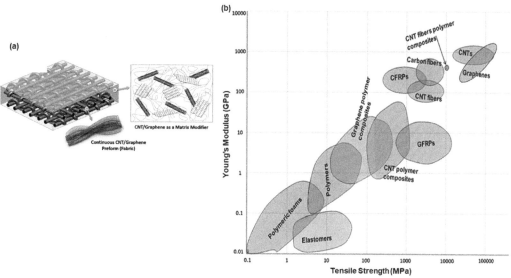

Figure 5.2 (a) A schematic of the CNT or graphene polymer composite, consisting of continuous CNT fiber preform (fabric) in a polymer matrix and chemically modified CNT or graphene as matrix modifiers. (Reproduced from the permission Valorosi, F. et al., (2020) Graphene and related materials in hierarchical fiber composites: Production techniques and key industrial benefits. Composites Science and Technology, 185, 107848-15) (b) Typical plot of Young's modulus against tensile strength with the mechanical properties of conventional polymer composites [33].

Polymer composites usually connect with the matrix-reinforcement interface. Though, it is also necessary to consider the van der Waals bonding between the walls of carbon nanotubes. The van der Waals bonding is relatively weak compared with the strong covalent bonding within the layers. However, in the MWNTs composites, the ability to reinforce is therefore limited by easy shearing between the walls or layers, respectively. Therefore, it is quite possible to track internal stress transfer between the walls of carbon nanotubes by the stress-induced Raman band shifts. Imperfect stress transfer is manifested as Raman band broadening during deformation and a lower Raman band shift rate compared to the single walled or monolayer material. By comparing the Raman band shifts under stress to SWNTs and MWNTs, the nanocomposites inter wall stress transfer efficiency can be accessed [33].

When considering the composite ability of the carbon nanotubes, it is necessary to discuss their functionalization behavior with different types of the materials. Therefore the extensive characteristics of carbon nanotubes functionalization with different types of materials can be briefly described as: carbon nanotubes tubes usually yield a mixture of various diameters and chiralities of nanotubes that are normally connected to their contamination with metallic and amorphous impurities. The post-synthesis chemical processing protocols that could purify tubes and possibly separate individual tubes in terms of diameter and chirality to utilize their different reactivities under viable routes, may be favorable for electronic and mechanical properties of these materials [34–37]. Moreover, the full potential CNTs reinforcements could be weak due to poor interfacial van der Waals interactions. Therefore, the nature of the dispersion of the CNTs is rather different than other conventional fillers, such as spherical particles and carbon fibers, due to the small diameter of CNTs within the nanometer range while possessing high aspect ratio (>1000) and large surface area. The surface modifications of the CNTs could minimize this issue; therefore, different approaches generally divided into chemical (covalent) and physical (non-covalent) functionalization as interactions between active materials and CNTs, as illustrated in Fig. 5.3(a,b,c,d).

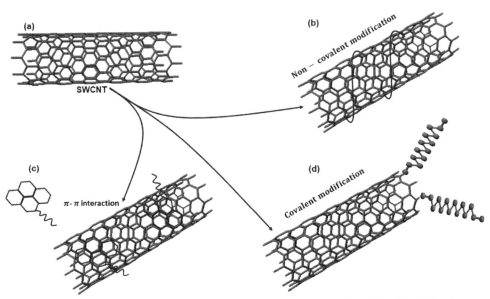

Figure 5.3 Functionalization of SWNT: **(a)** SWCNT, **(b)** non-covalent sidewall modifications, **(c)** surfactants $\pi-\pi$ interaction functionalization, **(d)** covalent modifications.

(i) *Covalent Functionalization*

In this kind of functionalization nanotubes end caps have a tendency to compose highly curved fullerene-like hemispheres, they are highly reactive as compared to the side walls [38, 39]. Generally sidewalls themselves contain defect sites like pentagonheptagon pairs, sp^3-hybrideized defects and vacancies in the nanotube, as shown in Fig. 5.4 [38].

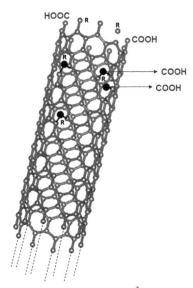

Figure 5.4 Typical defects schematic for a SWNTs, here sp^3-hybrideized defects (R=H and OH) leaves a hole lined with –COOH groups under oxidative defect condition, the SWNTs end terminated with COOH groups [38].

Usually chemical functionalization is based on the covalent bond of functional groups on the carbon form of CNTs. It can be formed from the end caps of nanotubes or to the sidewalls (i.e. defects). The direct covalent sidewall functionalization results in a change in hybridization sp^2 to sp^3 as well as a loss of p-conjugation layer. This process predominantly governs from high chemical reactive molecules. On other hand, in the highly curved fullerene-like hemispheres approach, the fluorination of CNTs is more relevant for the initial investigation of the covalent functionalization owing to the expectation that sidewalls could exist in a inert environment [40, 41]. However, it was found that the fluorinated CNTs C–F bonds are weaker than those in alkyl fluorides [42], this could provide substitution sites for additional functionalization [43]. In order to continue this, successful replacement of the fluorine atoms by amino, alkyl and hydroxyl groups have been achieved [44]. Similarly, many other methods have been successfully used such as, cycloaddition, Diels-Alder reaction, carbine, nitrene addition, chlorination, bromination, hydrogenation and azomethineylides [45–48].

Another method most frequently used is the defect functionalization of CNTs. In which the intrinsic defects are supplemented by oxidative damage to the nanotube framework by strong acids, as a consequence, creating functionalized holes with oxygenated functional groups [49]. The specific CNTs treatment with strong acid such as HNO_3, H_2SO_4 or a mixture of them [50, 51] or with the strong oxidants like $KMnO_4$ [52], ozone [53], reactive plasma [54, 55] are adopted to open the nanotubes. It therefore, generates oxygenated functional groups such as, carboxylic acid, ketone, alcohol and ester groups. This can serve as a tether for different types of chemical moieties on the ends and defect sites of these nanotubes. These kinds of functional groups have a rich chemistry as well as CNTs that can be used as precursors for further chemical reactions, like silanation [55], polymer grafting [56, 57], esterification [58], thiolation [51], and a few biomolecules [59]. The covalent functionalized CNTs also have a good advantage to their solubility in various organic solvents owing to the fact that CNTs can possess many functional groups such as polar or non-polar groups, as shown in Fig. 5.5.

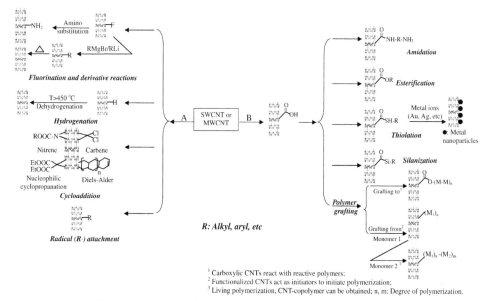

However, the CNTs functionalization methods discussed above have two main drawbacks. During the functionalization reaction, especially along with damaging ultrasonication process, a large number of defects are inevitably created on the CNT sidewalls. In extreme cases CNTs may fragment into smaller pieces, more specifically for the sp^2 to sp^3 carbon hybridization of CNTs. Such damaging effects results in the severe degradation in mechanical properties of CNTs as well as disruption of π electron bonds in nanotubes. The disruption of π electrons can directly relate to transport properties of CNTs because defect sites scatter electrons and phonons. This is directly responsible for the electrical and thermal conductions of CNTs. Another one is that the concentrated acids or strong oxidants are often used for CNT functionalization, these are environmentally unfriendly. As a consequence several efforts have been put forwarded to develop methods which are convenient to use as well as of low cost and less damage to CNT structure.

(ii) *Non-covalent Functionalization*

Non-covalent functionalization has an advantage in terms of no distraction of conjugated system of the CNTs sidewalls. This means that it does not affect the final structural properties of the material. The non-covalent functionalization can also be considered as an alternative approach to tune interfacial properties of nanotubes. The non-covalently CNTs functionalized aromatic compounds, surfactants and polymers use π–π stacking or hydrophobic interactions in the most part. In such non-covalent modifications CNTs can preserve their desired properties; however, improvement in their solubilities could be exceptional. It could be interpreted as, the aromatic small molecule absorption, polymer wrapping, surfactants, biopolymers and endohedral approaches.

The aromatic molecules, like pyrene, porphyrin and their derivatives may interact with the sidewalls of CNTs. In the literal meaning: π–π stacking interactions. This is opening the way for the non-covalent functionalization of CNTs, asdepicted in Fig. 5.6(a,b,c). A general approach for the non-covalent functionalization of CNTs sidewalls and immobilization of biological molecules to CNTs having a high degree of control has been demonstrated [60]. Further, CNTs/ FET devices functionalized non-covalently with a zinc porphyrin derivative can be fabricated to detect directly a photo induced electron transfer [61]. Moreover, CdSe–CNTs hybrids through the self-assembling pyrene-functionalized CdSe (pyrene/CdSe) nanoparticles on the surfaces of the CNTs were prepared [62].

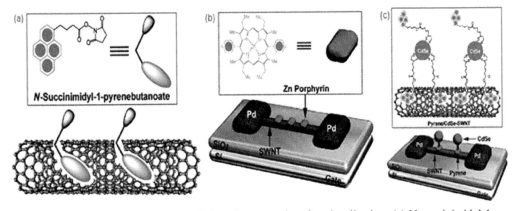

Figure 5.6   Aromatic small-molecule based non-covalent functionalization: **(a)** N-succinimidyl-1-pyrenebu-tanoate coated CNTs; **(b)** zinc porphyrin-coated CNTs; **(c)** pyrene/CdSe coated CNTs. (Reprint permission from Zhao, Y.L., Stoddart, J.F. (2009) Noncovalent functionalization of single-walled carbon nanotubes. Accounts of Chemical Research, 42, 1161–1171, Copyright © 2009, American Chemical Society)

Specifically the conjugated polymers can serve as excellent wrapping materials for the non-covalent functionalization of CNTs. This could directly relate to π–π stacking and van der Waals interactions between the conjugated polymer chains that contain aromatic rings as well as surfaces of CNTs, as illustrated in Fig. 5.7 [62–66]. There are some organic-soluble conjugated such as poly(*m*-phenylenevinylene)-co-(2,5-dioctoxy-*p*phenylene) vinylene (PmPV), poly(2,6-

pyridinlenevinylene)-co-(2,5-dioctoxy-*p*-phenylene)vinylene (PPyPV), poly-(5-alkoxy-*m*-phenylenevinylene)-co-(2,5-dioctoxy-*p*-phenylene)-vinylene (PAmPV), stilbene-like dendrimers, that have also been investigated for the non-covalent functionalization of CNTs [62–64, 67].

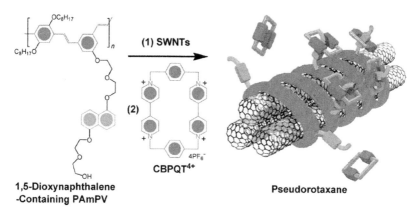

1,5-Dioxynaphthalene -Containing PAmPV

Pseudorotaxane

Figure 5.7 The side arms of the 1,5-dioxynaphthalene containing PAmPV-decorated CNTs hybrids associate with cyclobis (paraquat-*p*-phenylene) (CBPQT^{4+}) rings.

(Reprint permission from Zhao, Y.L., Stoddart, J.F. (2009) Noncovalent functionalization of single-walled carbon nanotubes. Accounts of Chemical Research, 42, 1161–1171, Copyright © 2009, American Chemical Society)

Moreover, surfactants polymers can also be used to functionalize the CNTs, as depicted in Fig. 5.8. The physical adsorption of surfactant on the CNTs surface could lower the surface tension. It can effectively prevent the aggregation of CNTs. Moreover, surfactant treated CNTs can also overcome the van der Waals attraction due to electrostatic/steric repulsive forces. The process efficiency depends widely on the properties of surfactants, medium chemistry and polymer matrix. The relation between surfactants and CNTs can occur in the different ways such as; (i) non-ionic surfactants, like polyoxyethylene 8 lauryl or C12EO8, polyoxyethylene octylphenylether (Triton X-100) [68, 69], (ii) anionic surfactants, like sodium dodecylsulfate (SDS), sodium dodecylbenzenesulfonate (NaDDBS), poly(styrene sulfate) (PSS) [70, 71] (iii) cationic surfactants, such as, dodecyl tri-methyl ammoniumbromide (DTAB), cetyltrimethylammounium 4-vinylbenzoate [72, 73]. Surfactants could be efficient for the solubilization of CNTs, as are known to be permeable as plasma membranes. Since surfactants are toxic for biological applications, their stabilized CNTs complexes potential use are limited for the biomedical applications such as, proteins, enzymes, DNA, etc. [74]. However, solubilization of CNTs with biological components is significantly more appropriate towards integration for new types of material with living systems, as shown in Fig. 5.9. Biomacromolecules under the non-covalent functionalization of CNTs can be included from simply saccharides and polysaccharides [75–78]. Although for a vast number of biomaterials such as *n*-decyl-*β*-Dmaltoside, *γ*-cyclodextrin, *η*-cyclodextrin, chitosan, pullulan

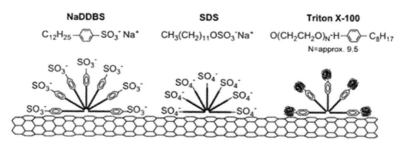

Figure 5.8 Schematic representation of surfactants to adsorb onto CNTs surface.

(Reprint permission from, Islam et al., 2003 High weight fraction surfactant solubilization of single-wall carbon nanotubes in water. Nano Letters, 3, 269–273, © 2003 American Chemical Society)

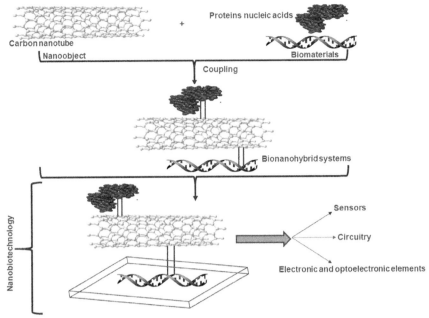

Figure 5.9 Conceptual generation of biomolecules-CNTs conjugates and their schematic to yield functional devices.

and phospholipid-dextran have been used in the non-covalent functionalization of CNTs [73, 79–83]. These materials have been preferred for use because of the advantage of the saccharides and polysaccharides exhibiting non-light absorption in UV-vis wavelength region. Therefore, the hybrids CNTs can also be characterized by using photochemical experiments. Thus, saccharide and polysaccharide-coated CNTs hybrids are biocompatible and may be applied for various medicinal purposes.

(iii) *Functionalization of Carbon Nanotubes from the Alternative Routes*
There are several methods to enhance the interaction between CNTs and the reacting materials, two routes have attracted much attention: namely covalent and non-covalent functionalization. Both methods have their advantages and disadvantages as discussed above. Though, traditional covalent functionalization strategy of CNT is the most frequently used approach through the chemical acid oxidation treatment. But the amount of defects generated during the functionalization hinder the intrinsic mobility of carriers along CNTs, therefore, it could not be considered as a model approach. Additionally, using this approach to functionalize nanotube surfaces not only identifies carboxylic acid groups but also leaves behind detrimental structures.

Thus harming their potential in terms of practical applications as well as compromising the mechanical properties of the nanotubes. Considering these key issues a general rule comes into picture that can alleviate these problems. The effective functionalization method should induce high density and homogenous surface functional groups, this could enhance the compatibility between CNTs and the foreign matrix. Under the allowed direct grafting, a little or no structural damage to the CNTs occurs.

Baek et al. have taken this issue and presented their view for an efficient route to covalently functionalize CNTs via direct Friedel-Crafts acylation technique (Fig. 5.10) [84–88]. This novel covalent grafting of the nanotubes may be a promising strategy to improve nanotube dispersion as well as to create microscopic interlinks. Hence, this kind of surface functionalization is not only limited to enhance the reactivity, but can also improve the specificity and provide an avenue for further chemical modification of CNTs. Hence, this approach has made remarkable achievement to enhance the functionalities of CNT-nanocomposites, which are usually not achievable from

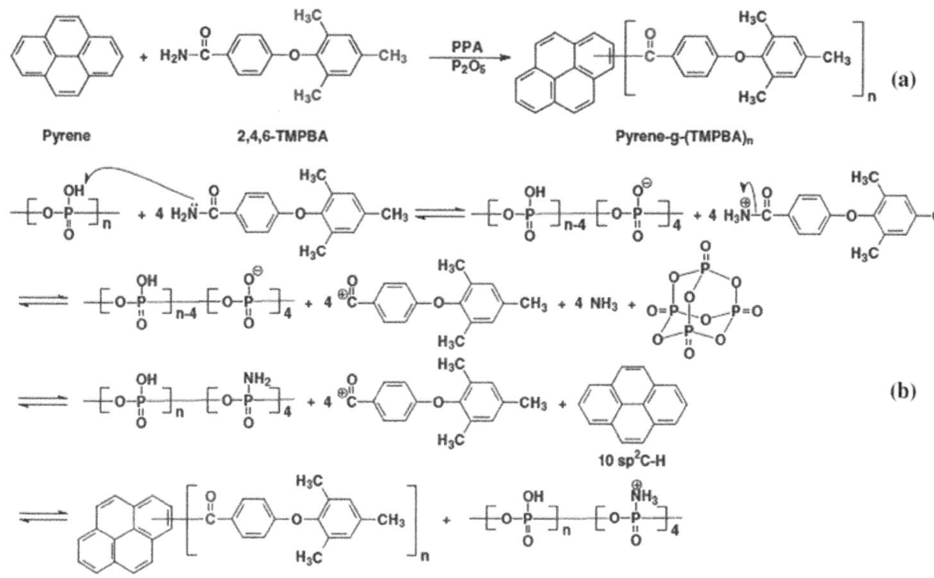

Figure 5.10 Friedel-Crafts acylation reaction of pyrene as a miniature graphene and organicmaterial in poly(phosphoric acid)/phosphorous pentoxide medium. (Reprint permission from Jeon et al., (2010) Edge-Functionalization of pyrene as a miniature graphene via friedel–crafts acylation reaction in poly (Phosphoric Acid). Nanoscale Research Letters, 5, 1686, copyright @ Springerlink).

individual components. This strategy can also have a less-destruction (or non-destructive) reaction mechanism for the efficient dispersion and functionalization of carbon nanomaterials. Therefore, the CNT damage under severe chemical treatments including oxidation and sonication can be avoided to a large extent. Highly enhanced properties can be achieved from improved dispersion stability along with chemical affinity of the matrices.

Nanocrystalline Chalcogenides CNTs Composites

These days much effort has been devoted to explore novel strategies that can alter the physical properties of carbon nanotubes (CNTs) by surface modification with organic, inorganic and biological species [89–97]. Such functional CNT-based composites could offer good prospects and opportunities for new applications in various areas. Linking semiconductor nanocrystals to CNTs has emerged as an active field [93, 98, 99]. The electronic interaction between CNTs and the external active semiconductor layer is supposed to be of equal importance because they play a crucial role in constructing optoelectronic devices [99]. Various types of nanocrystals, semi conducting metal chalcogenide nanocrystals have been extensively explored due to their quantum confinement effects and size- and shape-dependent characteristics. This kind of semiconductor nanocrystals can be applied in many different technological areas, including biological labeling and diagnostics, light emitting diodes, photovoltaic devices and lasers [100–104]. Many semiconducting nanocrystals of chalcogen elements with various compositions and shapes have been synthesized, as an important group II-VI semiconductor [105–107]. The functionalizing of CNTs with chalcogens nanocrystals can not only provide the advantages of combinations but also may result in properties that have potential applications in nanoscale electronic devices [93].

Since the carbon nanotubes (CNTs) have unique properties such as a high ratio, low mass, flexibility, a high mechanical strength and high electrical conductivity [108, 109], that offer various application potentials to sensors [110], electrocatalysts [111], catalysts [112], and energy conversion [113–116]. Therefore, functions of the semiconductor nanoparticles can be easily modified by CNTs to facilitate the electron transfer. Recently this kind of semiconductor material, chalcogenide has proven to be efficient light-driven photo-catalysts for H_2 production from water splitting [117]. Therefore, considerable efforts have been made to improve visible light response of binary or ternary sulfide solid solution photocatalysts [118–122], which may be more efficient and more stable, compared to their single-constituent chalcogenide [123]. Such sulfide solid composite solutions involve the injection of photogenerated electrons from one semiconductor into the lower lying conduction band of another one. Therefore , it may result in the lower recombination rate of photo-induced electron hole pairs [124]. These kinds of sulfide solid composite systems can also have tuneable absorption properties in the visible range of the solar spectrum, such as $Zn_xCd_{1-x}S$-based solid solutions, one of the most efficient visible light photocatalysts [120, 125–128].

Hence, water splitting is an attractive and promising novel property in such composite materials that can offer clean, economical and environmentally friendly conversion of solar energy into H_2 [129–131]. In contrast to the conventional wide band-gap semiconductors that can solely absorb the UV light and greatly restrict the practical applications, more recently the ternary metal sulfide $ZnIn_2S_4$ has been introduced, with the hope that it is an eco-friendly and chemically stable visible-light-driven photocatalyst with high activity of H_2 evolution as well as photocatalytic degradation of contaminants [132–139]. However, its major practical problem is, the photocatalytic activity of $ZnIn_2S_4$ should be improved. To resolve this issue, several approaches have been proposed to modify the visible-light driven photocatalysts with the proper band structures, it was found that the issue of photogenerated charge separation is the key factor that strongly affects the efficiency of the photocatalytic water-splitting process [140–142]. Therefore, to increase the utilization rate of the photogenerated charges and get high photocatalytic water-splitting activities, the photogenerated charges should be efficiently separated to avoid bulk/surface charge recombination as well as transfer to separated active sites on the surface of the photocatalysts [143–146]. To overcome these problems, various composite materials have been introduced by selectively coupling $ZnIn_2S_4$ nanomaterials with other semiconductors (TiO_2, CdSe, ZnS, MoS_2, NiS, and so on) or other noble metals [147–152]. More specifically, carbon-based materials, like active carbon, carbon nanotubes, have been widely introduced due to their excellent conductivity and extraordinary chemical stability [153–156]. Various nanostructured carbon materials such as two-dimensional layered materials and one-dimensional carbon material (carbon nanotubes and carbon nanofiber) can be promising materials.

Many attempts have been made to construct one- and two-dimensional nanocomposites by deposition of functional inorganic nanomaterials on the carbon backbone, with the random mixture of $ZnIn_2S_4$ and other semiconductors. However, their relatively weak interaction between inorganic species and the carbon backbone are usually not consistent with the desired uniformity and distribution of the inorganic component on the surface of the carbon backbone [154–157]. Therefore, it remains challenging to design new desirable strategies to fabricate well-defined structures by growing low-dimensional nanostructures of functional materials on the carbon backbone. The in situ controlled growth approach can offer constructing useful composites [158–160]. Using this kind of synthetic process, the effective interfacial contact and strong interaction between two components of the composite can be formed. This may lead to an enhanced photogenerated charge transfer and separation. The nanosheets ultrathin structures have been significantly demonstrated with improved photocatalytic activities. The hierarchical structure of the composites can be due to the large specific surface areas and the enhanced photogenerated charge carrier transfer from the interior to the surface of photocatalysts that participate in photocatalytic reactions [160, 161]. This is a notable example of the combination with their advantages. Innovations are not limited to this, and there are several reports on this topic, such as one-dimensional hierarchical structures of Ni_3S_2 nanosheets on carbon nanotube backbones, which can significantly improve supercapacitors and photocatalytic H_2-production performance [162]. With these, the electrochemical properties of one-dimensional carbon nanofibers (CNFs) and carbon nanotubes have a similar trend. However, the carbon nanofibers can also be easily prepared from electrospinning technology that is more appropriate to fulfill the requirements of the low cost as compared to carbon nanotubes [161].

Conversion materials like metal selenides have also received much attention due to their exceptional narrowband gap semiconducting properties as well as relatively high theoretical specific capacity [163–167]. Moreover, Na^+/K^+ elements atoms/molecules can be more easily inserted/extracted from metal selenides compared to metal oxides/sulfides/phosphides due to their much larger sodium/potassium ion insertion interstitial sizes [168]. They have also been considered as a promising candidate for electrodes due to the rapid growing energy demands. But the major hurdle to enhance their battery application is their relatively huge volume effect and poor electronic conductivity on sodiation/potassiation. To overcome the volume change and stable structure, the porous/hollow structure has been considered as one of the most effective strategies due to their favorable conditions like dynamically open structures and large surface areas [169–173]. The Prussian blue analogs with the kind of metal–organic frameworks consisting of metal centers and organic linkers on available self templates of porous materials can be suitable approaches to make these kinds of composites [174, 175]. Due to their advantages such as metal-organic frameworks, they can confine and in situ generate small-sized metal particles that can act as precursors to design metal oxides, phosphides and chalcogenides. Additionally, the metal–organic frameworks derived materials can possess hierarchical internal pore structures that are constructed from the micro/nanostructures with the advantage that organic linkers would be carbonized after the annealing process. Such a porous micro-sized material with nanosized characteristics can also avoid the agglomeration of pure nano-sized material; therefore, they can reveal high surface energy [176–179]. More specifically, when the metal-based nanoparticles are in situ coated by graphitized carbon matrix formed from organic frameworks. This could enhance the conductivity as well as buffering volume variation. Considering these factors, it is believed that metal–organic frameworks can be ideal precursors for the porous hierarchically structured metal compounds.

Moreover, to resolve the issue of low conductivity, another high mechanical tenacity and superior electrical conductivity material usually made as an additive to fabricate composite electrodes [180, 181]. Specifically, when CNTs are used to insert/adhere into active particle materials to make conductive network structures. This could have advantages to enhance both reaction kinetics and electrical conductivity of the whole electrode. As an example, fabrication of highly conductive Mn/Fe-involved bimetal selenide with the combined advantages in both hierarchical porous structures and the conductive composite when CNTs were used as an conductive additive. This ultimately enhanced the cyclic stability and rate capability. However, compared to monometallic selenides, the bimetallic compounds may have higher electronic conductivity due to their increased surface area and improved electrochemical properties. This could be due to synergistic interaction between two different metal atoms [182]. The bimetallic compounds can also improve the volumetric effects due to separation into the monometallic compounds on sodiation/desodiation as well as potassiation/depotassiation processes. Their substantial electrochemical activity can also result in abundant redox-active sites owing to the coexistence of two atoms and higher electronic conductivity under the covalent interaction between two metal atoms. This has been confirmed by the study of porous Mn–Fe–Se adhered/insertion and interlaced of CNTs (expressed as Mn–Fe–Se/CNTs) with a high specific surface area via a simple chemical precipitation approach and a subsequent one-step carbonization-selenization of Mn–Fe PBA precursor process [183].

In a similar manner Bismuth selenide (Bi_2Se_3) are also identified as potential materials due to their ability to make a type of A_2B_3 (A = Sb; B = Se, Te) chalcogenide structure materials. In which they can share the same layered rhombohedral crystal structure [184, 185]. Usually Bi_2Se_3 nanomaterials synthesize at a high temperature, however, they can also be incorporated from the facile wet-chemical approach in the form of ultrathin nanosheets [186–188]. The advantage of ultrathin nanosheets fabrication is that they can form one-dimensional nanostructures, which is appealing for research due to their unique properties. Owing to their excellent physical and chemical properties, the transition metal chalcogenides such as, ZnS, CuS, Bi_2S_3, MoS_2, CdS, etc. have been widely incorporated [189–194]. Copper sulfides have also attracted much attention due to their variations in stoichiometric compositions, complex structures, nanocrystal morphologies, valence states, different unique properties and their potential applications in many areas [5]. Their stoichiometric composition can vary in a wide range from CuS_2 copper-deficient side to Cu_2S copper-rich side, including CuS_2, Cu_7S_4, $Cu_{1.8}S$, $Cu_{1.94}S$, $Cu_{1.96}S$, and Cu_2S [195–196]. Copper sulfides also have different compositions such as important p-type semiconductors that exhibit almost ideal solar control characteristics and fast-ion conduction at high temperatures [197, 198]. Copper sulfides can be applicable in solar cell devices, photothermal conversion, coatings for microwave shields, solar control, etc. [199]. Covellite CuS has metallic conductivity and it can transform into a superconductor at 1.6 K as well as be used as a cathode material in lithium rechargeable batteries [200, 201]. It can also form various morphologies of copper sulfides including nanoparticles, nanowires, nanovesicles, nanodisks, micrometerscale hierarchical tubular structures, etc. [202]. They have advantages to synthesize at a large scale in the form of copper sulfide-core/carbon-shell cables and spheres composites via a simple, mild, and effective hydrothermal route.

Polycrystalline Chalcogenides CNTs Composites

CNTs rigidity, chemical inertness and strong π bonds interactions in pure form do not allow it to dissolve or disperse in common organic solvents/or polymer/or inorganic matrices. To solve this problem, several approaches have been developed, such as surfactant assisted dispersion [203], high power sonication [204], polymer wrapping [205], as well as surface modification due to inorganic coating [206]. In this regard, the hybrid nanocomposite materials synergetic behaviors can be considered as one of the impassive approaches owing to interactions between organic and inorganic components [207, 208]. This approach can have a large number of potential applications, mainly in electronic of nanodevices, gas sensing catalysis. Such as the interesting chalcogenide nanocomposites MoS_2, WS_2, NbS_2, etc., which have closed-cage structures to carbon fullerenes [209]. These two-dimensional materials are also known as Inorganic Fullerene-like (IF) [210]. IF nanoparticles may have superior properties such as, high modulus and low friction coefficients along with their small spherical size, closed structure and chemical inertness [211]. In the past such materials have been used as favorable solid lubricants under severe conditions. In addition to this, the incorporation of these inorganic fillers into various matrices [212, 213], including epoxy, polypropylene, and poly(phenylene sulphide), have also been considered to improve the thermal, mechanical and tribiological properties of the nanocomposites.

In this manner, the poly crystalline chalcogenide CNTs composite have also come into the picture, a few reports have been published from different authors selecting a two steps synthesis process, first they synthesized the chalcogenide alloys and later added certain amounts of CNTs and reheated the composite materials [214, 215]. Particularly, Upadhyay et al. reported a detailed work on the polycrystalline structure of $Se_{80}Te_{16}Cu_4$-CNTs composite materials by demonstrating their structural variations with different annealing temperature in the range of 340–380 K. They demonstrated that the polycrystalline XRD diffraction patterns due to the change in CuSe phase of the monoclinic phase of selenium. They also verified the polycrystalline from the TEM microstructure and SAED patterns. A clear morphological difference in between the pristine, 3 and 5 wt% CNTs and their above said temperature range were shown. They demonstrated that spheroidal particles formation with the increasing temperature in the temperature range of 340–360 K and above this the grain (crystal) size decreases. This is a direct evidence of increase in reaction rate and reduction in voids [215, 216]. It has also been noticed that the thermal conductivity and micro hardness of the annealed composite materials are enhanced as compared to pristine composition. Such enhancement can be related to an increase in crystallinity and a decrease in density of phonon scattering sites. Hence, subsequent graphitization of the CNTs can also improve the percolation of the CNTs composites [217–219].

Under a similar two steps synthesis process, Kurochka and Melnikova investigated the variation in electrical properties at low annealing temperatures (10 to 300 K). According to them, electrical properties of $AgGe_{1+x}As_{1-x}(S+CNT)_3$ systems composite materials vary due to the change in their structural parameters, i.e. increase in the systems crystallinity [220]. Moreover, the addition of CNT in the $AgGe_{1+x}As_{1-x}S_{3x}$ systems leads to noticeable changes in the electrical properties of materials, such as increase in electronic and ionic conductivity components as well as micro hardness, density and strength characteristics of the composite materials [220]. Hence, the polycrystalline structures of the chalcogenide-CNTs composites have shown a common trend to enhance the thermal, electrical, mechanical, density and strength properties with different annealing temperatures. However, very limited experimental data is available on this topic, therefore, it may not be considered as a general rule for such systems owing to the variable behavior of the chalcogenide-CNTs materials. Hence, a serious effort is needed to explore such composites polycrystalline structures.

Chalcogenide Glass–SWCNTs Nanocomposites

Single-walled carbon nanotubes (SWCNTs) have been intensively explored owing to their unique structure-dependent electronic and mechanical properties. They also have the potential ability as catalyst supports in heterogeneous configuration, such as high strength engineering fibers and the molecular wires for the next generation of electronics devices [221]. As for molecular wires, their long electron mean free paths and their ballistic transport properties are the most significant characteristics. It depends on helicity and diameter of the carbon nanotubes therefore; they are classified as either metallic or semiconducting. It is also supposed that the band gap of semiconducting nanotubes may decrease with increasing tubes diameter. This can be directly correlated to the well-established opinion, that the SWCNTs single defect can change the structure of the tube as well as characteristic from metallic to semiconducting.

To alter the carbon nanotubes electronic characteristics (metallic to semi metallic) chemical functionalization can be considered as one of the promising approaches. Under this functionalization, a composite system can form a moiety that the intrinsic properties may be configurable electronically. One of these kinds of structures is semiconductor nanocrystals like CdS and CdSe, (they are also known as quantum dots) [222]. Such systems have strongly sized dependent optical and electrical properties. Additionally, they also have a high luminescence yield as well as adjusting emission and absorption wavelengths ability by selecting the size of the nanocrystal. This makes quantum dots attractive for constructing optoelectronic devices with tailored properties [223]. Therefore, SWCNTs can covalently join to the CdSe semiconductor nanocrystals by short chain organic molecule linkers. The behavior of the organic capping groups on the nanocrystal surface as well as the organic bifunctional linkers can be directly correlated to the modulation of the interactions between the nanotubes and the nanocrystals with the implication of self-assembly. Such types of formed composites can be useful for numerous diverse applications [221, 224].

In a similar way, several others chalcogenide binary and ternary compositions with SWCNTs have been demonstrated as nanocomposite materials in the last few years, such as BN, WS_2, MoS_2, Cu (In,Ga) Se_2, etc. [225, 226]. Each of the reports has shown improved physical properties such as thermal, optical and electrical for these composites materials. However, most of the reports have presented the chalcogenides-SWCNTs composites structures in nanocrystalline form.

However, it is still challenging to synthesize chalcogenide glass-SWCNTs composites, owing to unique physical properties of SWCNTs, such as carbon–carbon bonding nanotubes, high specific surface areas, high Young moduli up to 1,800 GPa that is 100 times stronger than steel, maximum tensile strength (upto 30 GPa), high stiffness [227]. Such strong physical properties of the SWCNTs do not allow breaking their bonds in the low dimension fullerene chalcogenide glassy system. Therefore, it is believed that chalcogenide glass- SWCNTs composite cannot be formed in normal manner. To the best of our knowledge, very recently Kurochka et al. [227] successfully synthesized the chalcogenide glass-SWCNTs composite adopting the most frequently used melt quenched technique [228]. According to this report they made the $AgGe_{1.6}As_{0.4}$ $(S + SWCNT)_3$ glassy composites and investigated their properties as prepared as well as irradiated specimens. Although they clearly mentioned that the synthesized composite materials overall characteristics are very close to the glassy behavior but are not completely glass [228]. The structural modification demonstrated with the help of morphological analysis and the compositions elemental concentration were confirmed from the EDS pattern interpretation. They noticed that a clear morphological distinction between the $Ge_{1+x}As_{1-x}S_3$ and $AgGe_{1+x}As_{1-x}(S + SWCNT)_3$ as well as their protons and deuterons irradiated samples. The bonds formations in these composites have been interpreted with the help of the Raman patterns.

The irradiated $AgGe_{1.6}As_{0.4}(S + SWCNT)_3$ composites Raman spectral changes could also be associated with the transformation of ethane-like units and edge-sharing as well as corner-sharing structural units. As the glass matrix containing ethane-like and edge-sharing structural units may have a higher free energy than only corner-sharing units [50]. Therefore, the transformation occurs because the entropy of the glass tends to reach the magnitude characteristic of the supercooled liquid equilibrium state. As a consequence, an increase in the local ordering of the glass network may be possible [229]. The changes in the local atomic structure of irradiated $AgGe_{1.6}As_{0.4}$ $(S + SWCNT)_3$ as shown in Fig. 5.11.

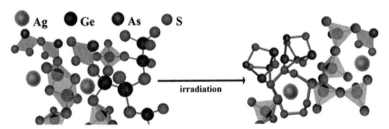

Figure 5.11 Schematic of the possible modifications in the fragment of layered atomic structure of the glassy matrix for the $AgGe_{1.6}As_{0.4}(S+CNT)_3$.
(Reprint permission from Kurochka, et al. (2019) Irradiation effect on the structural and electrical properties of the glassy AgeGeeAseS composite material containing carbon nanotubes. Solid State Ionics, 341, 115026)

Moreover, the mode of irradiation can also influence the structural transformation as well as the electrical properties of such composite materials. The changes in the material physical properties can be directly related to the irradiation process which is mainly connected with alternation in the atomic structure of the material than the embedding of additional charge carriers. As an example, under the deuterium irradiation, high energy beam of deuterons interacts with $AgGe_{1.6}As_{0.4}$ $(S+SWCNT)_3$ composite materials, and as a consequence the surface of the material may be destroyed due to thermal heating during the irradiation process. Therefore, some amounts of the Ge and As atoms are sputtered from the surface of the materials. Subsequently, S atoms and Ag^+ ions can push into the interior of the specimen via a collision cascade. The high energy interaction with the solid can be lost rapidly due to interactions with neighboring atoms. Therefore, most parts of the metallic element silver after the irradiating beam impact may lie slightly below the depth of interaction. This action can lead to the excitation of atoms as well as the formation of ions; therefore, ionization losses can be significant. This can also impact on the concentration of silver ions due to the incident side of the beam of heavy particles that may be larger than the opposite side, hence polarization occurs within the composite materials. Therefore, fraction of the ionic conductivity component in the total conductivity (electronic and ionic) of materials may remain practically unchanged. While, in the case of the deuteron beam irradiation of these composite materials no change occurs in the ionic conductivity component, therefore, the value of the electronic conductivity component decreases slightly from $(6.3 \pm 0.1) \times 10^{-11}$ S/m (for the non-irradiated sample) to $(2.3 \pm 0.5) \times 10^{-11}$ S/m [228]. This action can be because a fraction of the protons can interact with the material surface and knock out the electrons, therefore, subsequently undergoing a neutral state as per the reaction $2H^+ + 2e^- \rightarrow H_2(g)$ [230]. As a consequence, the electronic conductivity can be reduced. Additionally, a decrease may occur in the magnitude of the electronic conductivity component owing to the rearrangement of the atomic structure of the composite structure after irradiation [228]. Hence, in such composite materials fabrication the used techniques or methods can play an important role in their physical properties investigations.

Chalcogenides Glass–MWCNTs Nanocomposites

While chalcogenide materials have provided many unquestionable advantages, such as, infrared transmission and detection, threshold and memory switching, optical fibers, functional elements in integrated-optic circuits, non-linear optics, holographic and memory storage media, chemical and bio-sensors and photovoltaics, etc. [231, 232]. However, they are not free from a few irrefutable drawbacks, such as, low edging, low electrical conductivity, stability, etc., compared to their counterpart technological materials like nanomaterials [232, 233]. For more than five decades, several investigators have made an effort to resolve this problem by demonstrating the scientific and technological outcomes. A vast number of reports address this key issue that can minimize efficiency and fabricate these kinds of composite materials including metallic nano phase and rare earth composites [234]. In this order, the metal-chalcogenide alloys have been widely investigated in recent years due to their unique features such as, high thermal stability and structural complexity that possibly can produce nano phase helicharical structure [235–237]. However, in continuing this great effort, remarkable improvements have been made but owing to different kinds of chalcogenide composites limitations their working performances on such devices could not reach the desired demanding level in the modern world. Therefore, to fulfill future demands technologies based on such materials, more effort is required to make advanced chalcogenide composites by exploring various types of existing materials such as chalcogenides – nanocomposites.

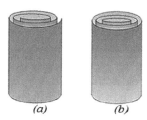

(a)   *(b)*

Figure 5.12 A schematic illustration of 'Swiss roll' and 'Russian doll' models for multiwalled nanotubes.

In order to explore various suitable materials for chalcogenides the MWCNTs can also be considered as a potential material. This is due to MWCNTs structured layer in either scrolls-like, 'Swiss-roll', or 'Russian doll' or nested, with the arrangement of discrete tubes as shown in Fig. 5.12. In addition, MWCNTs structure can also consist of a mixture of these two arrangements. Their structural relationship between successive cylinders has been extensively studied by Zhang and colleagues and by Reznik et al. [238]. Therefore, MWCNTs overall structure is unlike ABAB stacking of single-crystal graphite in which cylindrical carbon nanotubes, except possibly in small areas. However, very large numbers of possible cylindrical graphene structures can exist. In which theoretically all nanotubes larger than the archetypal (5, 5) and (9, 0) tubes can be capped.

The number of possible caps can grow rapidly with the increasing diameter. The capping of carbon nanotubes can be from various approaches considering their suitability for a particular tube [238]. Similar to fullerenes, all capped nanotubes follow the Euler's law. Therefore, a hexagonal lattice with any size or shape possesses a closed structure due to the inclusion of precisely 12 pentagons. Thus, any nanotube cap should contain six pentagons and only consider the strain that may be isolated from each other (excluding for the moment, caps containing heptagons).

Where the smallest tubes may be capped with isolated pentagons with two archetypal tubes as depicted in Fig. 5.13. In which for each of these there can be possibly one cap corresponding to the C_{60} molecule that divide in two different ways.

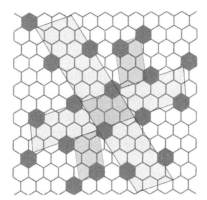

Figure 5.13   A projection map of the caped carbon nanotube.

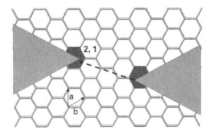

Figure 5.14   Vector connection of two pentagonal defects.

To consider this, Fujita et al. calculated the number of possible caps for the MWCNTs larger than those by using a method that was based on 'projection mapping'. This conclusion led to a honeycomb network that could be folded to form a given fullerene or nanotube. These kinds of pentagons may be constructed by removing a 60° triangular segment of lattice, this leads to the formation of a conical defect that is known as a 60° positive wedge disclination. Therefore, an icosahedral fullerene can be fully specified by the vector that connects two adjacent pentagons. Examining the icosahedral fullerene C_{140} in which the defined vector can be seen in Fig. 5.14. Further the defects can form a regular triangular array.

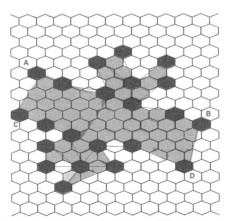

Figure 5.15   Schematic of projection map for the two different ways capped chiral nanotubes.

Therefore, fullerene can be formed due to the removal of the non-shaded part of the lattice and superimposing rings with the same numbers. Additionally, the nanotube can be capped at each end with one-half of a C_{140} molecule. To succeed the mapping, simply follows the two lines AC and BD. This could result in tube chiral shape associated with the vector (10, 5). Fujita et al., demonstrated a general icosahedral fullerene with the help of indices (nf, mf), by dividing half in a direction perpendicular to one of the five-fold axes, that can cap a nanotube having indices (5nf, 5mf). This leads to a series that is called 'magic number' icosahedral fullerenes, C_{60}, C_{240}, C_{540}, ..., that may possess indices (1, 1), (2, 2), (3, 3) ..., through the bisected cap of the armchair tubes series associated with the vectors (5, 5), (10, 10), (15, 15) and so on. Hence, bisected in a direction perpendicular to one of the three-fold axes, such fullerenes can cap the tubes (9, 0), (18, 0), (27, 0), etc. Therefore, most of the nanotubes larger than the (5, 5) and (9, 0) tubes (with one exception) can be capped in more than one way, as shown in Fig. 5.15. This indicates that there are two different ways of capping the chiral nanotube according to the defined vector (7, 5). Further it was demonstrated that there are 13 possible ways to cap the nanotubes. According to the theory of the nanotube, capping can also occur up to 3 nm in diameter. The isolated-pentagon can also cap in the form of (9, 0) and (10, 0) as depicted in Fig. 5.16. Thus in general, with this approach, it can produce even larger numbers that can possibly be capped. Theoretically 39 possible ways isolated-pentagon can be capped with the (9, 0), (10, 0) and (11, 0) nanotubes configuration, but experimentally it is limited only up to 21. Hence, the number of caps may be huge as the diameter increases. Moreover, the number of caps including adjacent pentagons varied up to with $d^{7.8}$ [238]. For those that can fulfill the isolated pentagon rule, the number of caps was lesser for small diameters; however, the power-law behavior may occur for larger diameters. In the case of large tube diameters, the fraction of caps with adjacent pentagons may be negligible. With the remarkable note, a fewer caps for armchair and zigzag than for chiral nanotubes was found due to higher symmetry of the achiral tubes. Further Reich et al. also demonstrated that a given nanotube can have thousands of distinct caps which is quite the opposite for the inverse problem. Therefore a given cap only fits to one particular nanotube.

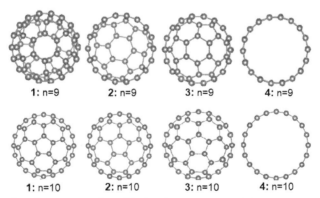

Figure 5.16 Atomic representation of isolated-pentagon caped for (9, 0) and (10, 0) tubes.

Additionally, with these special features of two-dimensional MWCNTs cylinder each carbon atom can also precisely connect with three neighbor atoms. In which the lattice of MWCNTs molecules predominantly consists of hexagons with a number of additional pentagons or heptagons within the structure [95]. This allows MWCNTs to incorporate inside the inorganic matrix to modify the properties of the special polymeric materials including structural, optical, electrical conductivity and mechanical enhancement. Moreover this kind of structural modification of the MWCNT might be the consequence of reduction in interfacial energy between the MWCNTs side wall surface and complex alloy constitutes [239]. The first successful effort was made by Singh in 2013, adopting a single step melt quenched approach [240] to consider these favorable conditions of MWCNTs to make composites with chalcogenides under the glassy configuration. He initially demonstrated in a report on the Se_{96}–Zn_2–Sb_2 + 0.05% MWCNTs composites by showing the chalcogenide – MWCNTs composites under the glassy system by using the schematic process as shown in Fig. 5.17.

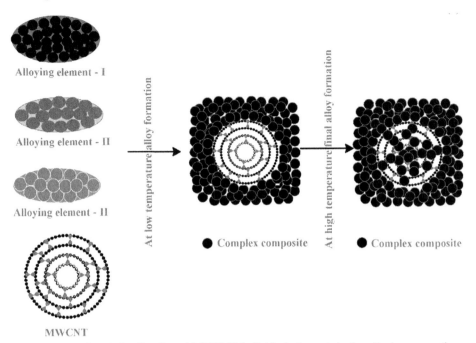

Figure 5.17   Step I: Se, Zn, Sb and MWCNT individual elements in the alloying proportion,
Step II: Se–Zn–Sb alloy formation in which unaffected MWCNT are present,
Step III: inclusion of the Se–Zn–Sb alloy constituent into the sidewall surfaces of the MWCNT.

This report explored the composite glassy structure with the demonstration of diffuse MWCNTs structure in a low dimension chalcogenide configuration. Additionally, the composite's exceptional infrared transmission percentage level was not achieved as those in the individual chalcogenide glassy alloys. Hence with this unique innovative research work it was established that small amounts of MWCNTs can diffuse in low dimension chalcogenide glassy system, although earlier it was theoretically believed that diffusion of MWCNTs was almost impossible [240]. In order to explore this novel concept with other chalcogenide alloys under the glassy system, Singh also extended a study extensively and reported several research outcomes exploring different physical properties [240–244]. Later different authors also contributed by demonstrating their research outcomes in this newly established area, by making the chalcogenide – carbon nanotubes composites with the different glassy configurations [220, 245–251]. Most of the reports were based on a single step synthesis process, while a few of them adopted the two steps synthetic process to make the composite materials [240–244–251]. This is due to the fact that the single step melt quench synthesis process of chalcogenide – carbon nanotubes are more complex than the two steps process.

STRUCTURAL MODIFICATIONS

In view of various concepts discussed, it can be said that, a composite configuration simply reflects the intrinsic properties of the individual component of the alloy, those discrete in size, shape as well as with a remarkable structural property relationship under weak and strong interactions between their components [95]. In the growing chalcogenide composites science area including their structural interpretation has been considered as one of the most significant parameters to develop an understanding about it. In which flexible fullerene like structural properties can be considered as the backbone of the composites structure with other compounds, a schematic of the fullerene like of Se is illustrated in Fig 5.18 [252, 253]. More specifically, the structures of the chalcogenide-MWCNTs composites could be interpreted as the following:

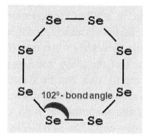

Figure 5.18   Selenium fullerene like chain structure.

A large number of excitonic levels in carbon MWCNTs are extremely important to form composites with low dimension materials. The existence of impurities in MWCNTs can allow in forming states within the band gap of semiconducting materials as well as it may also generate the electronic transitions from the valence band to impurity levels; as a consequence, the total photonic transitions would be modified in a semiconductor. Modification in MWCNTs low energy π-bond electron states can also play an important role in the reconstruction of electronic structure composites. Owing to this the π-bonds of the MWCNTs can produce plasmon resonance within successive layers [254]. Under the successive layers producing π-plasmon is the collective excitation of the π bond electrons that depend on the surface plasmons of the MWCNTs. Hence, the transition is allowed in between the π and $\pi*$ energy bands at the same cutting line corresponding to initial and final states [239]. In addition the existence of plasmon resonance within the successive layers can modify the MWCNTs structure to a large extent and possibly their low energy $\pi - \pi$ may break at even a lower temperature during the composite melt quenched synthesis process. Hence, the possibility of the thermal quantization process occurring is abundant in such types of composites, whereas, under the low dimension alloys, a strong and stiff material with weaker bonds can diffuse or react to form a new structure. Adopting the novel assumption of '*thermonic energy tunneling effect*', it is possible to break or diffuse strong MWCNTs (or CNTs) lower energy bonds and other weaker bonds at even below their melting points under a continuous excess heating environment. Singh successfully fabricated the composite materials. Here it should be noted that to make such low dimension alloys - MCNTs composites structures, the alloys individual element interactions with MWCNTs at the specific temperature under the molten form cannot be ignored. Such metallic elements usually have a strong tendency to react with the MWCNTs (or CNTs), semi metallic elements have moderate reactive behavior, however, non-metals have unreactive behavior with them. To make this kind of composite materials, the choice of alloying elements can also play a crucial role.

Modifications in Surface Morphologies

SZS-MWCNTs Glassy Composite

To verify these concepts, Singh initially demonstrated two different chalocogenide alloys compositions to make the MWCNTs composites by selecting the different groups of alloying elements [240]. His first novel work with $Se_{96}-Zn_2-Sb_2$ and $Se_{96}-Zn_2-Sb_2 + 0.05\%$ MWCNTs (SZS, SZS–MWCNTs) demonstrated the structural properties with the XRD patterns, FESEM morphologies, EDS patterns and EDS elemental mapping for the both Se_{96}–Zn_2–Sb_2 glassy as well as the Se_{96}–Zn_2–Sb_2 + 0.05% MWCNTs composite, as illustrated in Fig. 5.19(a-l). The XRD patterns of SZS–MWCNTs materials has clearly indicated that the overall amorphous behavior of this composite, by showing no well-developed of any crystalline peak in the pattern. Hence, the crystallographic structure of chalcogenide alloy SZS and its SZS–MWCNTs has a random atomic arrangement in an intrinsic complex glassy structure. This is a direct experimental evidence of the loss of structural configuration of MWCNTs in the low dimension chalcogenide SZS system. Moreover, as discussed that at around ± 1000 °C temperature, in MWCNTs only weaker bonds can be usually influenced, so query remains about the stronger σ bonds of the MWCNTs, whether it has completely dissolved or partially affected or stand in the original form within the complex configuration. While the XRD patterns of the alloy and composite explain that there is no place for a third query, however, the remaining two can be considered for further investigation. The microscopic analysis of these materials can help in answering more questions appropriately, therefore, it is worth interpreting these materials at high resolution surface morphologies, as depicted in Fig. 5.19(b, c). The SZS alloy and SZS–MWCNTs composite surface morphology has a significant distinct surface morphologies, specifically the composite has a clear signature of partial diffusion of MWCNTs in a SZS-amorphous scheme. This literally means that strong σ bonds of the MWCNTs have not completely diffused in the complex SZS configuration, while one seeks to preserve the backbone of MWCNTs in an amorphous glassy system. Adopting the basic idea under such a structural configuration, the composite properties can be changed and their devices overall performance may be boosted. Through the existing compressed additional MWCNTs high efficient (conducting) channel in the low dimension chalcogenide regime. Moreover, question remains unsolved whether the constituents of SZS alloy could enter the compressed MWCNTs or not, this is a crucial parameter to define developed SZS–MWCNTs material in composite form, consequently, to explore their enhanced physical properties and performances. The EDS mapping outcome of the SZS alloy and SZS–MWCNTs could efficiently describe the inclusion of alloying elements [242]. The EDS mappings of the both materials are illustrated in Fig. 5.19(d-f, h-k). The homogeneous distribution of the alloying elements in EDS mapping, in terms of their (Se,

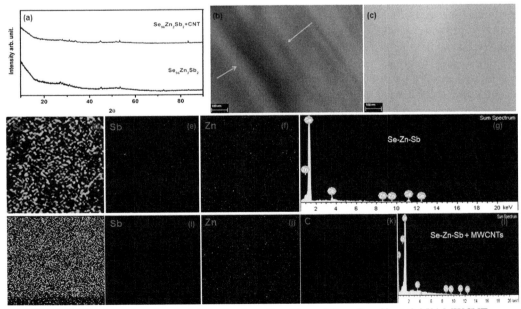

Figure 5.19(a, b, c, d, e, f, g, h, i, j, k, l) Se_{96}–Zn_2–Sb_2 and Se_{96}–Zn_2–Sb_2 + 0.05% MWCNTs composites, XRD patterns, FESEM morphologies, EDS elemental mapping, EDS patterns.

Sb and Zn) compositional amounts has revealed the homogenous mixing throughout the SZS glassy configuration. However, SZS–MWCNTs configuration elemental mapping also has an almost homogeneous distribution of the alloying elements particles throughout the configuration, including carbon particles. Thus, the inclusion of the SZS alloying elements was confirmed with the help of elemental mapping of the alloy as well as composite. To confirm the presence of amounts of each element in the formed configurations the EDS patterns results has also been studied (See Fig. 5.19(g, i)). The interpreted EDS patterns of the SZS and SZS–MWCNTs are also in accordance with the elemental mapping outcomes [242]. Hence, with this initial work it was established that the structure of the SZS can be modified from the incorporation of MWCNTs overall in a glassy system.

GTS-MWCNTs Glassy Composite

In order to extend the of investigation with other chalcogenide alloys – carbon nanotubes, Singh et al. reported the exceptional finding on $Se_{55}Te_{25}Ge_{20}$ (GTS), $Se_{55}Te_{25}Ge_{20}$ (GTS) – MWCNTs [244]. With this it was established that chalcogenide–MWCNTs can be fabricated under the glassy system with various chalcogenide alloys by selecting their suitable constituents. This study demonstrated that Ge–Te chalcogenide alloy could be considered as potential material for the non-volatile Phase Change Random Access Memory (PCRAM) application [255]. Ge–Te has a high crystallization (189°C) and melting (700°C) temperatures possessing a large crystallization time and a high RESET current [256, 257]. This weakness can be improved by making their composites through structural modification.

Since in two dimensional multi-walled carbon nanotubes (MWCNTs) each carbon atom cylinders is precisely connected with three neighbors. The MWCNTs lattice molecules consist of hexagons with a number of additional pentagons or heptagons in the structure [95]. Therefore, incorporation of MWCNTs into an inorganic matrix can modify special polymeric material (chalcogenide) chains and rings. As a consequence, physical properties of the formed chalcogenide material due to the structural modification of the MWCNTs result in the reduction in their interfacial energy between MWCNT side wall surfaces from the inclusion of complex alloy constituents. However, MWCNTs can have a large number of molecules of the carbon atoms under the strongly bounded flat configuration with the honeycomb lattice site structure. In their typical structure, out of four outer electrons three electrons are strongly bonded with three neighboring atoms. The structural modifications of GTS – MWCNTs nanocomposite with respect to GTS can be interpreted with the XRD pattern, FESEM surface morphologies are depicted in Fig. 5.20(a, b, c).

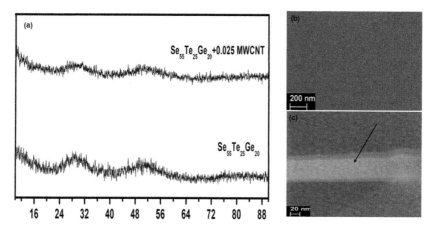

Figure 5.20(a, b, c) XRD patterns and FESEM surface morphologies of $Se_{55}Te_{25}Ge_{20}$ (GTS), and $Se_{55}Te_{25}Ge_{20}$ (GTS) + MWCNTs.

According to the XRD patterns of these compositions materials, no sharp crystalline peak exists, therefore, overall their crystallographic structure are amorphous. Moreover FESEM surface morphological interpretation demonstrates that a clear distinct picture by showing the existence of diffused state MWCNTs under the glassy system. This is another evidence that deformation of MWCNTs is in the chalcogenide glassy configuration [243]. A further query comes to mind that if MWCNTs do exist in GTS under the glassy system, then one needs to know which kind of GTS constituents distributions within the configuration as well as the composition elemental amounts in a proper proportion. These two answers could provide an interpretation of EDS elemental mapping and EDS patterns of the MWCNTs + GTS composite. As EDS elemental mapping depicted in Fig. 5.21(a, b, c, d, e, f, g, h, i), a dense elemental distribution mapping of the Se (see Fig. 5.21(e)) can be clearly visualized with less inclusions within the MWCNTs regime, by showing a narrow strip in the morphology, whereas the elemental distribution mapping of the element Te has exhibited (see Fig. 5.21(g)) comparatively higher inclusion ability in the MWCNTs within the strip. Subsequently, a homogeneous elemental distribution mapping of the Ge (see Fig. 5.21 (f)) demonstrates that the semiconducting material can have a higher order inclusion ability than amorphous semiconducting materials (Se and Te). This may increase the metallicity of the material. Moreover, an uneven scattered distribution mapping of the element carbon atoms throughout the surface area (see Fig. 5.21(d)) can be correlated to the loss of the MWCNTs periodic structure within the complex GTS composite. Further, the EDS pattern of the GTS- MWCNTs composite elemental amounts has provided the evidence about their presence in an appropriate manner (see Fig. 5.21(h)). Hence, the EDS mapping and their EDS pattern (see Fig. 5.21(d, i)) interpretation has clearly established that diffusion of the MWCNTs in GTS complex glassy configuration with their appropriate elemental concentrations [243].

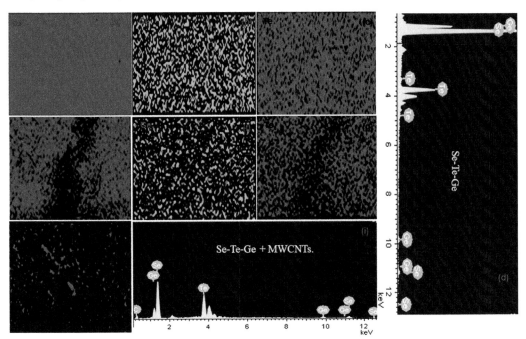

Figure 5.21(a, b, c, d, e, f, g, h, i)　EDS elemental mapping and EDS patterns of GTS + MWCNTs composite and GTS alloy.

This newly identified and established field research studies are not limited to these exceptional introductory reports to explore chalcogenide–MWCNTs composites under the glassy system. Various ingestions synthetic process either single step process or two steps such as demonstrated by Sen et al., the thermal and mechanical properties of CNT–$Se_{90-x}Te_{10}Ag_x$ ($x = 0$, 5 and 10) glassy composites by adopting two step synthetic process, while Ganaie at al. investigated the structural, electrical and dielectric properties of doped CNT using the two steps synthetic process [245, 258]. Moreover, Upadhyay et al. and Jaiswal et al. [247–251] also did extensive research adopting the two step synthetic process and reported such composite materials structural, optical and electrical properties. Hence, a quest for investigations of chalcogenide- carbon composites under a glassy system is desired to develop a better understanding of these materials.

Modifications in Raman Spectroscopic

Raman spectroscopic analysis of materials can provide valuable analysis about their structures. Using Raman spectroscopy, the characteristics of amorphous materials structures can also be accessed. Owing to this Raman spectroscopy can efficiently detect signals even it comes from small amounts of crystallites. Therefore, it requires very high sensitivity for the local ordering in the structure. Hence, weak Raman spectroscopic signals could correlate to the amorphous structure of the materials. According to well-established facts the Raman spectra of amorphous materials are similar to crystalline materials possessing a distinguishable peak broadening in their relative peak position [259]. Considering these well described facts of Raman spectroscopy, Singh et al., also studied the SZS alloy and SZS–MWCNTs composite in their initial report [240]. The outcome of this study demonstrated that a broad Raman active Se–Se peak in the region between 220 to 270 cm^{-1} in which a sharp position at 252 cm^{-1} for the SZS alloy, as illustrated in Fig. 5.22(a). The demonstrated Se–Se peak position for the SZS composition is in consistence with the pure Se–Sb alloy peak value of 250 cm^{-1} that arises due to Se_8 ring structure of the material [260, 261]. It was noticed that Se–Sb Raman low phonon band peak in SZS alloy found at 140 cm^{-1}, and their sharp band edge was described at 107 cm^{-1}, that is in consistence with other reports [262]. Similarly, the SZS–MWCNTs composite has also exhibited Se–Se, Se–Sb and Zn–Sb peaks at 236, 140 and 107 cm^{-1} and possess mixed Raman G and D modes low energy phonon peak at ~188 cm^{-1}, as depicted in Fig. 5.22. In this a broad diffuse D and G bands active peaks, at 1344 and 1574 cm^{-1} and a defect diffuse two-dimensional band has been recognized at 2699 cm^{-1}, this has been correlated to E_2g Raman active mode.

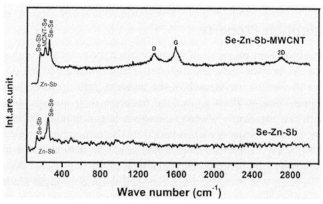

Figure 5.22   Raman spectrum of Se_{96}–Zn_2–Sb_2 alloy and Se_{96}–Zn_2–Sb_2 + 0.05% MWCNTs composite.

Moreover in the extensive study he also demonstrated the GTS alloy and GTS– MWCNTs composite. This study also demonstrated the diffusion of MWCNTs in GTS complex configuration by showing the Raman spectroscopic analysis in the wave number range upto 3000 cm^{-1}. In which material Se–Te–Ge exhibited a broad Raman peak single at ~196 cm^{-1} [263–265]. The appearance of a broad Raman peak could be correlated to such materials in a random large scale ($\leq$ 20Å) atomic distribution within the configurations. Owing to amorphous GeSe and GeTe phases coexistence during the melt quench process, the predominantly A_1 vibrational mode of GeSe and GeTe$_2$ may be involved, as illustrated in Fig. 5.23 [266]. In which the individual Raman strong peaks at 300, 121 and 237 cm^{-1} of the elements Ge, Te and Se are absent [255]. Additionally, this Raman spectroscopic interpretation has also demonstrated that the homogeneous diffusion of the strong carbon Raman D, G and two-dimensional peaks at 1353, 1582 and 2708 cm^{-1} for the GTS–MWCNTs composite, as depicted in Fig. 5.23. However, the weak MWCNTs Raman peak at ~191 cm^{-1} appeared to be the combination of GeTeSeC single broad peak. This can be correlated to the dispersion of sp^2 carbon nanotubes phonon modes A, E_1 and E_2 symmetry [262, 267] due to the splitting of doubly degenerate 2D peak into non-degenerate mode. Hence the intrinsic structure of the MWCNTs has been modified in a GTS glassy configuration.

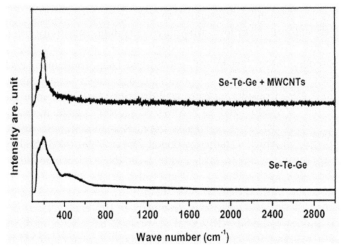

Figure 5.23 Raman spectrum of GTS alloy and GTS–MWCNTs composite.

OPTICAL PROPERTIES

UV/Visible Absorption Property

Usually UV/Visible absorption property of amorphous semiconductors is described in terms of their interaction between incident light and material. The absorbance of incident light depends on the material behavior as well as wavelength of the incident light. In the case of incident photons energy below the optical energy band gap of the material, their absorption would be relatively low. As well as with the increasing photons energy a larger number of electrons may absorb the incident photons, as a consequence absorbance of the respective wavelength photons could rapidly increase [268]. Such a corresponding profile could be demonstrated in term of Urbach's formula [252]. Additionally, the band tail encroachment in the gap region due to negative optical absorption correlation energy at the edge has been well defined in these kinds of materials. The band tail encroachment in glassy amorphous can be directly related to the existing charge defects within the configuration [269]. Considering these basic optical properties of glassy amorphous materials various investigators have extended the view of UV/Visible light absorption over a similar configuration such as, carbon nanotubes [252, 253, 270].

In particular, SZS glassy alloy and SZS–MWCNTs chalcogenide composite under the glassy configuration have also been reported [242]. According to this demonstration UV/Visible absorption spectrum was reordered in the wave length range 200 to 1000 nm, in which broad and small absorbance is noticed in the wave length range 350–950 nm. As the well-established UV/Visible spectrum analysis concept of amorphous materials, the last peak of the higher wave length side peak is considered in the wave length range 600–850 nm, as illustrated in the Fig. 5.24(a, b). A sharp absorption peak was also reordered for the SZS–MWCNT composite. The absorbance percentage was recorded higher for SZS–MWCNT composite than SZS alloy. Using the well-known Tauc plots, SZS alloy and SZS–MWCNTs optical energy bands were evaluated 1.37, 1.39 eV, which is in the range of well defined semiconducting materials (≥ 3 eV) [240]. This study indicated a marginal enhancement in optical energy band for the composite. This is inconsistent with the expectation of theoretical simulations on possible chalcogenide-nanotubes composite materials, according to the expectation that small amounts of MWCNTs incorporation could drastically enhance absorbance of such materials, this experimental finding contradicts the other one [240]. Further, in the extended study SZS alloy and SZS–MWCNTs other key optical parameters such as extinction coefficient, refractive index, real and imaginary dielectric constants have also been reported.

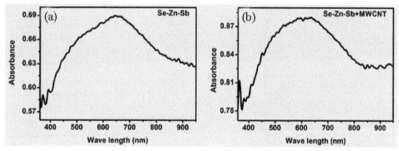

Figure 5.24(a, b)   UV/Visible absorption spectrum of SZS alloy and SZS–MWCNTs composite.

Moreover, to recognize the optical behavior MWCNTs composite within the glassy configuration, the GTS alloy and GTS-MWCNTs composite UV/Visible absorption have been investigated [244]. This study clearly demonstrated that the existence of sharp absorbance within the wavelength range 380 to 1050 nm, as illustrated in Fig. 5.25(a, b). The GTS alloy broad UV/Visible absorption peak has a relatively sharp decline band tail edge compared to GTS–MWCNTs composite broad absorption reduced band tail. The optical energy band gaps of these materials have been evaluated in the wavelength range of 600 to 900 nm by using the well-known Tauc relationship. The evaluated optical energy band gaps of the GTS alloy and GTS–MWCNTs composite are 1.26, and 1.30 eV, respectively. This study also has a similar trend like SZS–MWCNTs composite material with marginally improved behavior but not in a drastically changed manner. By means of the optical behavior of the chalcogenide–carbon nanotubes composite materials depend on the chalcogenide alloys constituents. By increasing the metallicity of the chalcogenide alloys the chalcogenide–carbon nanotubes composites absorbance could be in an increasing order. However, before reaching a conclusion, more data on chalcogenide–nanotube composites under glassy configuration is required.

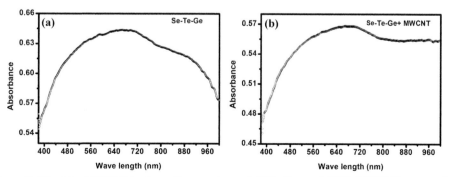

Figure 5.25(a, b)   UV/Visible absorption spectrum of GTS alloy and GTS–MWCNTs composite.

The marginal enhancement of optical absorbance of the chalcogenide–carbon nanotubes composite materials could be correlated to their structural modifications. From this view, the formation of glassy configuration with the existence of diffused MWCNTs occurs due to the bonding angles of the major chalcogenide constituent (Se) and MWCNTs that may be changed and overall formed composite single alloy is different from them. Hopefully newly formed composite materials bonding angle would lie between these two governing constituents, as illustrated in Fig. 5.26. However, to define the correct bonding angle of the various chalcogenide–MWCNTs composites materials a rigorous theoretical description is required, which is eagerly awaited. Moreover, the bond angle modification of the composite could also permit formation of additional sub energy levels within the forbidden gap, possibly closer as compared to base chalcogenide glassy configuration, as depicted in Fig. 5.26. This may lead to the enhancement in optical properties of chalcogenide–MWCNTs composite materials.

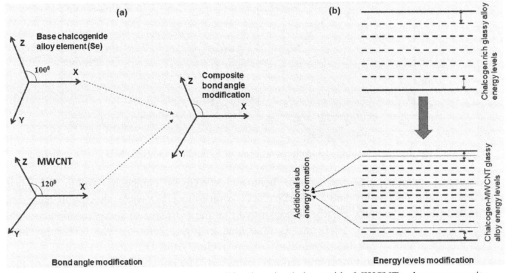

Figure 5.26   A schematic for structural modifications in chalcogenide–MWCNTs glassy composites.

Infrared (IR) Interpretation

Usually IR properties for different kinds of materials can provide valuable information about the impure atoms and their chemical bonding in the spectral ranges of 0.5–7, 0.8–12 and 1.2–16 μm [271]. In chalcogenide glassy materials mid IR (3–25 μm wavelength, corresponding wave number: 3333 to 400 cm^{-1}) transparency is technologically important as it can cover the atmospheric windows the 3–5 and 8–12 μm wavelength regions along with molecular fingerprints [272]. Specifically the IR optics utility of chalcogenide glasses depends on the restrictions of vibrational bands absorption that governs through the incorporation of foreign impurities. Hence, structural modifications of the chalcogenide glasses could also influence the IR optical property of the material. In this selection of chemical composition could play a crucial role in achieving higher order IR transmitting materials [273]. Therefore, it is worth to noting the IR property of chalcogenide–nanotubes composite systems. Singh also demonstrated the FT–IR transmission property of the SZS alloy and SZS–MWCNTs glassy composite materials in the wave number range upto 10000 cm^{-1}, as depicted in Fig. 5.27(a). This outcome has shown an impressive enhancement in transparency percentage in this range for the SZS–MWCNTs as compared to SZS parent alloy. This enhancement of IR transparency percentage could be due to the diffusion of elemental covalent bonds of metalloid Sb and metallic Zn in non-metal Se chains and rings, which has appeared in SZS alloy as well as SZS–MWCNT glassy composite.

Moreover the FT–IR characteristic has also been examined for the GTS glassy alloy and GTS–MWCNTs composite under the glassy system, as illustrated in Fig. 5.27(b) [244]. This study has shown a similar result with a significant IR transparency enhancement for the GTS–MWCNTs composite material as compared to GTS glassy alloy, as depicted in Fig. 5.27(b). By showing the non-existence sharp absorption peak (except a very weak absorption band at 3450 cm^{-1}) for the GTS alloy configuration that could be related to the elements Ge and Te covalent bond diffusion in the host Se chains and rings. While the GTS–MWCNTs composite IR transparency throughout the spectral range exhibited a few very weak absorption bands at 790, 1090, 1635, 2350, 3450 cm^{-1}, and having a relatively higher transmission percentage than the parent configuration. Here it is worth noting that between the FT–IR transmission profiles of SZS–MWCNTs and GTS–MWCNTs composites there exists several very weak absorption bands, this could be interpreted by increasing metallacity of chalcogenide composite with which their configuration may have a strong diffused structure in which distinct micro phase formations can prevail under the glassy structure.

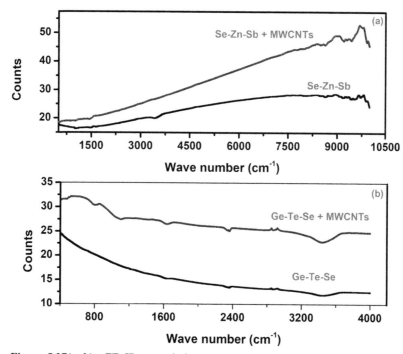

Figure 5.27(a, b)   FT–IR transmission spectra for the SZS, SZS–MWCNTs and GTS, GTS–MWCNTs materials.

The mechanism of IR could be understood in a simple way as shown in a schematic model for the chalcogenide–nanotube composite under the glassy configuration illustrated in Fig. 5.28. When a suitable light occurs on the materials, then molecules of the glassy system absorb the incoming light (Step-I), subsequently it undergoes vibrations or bounded excitation (Step-II). They may continuously vibrate along their original position and radiate the energy in the form of IR energy (Step-II) in each relaxing vibration cycle. This vibration process is continuous throughout the employing light time period and recorded from a spectrometer to build the IR pattern of the material (Step-III). Since the chalcogenide–carbon nanotubes composites have more complex structures than the parent chalcogenide glassy alloy, they can build a higher order IR transparency percent profile, as can be noticed in the SZS, SZS–MWCNTs, GTS, GTS–MWCNTs experimental demonstration.

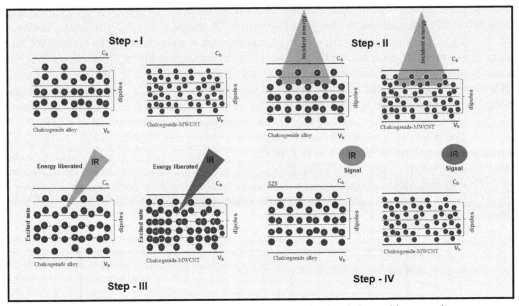

Figure 5.28   Schematic model for IR mechanism for the chalcogenide–nanotubes composite under the glassy system.

MECHANICAL PROPERTIES

Though, according to the theoretical simulation predictions, incorporation of CNTs within the chalcogenide glassy configuration can drastically enhance the mechanical properties of the systems [274, 275]. Yet according to our knowledge there is no report on mechanical properties on single step synthesized chalcogenide–CNTs composite glassy configuration. Although some reports on two steps (or the reheating process) synthesized chalcogenide–CNTs composite materials mechanical properties have been described [247]. As per interpretations of $Se_{85}Te_{10}Ag_5$–MWCNTs glassy composites, the micro hardness of the composite materials is drastically increased as compared to their parent glassy alloy. They also demonstrated that the enhancement in hardness of MWCNTs glassy composites may be due to barely deformable covalent C–C bonds within the parent glassy configurations. Moreover the Vickers indentation on the polished sample of the chalcogenide–MWCNTs composites systems providing visualized evidence in consistence of the enhancement of the micro hardness were also studied. However, there is no direct single step synthesized result on chalcogenide–nanotubes composites within the glassy configuration, therefore, it is expected that this kind of composition composites micro hardness and other mechanical property parameters could be slightly higher than two steps synthesized processed materials, owing to the fact that reheating can reduce the micro hardness of the materials. Nonetheless, this is a matter of further extensive research.

ELECTRICAL PROPERTIES

Similar to the mechanical properties, there is no electrical properties report on single step synthesized chalcogenide–CNTs glassy composite materials. However, some descriptions have been published with the distinct chalcogenide compositions CNTs composites fabricated from the reheating process, such as $CNTs–AgAsS_2$, $(Se_{85}Te_{10}Ag_5)_{100-x}(CNT)_x$, $AgGe_{1+x}As_{1-x}(S+CNT)_3$, $(CNT)_x$ $(Cu_5Se_{75}Ge_{10}In_{10})_{100-x}$ [214, 220, 247, 251]. Almost all experimental investigations of the electrical conductivities of chalcogenide–CNTs composites have demonstrated that a sharp enhancement in their conductance compared to their parent chalcogenide alloy. Using the I-V characteristics of the chalcogenide–CNTs composite, materials electrical conductivity can be evaluated from the following relationship:

$$\sigma = \frac{1}{\rho} = \left(\frac{1}{R}\right) \times \left(\frac{T}{A}\right) \tag{5.1}$$

Here σ and ρ are the conductivity and resistivity, R is the Ridbarg constant, T and A are the temperature and area.

The room temperature significant enhancement in electrical conductivity of the chalcogenide–CNTs glassy composites could be correlated to the ionic and electronic conductivity mechanism due to the existence of CNTs within the glassy configuration. It favored the formation of CNT–CNT continuous interconnected conductive network throughout the host polymeric glassy network, also known as percolation network [276]. Therefore, it could be considered that CNT high electronic conductive (10^6 S/cm) and their excellent mobility (10^5 cm^2/V/s) make a percolation path by the CNT in a glassy complex composite [214, 277, 278]. As a consequence, an effective conducting channel could be available for the electronic conduction through outer surfaces of MWCNTs [20] in a glassy composite. Thus the overall electrical conductivity can be increased due to conduction through CNT at room temperature.

On the other hand, change in σ_{dc} with temperature could be expressed with the help of variation of log σ_{dc} versus $1000/T$, it was found to be a straight line for the thermally activated mechanism. The activation energy and σ_{dc} conductivity can be calculated using the following Arrhenius formula:

$$\sigma_{dc} = \sigma_0 \exp\left(\frac{\Delta E_{dc}}{KT}\right) \tag{5.2}$$

Here σ_{dc}, σ_0, ΔE_{dc} and K are the dc conductivity, pre-exponential factor, activation energy and Boltzmann constant, respectively. It is worth noting that with the increasing CNT wt%, the dc conductivity can increase up to 10^{-5} ohm^{-1} m^{-1} to 10^{-3} ohm^{-1} m^{-1} at 304 K for the MWCNTs/Cu$_5$Se$_{75}$Ge$_{10}$In$_{10}$ composite [251]. This enhancement is approximately 100 times in their electrical conductivity owing to its outstanding mobility and highly conducting nature of CNTs. Therefore, the loss in E_{dc} with the increase in frequency may be due to the electronic jump between localized states with frequency.

Hence, the chalcogenide–MWCTs composites electrical properties could be modified largely in terms of a drastic enhancement in their electrical conductivity at both room temperature as well as with increasing temperature.

THERMAL PROPERTIES

In order to explore thermal properties, the chalcogenide–CNTs glassy nanocomposites were fabricated from the direct single step synthesis process. To know the thermal behavior of the chalcogenide–MWNCTs composite, Singh also demonstrated the Differential Scanning Calorimetry (DSC) thermo-grams to describe the crystallization kinetics [241]. He performed the experiment in view of deducing the higher order crystallization kinetics characteristics with the improved thermal stability parameters for the MWCNTs composite, subsequently, to provide another confirmative evidence about materials single phase homogeneous configuration within the glassy state. He demonstrated that the non-isothermal DSC measurements with SZS alloy and SZS–MWCNTs glassy composite within the temperature range from 300 K to 523 K, as depicted in Fig. 5.29. According to this study, the SZS alloy and SZS–MWCNTs behaviors have been defined as chalcogenide glassy materials by showing the well-known glass transition peak, crystallization peak and melting peaks of the materials. Further, by analyzing the DSC thermograms, the materials glass transition temperature (T_g), crystallization temperature (T_c) and melting temperature (T_m) have been defined. The defined values of the T_g, T_c, T_p (peak crystallization temperature) and T_m at different heating rates are listed in Table. 5.1. Using these experimental data, he defined their glass forming ability, activation energies of crystallizations at T_g, T_c T_p by using different approaches such as Hruby, Ozawa relation, Augis and Bennett and Takhor relationship. The SZS–MWCNTs composites evaluated values of the T_g, T_c, T_p at DSC different heat settings (5, 10, 15, 20 °C/min) were found to be lower than the parent glassy SZS alloy.

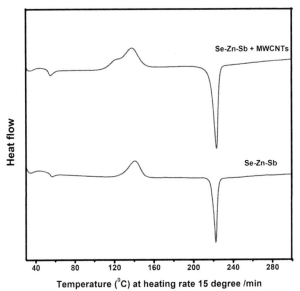

Figure 5.29   Non-isothermal DSC thermograms at heating rate 15 °C/min for the SZS alloy and SZE–MWCNTs composite.

Table 5.1   Crystallization kinetic parameters such as T_g, T_c, T_p and T_m at different DSC heating rates for the SZS alloy and SZS- MWCNT composite materials [241].

	Heating rates	T_g **(K)**	T_c **(K)**	T_p **(K)**	T_m **(K)**
Se–Zn–Sb	5	322	382	407	493
	10	323	387	411	494
	15	327	393	417	495
	20	328	397	433	495
Se–Zn–Sb+MWCNT	5	320	374	392	495
	10	322	381	402	496
	15	324	384	410	496
	20	326	390	414	497

This result is unexpected as almost every theoretical simulation predicted the possibility of small amounts of CNTs in chalcogenide glassy configurations could drastically enhance their thermal crystallization kinetics properties [241]. Corresponding activation energies at the T_g, T_c and T_p of SZS–MWCNTs composite were found to be greater than the parent SZS alloy, the existing activation energies values are listed in Table 5.2. The obtained crystallization parameter values of SZS–MWCNTs indicate their inferior kinetic properties. Moreover, the thermal stabilities of SZS alloy and SZS–MWCNTs composite have been defined by using two different approaches, i.e. Saad and Poulin and Hu et al. thermal stability criterion. These two thermal stabilities criteria results demonstrated that the lower value for the SZS–MWCNTs composites compared to SZS alloy. This is a major contradiction between the theoretical approach and experimental finding, as it is believed that any form of CNTs stiffness and thermal stability should be higher than inorganic alloys.

Table 5.2 Crystallization activation energies E_g, E_c, E_p for the SZS alloy and SZS–MWCNT composite materials [241]

Compositions	Activation energy E_g (KJ/mol), at T_g	Activation energy E_c (KJ/mol), at T_c	Activation energy E_p (KJ/mol), at T_p
Se-Zn-Sb	195.49	106.21	72.06
Se–Zn–Sb + MWCNT	207.53	118.21	79.89

This could be correlated to CNTs circular heat flow behavior during the synthesis of the composite material, as shown in Fig. 5.30. Due to such characteristics of heat flow and filling, the constituents of SZS alloys may produce additional heat within the MWCNTs side wall, therefore, unsaturated hydrogen like bonds as well as pentagons π bonds are affected more, however, it can preserve the CNT structure within the glassy due to partially affected stronger σ bonds.

Therefore this usual behavior is still open for future innovations, as per our view the properties of individual chalcogenide alloy depends on their alloying constituents and their reactivity with MWCNTs may be different in molten conditions. Thus, it could be said that crystalline kinetics and thermal stability may vary for different chalcogenide–MWCNTs composite materials. To make a conclusive view on thermal properties of the chalcogenide–MWCNTs composite materials, more experimental data is required.

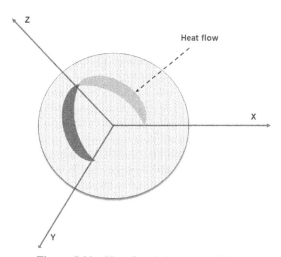

Figure 5.30 Heat flow behavior in CNTs.

CONCLUSIONS

The emergence of composite materials for future potential use in different scientific and technological applications, have described the recently introduced chalcogenide–MWCNTs composites by introducing the advantages of the composite materials in various forms. More specifically, focusing attention towards one of the potential research area chalcogenide -nanotubes composites materials under a glassy configuration. In order to introduce this growing scientific field, various types of chalcogenide composites materials including carbon nanotubes have been discussed. Subsequently, the CNTs flexibility towards different kinds of nanocomposites including low dimension chalcogenide alloys have been described in detail. Additionally, the bond sharing ability in terms of covalent functionalization, non-covalent functionalization and alternative routes functionalization of the CNTs have also been extensively reviewed. Considering the favorable conditions of chalcogenide alloys and carbon nanotubes for composite formation, various previous reports on nanocrystalline chalcogenide alloys–CNTs and polycrystalline chalcogenides CNTs composite materials with their key advantages have been studied. The possible deformation of MWCNTs in a chalcogenide glassy alloy, can drastically enhance the physical properties, such as thermal, optical and electrical properties, owing to the structural changes of MWCNTs. Further, the structural or configurational changes of MWCNTs as well as chalcogenide alloy owing to composites formation have been described. In this order, one step synthesized chalcogenide–MWCNTs glassy composite materials surface morphologies have been addressed. The variations in Raman spectrum peaks as well as their structural interpretation for the chalcogenide–MWCNTs glassy composites materials has also been taken into account. This also includes the optical properties variation in chalcogenide–MWCNTs glassy composites by demonstrating the UV/Visible absorption and infrared interpretation. Moreover, there is no individual report on mechanical and electrical properties on such materials synthesized from a direct single method, however, some reports are available for those based on the reheating synthesis or two step process. The outcomes of different studies on chalcogenide–CNTs composites materials have been demonstrated, a drastic enhancement in mechanical properties as well as in electrical conductivity are interpreted. In contrast to mechanical and electrical properties, thermal properties of single steps synthesized chalcogenide–MWCNTs glassy composite report is available, and has also been considered in this study. The outcomes of this study may help surmount a major hurdle for theoreticians and experimentalists to reconsider a generalized concept for prospective chalcogenide–MWCNTs glassy composites, owing to their lower thermal stability and other crystalline kinetics parameters for the specific composite. However, this field is fairly new and needs further in-depth and extensive investigations with such kinds of composite materials to make an indisputable view. This is due to the fact that chalcogenide alloys characteristics or behavior could vary with the selection of alloying constitutes.

References

1. Guo, Z., Wei, S., Shedd, B., Scaffaro, R., Pereira, T., Hahn, H.T. (2007) Particle surface engineering effect on the mechanical, optical and photoluminescent properties of Zno/vinyl-ester resin nanocomposites. J. Mater. Chem., 17, 806–813.

2. Zhu, J., Wei, S., Ryu, J., Sun, L., Luo, Z., Guo, Z. (2010) Magnetic epoxy resin nanocomposites reinforced with core−shell structured Fe@ Feo nanoparticles: Fabrication and property analysis. ACS Appl. Mater. Interfaces, 2, 2100–2107.

3. He, Q., Yuan, T., Wang, Y., Guleria, A., Wei, S., Zhang, G., Sun, L., Liu, J., Yu, J., Young, D.P. (2016) Manipulating the dimensional assembly pattern and crystalline structures of iron oxide nanostructures with a functional polyolefin. Nanoscale, 8, 1915–1920.

4. Hull, D., Clyne, T.W. (1996) An Introduction to Composite Materials, 2nd Ed. Cambridge University Press, ISBN 0 521 38190 8.

5. Mendeleev, D.I. (1864) Glass Production. Sankt, Petersburg.

6. Sosman, R.D. (1927) Properties of Silica. American Chemical Society Monograph Series No. 37, 37, Chemical Catalog Co. Inc., New York.

7. Tarasov, V.V. (1959) New Problems of Glass Physics. State Publishers on Construction Industry, Moscow.

8. Tarasov, V.V. (1979) Problems of Glass Physics. Stroiizdat Publishers, Moscow.

9. Zachariasen, W.H. (1932) The atomic arrangement in glass. J. Am. Chem. Soc., 54, 3841–3851.

10. Porai-Koshits, E.A. (1959). Glass structure. *In*: Bezborodov, M.A., Bobkova, N.N., Brekhovskikh, S.M., Ermolenko, N.I., Mazo, E.E., Porai-Koshits, E.A. (eds), Diagrammy stekloobrazuyushchikh sistem (Phase Diagrams of Glass-forming systems). Belarus. Politekh. Inst., Minsk, pp. 13–29.

11. Goodman, C.H.L. (1975) Strained mixed-cluster model for glass structure. Nature, 241, 370–372.

12. Uhlman, D.R. (1977) Glass formation. J. Non-Cryst. Solids, 25, 43–85.

13. Lucovsky, G. (1979) Selenium, the amorphous and liquid states. *In*: Gerlach E., Grosse P. (eds), The Physics of Selenium and Tellurium. Springer Series in Solid-State Sciences, vol 13. Springer, Berlin, Heidelberg, pp. 178–192.

14. Elliott, S.R. (1984) Physics of Amorphous Materials. Longman, London.

15. Minaev, V.S. (1991) Stekloobraznye Poluprovodnikovye Splavy (Vitreous Semiconductor Alloys). Metallurgy Publishers, Moscow.

16. Frankenheim, M.L. (1851) Kristallization und amorphie. J. Prakt. Chem., 54(7–8), 430–476.

17. Wang, R. and Merz, M.D. (1977) Polymorphic bonding and thermal stability of elemental non-crystalline solids. Phys. Status Solidi (a), 39, 697–703.

18. Blinov, L.N. (1985) On correlation between glass formation and polymorphism in chalcogenide systems. Abstracts of All-Union Conference "Vitreous Semiconductors", Leningrad, USSR, pp. 176–177.

19. Sugai, S. (1986) Two-directional photo-induced crystallization in $GeSe_2$ and $SiSe_2$ glasses. Phys. Rev. Lett., 57, 456–459.

20. Sugai, S. (1987) Stochastic random network model in Ge and Si chalcogenide glasses. Phys. Rev. B, 35, 1345–1361.

21. Mott, N.F. and Davis, E.A. (1979) Electron Processes in Non-Crystalline Materials. Clarendon Press, Oxford.

22. Feltz, A. (1983) Amorphe und Glasartige Anorganische Fest-korper. Akademie-Verlag, Berlin.

23. Lucovsky, G., Hayes, T.M. (1979) Short-range order in amorphous semiconductors. *In*: Brodsky, M.H. (ed), Amorphous Semiconductors. Springer, Berlin, Heidelberg, New York, pp. 268–310 (Russian Translation).

24. Leko, V.K. (1993) The structure of vitreous silica. Fiz. Khim. Stekla, 19, 673–715 (Glass Phys. Chem. (English Transl.), 19, 351–374).

25. Minaev, V.S. (1996) Polymorphic-crystalloid structure of glass. Fiz. Khim. Stekla, 22, 314–325 (Sov. J. Glass Phys. Chem. (Engl. Transl.), 22, 235–242).

26. Fairman, R., Ushkov, B. (2004) Semiconducting Chalcogenide Glass I: Glass Formation, Structure, and Stimulated Transformations in Chalcogenide Glasses. Elsevier, San Diego.

27. Kokorina, V. (1996) Glasses for Infrared Optics. CRC Press, Boca Raton, Fla.

28. Chiang, Y.S., Ing Jr., S.W. (1969) Studies on selenium–organic polymer interfaces. 13, 883–897 (1969).

29. Geim, A.K. (2009) Graphene: Status and prospects. Science, 324, 1530–1534.

30. Zang, J., Ryu, S., Pugno, N., Wang, Q., Tu, Q., Buehler, M.J., Zhao, X. (2013) Multifunctionality and control of the crumpling and unfolding of large-area graphene. Nat. Mater., 12, 321–325.

31. Ramanathan, T., Abdala, A.A., Stankovich, S., Dikin, D.A., Herrera-Alonso, M., Piner, R.D., Adamson, D.H., Schniepp, H.C., Chen, X., Ruoff, R.S., Nguyen, S.T., Aksay, I.A., Homme, R.K.P., Brinson, L.C. (2008) Functionalized graphene sheets for polymer nanocomposites. Nature Nanotechnology, 3, 327–331.

32. Song, S.H., Park, K.H., Kim, B.H., Choi, Y.W., Jun, G.H., Lee, D.J., Kong, B.S., Paik, K.W., Jeon, S. (2013) Enhanced thermal conductivity of epoxy–graphene composites by using non-oxidized graphene flakes with non-covalent functionalization. Advanced Materials, 25, 732–737.

33. Kinloch, I.A., Suhr, J., Lou, J., Young, R.J., Ajayan, P.M. (2018) Composites with carbon nanotubes and graphene: An outlook. Science, 362, 547–553.

34. Strano, M.S., Kyke, C.A., Usrey, M.L., Barone, P.W., Allen, M.J., Shan, H., Kittrell, C., Hauge, R.H., Tour, J.M., Smalley, R.E. (2003). Electronic structure control of singlewalled carbon nanotube functionalization. Science, 301, 1519–1522.

35. Li, H., Zhou, B., Lin, Y., Gu, L., Wang, W., Fernando K.A.S., Kumar, S., Allard, L.F., Sun, Y.P. (2004). Selective interactions of porphyrins with semiconducting single walled carbon nanotube. J. Am. Chem. Soc., 126, 1014–1015.

36. Niyogi, S., Hamon, M.A., Hu, H., Zhao, B., Bhowmik, P., Sen, R., Itkis, M.E., Haddon, R.C. (2002) Chemistry of single-walled carbon nanotubes. Acc. Chem. Res., 35, 1105–1113.

37. Tasis, D., Tagmatarchis, N., Georgakilas, V., Prato, M. (2003) Soluble carbon nanotubes. Chem. Eur. J., 9, 4000–4008.

38. Hirsch, A. (2002) Functionalization of single-walled carbon nanotubes. Angew. Chem., Int. Ed., 41, 1853–1859.

39. Sinnott, S.B. (2002) Chemical functionalization of carbon nanotubes. J. Nanosc. Nanotechnol., 2, 113–123.

40. Bianco, A., Prato, M., Kostarelos, K., Bianco, A. (2008) Functionalized carbon nanotubes in drug design and discovery. Acc. Chem. Res., 41, 60–68.

41. Mickelson, E.T., Huffman, C.B., Rinzler, A.G., Smalley, R.E., Hauge, R.H., Margrave, J.L. (1998) Fluorination of single-wall carbon nanotubes. Chem. Phys. Lett., 296, 188–194.

42. Kelly, K.F., Chiang, I.W., Mickelson, E.T., Hauge, R.H., Margrave, J.L., Wang, X., et al. (1993) Insight into the mechanism of sidewall functionalization of single-walled nanotubes: An STM study. Chem. Pys. Lett., 313, 445–450.

43. Touhara, H., Inahara, J., Mizuno, T., Yokoyama, Y., Okanao, S., Yanagiuch, K., et al. (2002) Fluorination of cup-stacked carbon nanotubes, structure and properties. Fluorine Chem., 114, 181–188.

44. Stevens, J.L., Huang, A.Y., Peng, H., Chiang, I.W., Khabashesku, V.N., Margrave, J.L. (2003) Sidewall amino functionalization of SWNTs through fluorination and subsequent reactions with terminal diamines. Nano Lett., 3, 331–336.

45. Hu, H., Zhao, B., Hamon, M.A., Kamaras, K., Itkis, M.E., Haddon, R.C. (2003). Sidewall functionalization of single-walled carbon nanotubes by addition of dichlorocarbene. J. Am. Chem. Soc., 125, 14893–14900.

46. Unger, E., Graham, A., Kreupl, F., Liebau, M., Hoenlein, W., (2002). Electrochemical functionalization of multi-walled carbon nanotubes for solvation and purification. Curr. Appl. Phys., 2, 107–111.

47. Kim, K.S., Bae, D.J., Kim, J.R., Park, K.A., Lim, S.C., Kim, J.J., Choi, W.B., Park, C.Y., Lee, Y.H. (2002) Modification of electronic structures of a carbon nanotube by hydrogen functionalization. Adv. Mater., 14, 1818–1821.

48. Tagmatarchis, N., Prato, M.J. (2004) Functionalization of carbon nanotubes via 1,3-dipolar cycloadditions. J. Mater. Chem., 14, 437–439.

49. Chen, J., Hamon, M.A., Hu, H., Chen, Y., Rao, A.M., Eklund, P.C., Haddon, R.C. (1998) Solution properties of single-walled carbon nanotubes. Science, 282, (5386), 95–98.

50. Esumi, K., Ishigami, M., Nakajima, A., Sawada, K., Honda, H. (1996) Chemical treatment of carbon nanotubes. Carbon, 34, 279–281.

51. Liu, J., Rinzler, A.G., Dai, H., Hafner, J.H., Bradley, R.K., Boul, P.J., et al. (1998) Fullerene pipes. Science, 280, 1253–1256.

52. Yu, R., Chen, L., Liu, Q., Lin, J., Tan, K.-L., Ng, S.C., Chan, H S.O., Xu, G.-Q., Hor, T.S.A. (1998) Platinum deposition on carbon nanotubes via chemical modification. Chemistry of Materials, 10, 718–722.

53. Sham, M.-L., Kim, J.-K., (2006) Surface functionalities of multi-wall carbon nanotubes after UV/ Ozone and TETA treatments. Carbon, 44, 768–777.

54. Wang, S.C., Chang, K.S., Yuan, C.J. (2009) Enhancement of electrochemical properties of screen-printed carbon electrodes by oxygen plasma treatment. Electrochimica Acta, 54, (21), 4937–4943.

55. Ma, P.C., Kim, J.K., Tang, B.Z. (2006) Functionalization of carbon nanotubes using a silane coupling agent. Carbon, 44, 3232–3238.

56. Sano, M., Kamino, A., Okamura, J., Shinkai, S. (2001). Self-organization of PEO-graft-single-walled carbon nanotubes in solutions and Langmuir–Blodgett films. Langmuir, 17, (17), 5125–5128.

57. Kong, H., Gao, C., Yan, D. (2003) Controlled functionalization of multiwalled carbon nanotubes by in situ atom transfer radical polymerization. J. Am. Chem. Soc., 126, 412–413.

58. Hamon, M.A., Hui, H., Bhowmik, P. (2002) Ester-functionalized soluble single-walled carbon nanotubes. Appl. Phys. A., 74, 333–338.

59. Coleman, J.N., Khan, U., Gun'ko, Y.K. (2006) Mechanical reinforcement of polymers using carbon nanotubes. Adv. Mater., 18, 689–706.

60. Chen, R.J., Zhang, Y., Wang, D., Dai, H. (2001) Noncovalent sidewall functionalization of single-walled carbon nanotubes for protein immobilization. J. Am. Chem. Soc., 123, 3838–3839.

61. Hecht, D.S., Ramirez, R.J.A., Briman, M., Artukovic, E., Chichak, K.S., Stoddart, J.F., Gruner, G. (2006) Bioinspired detection of light using a porphyrin-sensitized singlewall nanotube field effect transistor. Nano Lett., 6, 2031–2036.

62. Hu, L., Zhao, Y.-L., Ryu, K., Zhou, C., Stoddart, J.F., Gruner, G. (2008) Light-induced charge transfer in pyrene/CdSe-SWNT hybrids. Adv. Mater., 20, 939–946.

63. Star, A., Stoddart, J.F., Steuerman, D., Diehl, M., Boukai, A., Wong, E.W., Yang, X., Chung, S.-W., Choi, H., Heath, J.R. (2001) Preparation and properties of polymer-wrapped single-walled carbon nanotubes. Angew. Chem. Int. Ed., 40, 1721–1725.

64. Star, A., Liu, Y., Grant, K., Ridvan, L., Stoddart, J.F., Steuerman, D.W., Diehl, M.R., Boukai, A., Heath, J.R. (2003). Noncovalent side-wall functionalization of singlewalled carbon nanotubes. Macromolecules, 36, 553–560.

65. Star, A., Stoddart, J.F. (2002) Dispersion and solubilization of single-walled carbon nanotubes with a hyperbranched polymer. Macromolecules, 35, 7516–7520.

66. Cheng, F., Adronov, A. (2006) Noncovalent functionalization and solubilization of carbon nanotubes by using a conjugated Zn-porphyrin polymer. Chem. Eur. J., 12, 5053–5059.

67. Yi, W., Malkovskiy, A., Chu, Q., Sokolov, A.P., Colon, M.L., Meador, M., Pang, Y. (2008). Wrapping of single-walled carbon nanotubes by a π-conjugated polymer: The role of polymer conformation-controlled size selectivity. J. Phys. Chem. B ,112, 12263–12269.

68. Steuerman, D.W., Star, A., Narizaano, R., Choi, H., Ries, R.S., Nicolini, C., Stoddart, J.F., Heath, J.R. (2002) Interactions between conjugated polymers and singlewalled carbon nanotubes. J. Phys. Chem. B, 106, 3124–3130.

69. Gong, X., Liu, J., Baskaran, S., Voise, R.D., Young, J.S. (2000) Surfactant-assisted processing of carbon nanotube/Polymer composites. Chem. Mater., 12, 1049–1052.

70. Vaisman, L., Marom, G., Wagner, H.D. (2006) Dispersions of surface-modified carbon nanotubes in water-soluble and water-insoluble polymers. Adv. Funct. Mater., 16, 357–363.

71. Islam, M.F., Rojas, E., Bergey, D.M., Johnson, A.T., Yodh, A.G. (2003) High weight fraction surfactant solubilization of single-wall carbon nanotubes in water. Nano Lett., 3, 269–273.

72. Yu, J., Grossiord, N., Koning, C.E., Loos, J. (2007) Controlling the dispersion of multi-wall carbon nanotubes in aqueous surfactant solution. Carbon, 45(3), 618–623.

73. Whitsitt, E.A., Barron, A.R., (2003), Silica coated single walled carbon nanotubes. Nano Lett., 3, 775–778.

74. Kim, T.H., Doe, C., Kline, S.R., Choi, S.M. (2007) Water-redispersible isolated single-walled carbon nanotubes fabricated by in situ polymerization of micelles. Adv. Mater., 19, 929–933.

75. Klumpp, C., Kostarelos, K., Prato, M., Bianco, A. (2006) Functionalized carbon nanotubes as emerging nanovectors for the delivery of therapeutics. Biochimica et Biophysica Acta (BBA) - Biomembranes, 1758, 404–412.

76. Barone, P.W., Strano, M.S. (2006) Reversible control of carbon nanotube aggregation for a glucose affinity sensor. Angew. Chem., Int. Ed., 45, 8138–8141.

77. Chambers, G., Carroll, C., Farrell, G.F., Dalton, A.B., McNamara, M., in het Panhuis, M., Byrne, H.J. (2003) Characterization of the interaction of gamma cyclodextrin with single-walled carbon nanotubes. Nano Lett. 3, 843–846.

78. Ikeda, M., Hasegawa, T., Numata, M., Sugikawa, K., Sakurai, K., Fujiki, M., Shinkai, S. (2007) Instantaneous inclusion of a polynucleotide and hydrophobic guest molecules into a helical core of cationic 1,3-glucan polysaccharide. J. Am. Chem. Soc., 129, 3979–3988.

79. Cheng, F., Adronov, A. (2006) Noncovalent functionalization and solubilization of carbon nanotubes by using a conjugated Zn-porphyrin polymer. Chem. Eur. J., 12, 5053–5059.

80. Ishibashi, A., Nakashima, N. (2006) Individual dissolution of single-walled carbon nanotubes in aqueous solution of steroid or sugar compounds and their Raman and Near-IR spectral properties. Chem. Eur. J., 12, 7595–7602.

81. Dodziuk, H., Ejchart, A., Anczewski, W., Ueda, H., Krinichnaya, E., Dolgonos, G., Kutner, W. (2003) Water solubilization, determination of the number of different types of single-wall carbon nanotubes and their partial separation with respect to diameters by complexation with η-cyclodextrin. Chemical Communications, 8, 986–987.

82. Yan, L.Y., Poon, Y.F., Chan-Park, M.B., Chen, Y., Zhang, Q. (2008) Individually dispersing single-walled carbon nanotubes with novel neutral pH water-soluble chitosan derivatives. J. Phys. Chem. C, 112, 7579–7587.

83. Yan, L.Y., Poon, Y.F., Chan-Park, M.B., Chen, Y., Zhang, Q. (2008) Individually dispersing single-walled carbon nanotubes with novel neutral pH water-soluble chitosan derivatives. J. Phys. Chem. C, 112, 7579–7587.

84. Goodwin, A.P., Tabakman, S.M., Welsher, K., Sherlock, S.P., Prencipe, G., Dai, H. (2009) Phospholipid-dextran with a single coupling point: A useful amphiphile for functionalization of nanomaterials. J. Am. Chem. Soc., 131, 289–296.

85. Baek, J.-B., Lyons, C.B., Tan, L.-S. (2004). Covalent modification of vapour-grown carbon nanofibers via direct Friedel-Crafts acylation in polyphosphoric acid. J. Mater. Chem., 14(13), 2052–2056.

86. Lee, H.-J., Oh, S.-J., Choi, J.-Y., Kim, J.W., Han, J., Tan, L.-S., Baek, J.-B. (2005). In situ synthesis of Poly(ethylene terephthalate) (PET) in ethylene glycol containing terephthalic acid and functionalized multiwalled carbon nanotubes (MWNTs) as an approach to MWNT/PET nanocomposites. Chem. Mater. 17, 5057–5064.

87. Oh, S.-J., Lee, H.-J., Keum, D.-K., Lee, S.-W., Wang, D.H., Park, S.-Y., Tan, L.-S., Baek, J.-B. (2006) Multiwalled carbon nanotubes and nanofibers grafted with polyetherketones in mild and viscous polymeric acid. Polymer, 47, 1132–1140.

88. Choi, J.-Y., Han, S.-W., Huh, W.-S., Tan, L.-S., Baek, J.-B. (2007) In situ grafting of carboxylic acid-terminated hyperbranched poly(ether-ketone) to the surface of carbon nanotubes. Polymer, 48, 4034–4040.

89. Jeon, I.-Y., Choi, E.-K., Bae, S.-Y., Baek, J.-B., (2010). Edge-Functionalization of pyrene as a miniature graphene via friedel–crafts acylation reaction in Poly(Phosphoric Acid). Nanoscale Res. Lett., 5, 1686–1691.

90. Li, Q.W., Sun, B.Q., Kinloch, I.A., Zhi, D., Sirringhaus, H., Windle, A.H. (2006) Enhanced self-assembly of pyridine-capped CdSe nanocrystals on individual single-walled carbon nanotubes. Chem. Mater., 18, 164–168.

91. Shi, D., Lian, J., Wang, W., Liu, G., He, P., Dong, Z., Wang, L., Ewing, R.C. (2006) Luminescent carbon nanotubes by surface functionalization. AdV. Mater., 18, 189–193.

92. Du, J.M., Fu, L., Liu, Z.M., Han, B.X., Li, Z.H., Liu, Y.Q., Sun, Z.Y., Zhu, D.B. (2005) Facile route to synthesize multiwalled carbon nanotube/zinc sulfide heterostructures: Optical and electrical properties. J. Phys. Chem. B, 109, 12772.

93. Zhao, L.P., Gao, L. (2004) Coating multi-walled carbon nanotubes with zinc sulphide. J. Mater. Chem., 14, 1001.

94. Gu, F., Li, C., Wang, S. (2006) Solution-chemical synthesis of carbon nanotube/ZnS nanoparticle core/shell heterostructures. Inorganic Chemistry, 46, 13, 5343–5348.

95. Peng, X., Wong, S.S. (2009) Controlling nanocrystal density and location on carbon nanotube templates. Chem. Mater., 21, 682–694.

96. Eder, D. (2010) Carbon nanotube-inorganic hybrids. Chem. Rev., 110, 1348–1385.

97. Shou-ai, F., Jiang-hong, Z., Zhen-ping, Z. (2008) The manufacture of carbon nanotubes decorated with ZnS to enhance the ZnS photocatalytic activity. New Carbon Materials, 23, 228–234.

98. Huang, S., Li, S., He, Q., An, H., Xiao, L., Hou, L. (2019) Formation of CoTe2 embedded in nitrogen-doped carbon nanotubes-grafted polyhedrons with boosted electrocatalytic properties in dye-sensitized solar cells. Applied Surface Science, 476, 769–777.

99. Ravindran, S., Chaudhary, S., Colburn, B., Ozkan, M., Ozkan, C.S. (2003) Covalent coupling of quantum dots to multiwalled carbon nanotubes for electronic device applications. Nano Lett., 3, 447–453.

100. Cao, J., Sun, J.Z., Hong, H., Li, H.Y., Chen, H.Z., Wang, M. (2004) Carbon nanotube/CdS core–shell nanowires prepared by a simple room-temperature chemical reduction method. AdV. Mater., 16, 84–87.

101. Huang, J.X., Xie, Y., Li, B., Liu, Y., Qian, Y.T., Zhang, S.Y. (2000) In-situ source–template–interface reaction route to semiconductor CdS submicrometer hollow spheres. AdV. Mater., 12, 808–811.

102. Lee, S.M., Jun, Y.W., Cho, S.N., Cheon, J. (2002) Single-crystalline star-shaped nanocrystals and their evolution: Programming the geometry of nano-building blocks. J. Am. Chem. Soc., 124, 11244–11245.

103. Yu, S.H., Yoshimura, M. (2002) Shape and phase control of ZnS nanocrystals: Template fabrication of wurtzite ZnS single-crystal nanosheets and ZnO flake-like dendrites from a lamellar molecular precursor $ZnS \cdot (NH_2CH_2CH_2NH_2)_{0.5}$, AdV. Mater., 14, 296–300.

104. Ma, Y., Qi, L., Ma, J., Cheng, H., Shen, W. (2003) Synthesis of submicrometer-sized CdS hollow spheres in aqueous solutions of a triblock copolymer. Langmuir, 19, 9079–9085.

105. Sone, E.D., Zubarev, E.R., Stupp, S.I. (2002) Semiconductor nanohelices templated by supramolecular ribbons. Angew. Chem. Int. Ed., 41, 1705–1709.

106. Prevenslik, T.V. (2000) Acoustoluminescence and sonoluminescence. J. Lumin., 87–89, 1210–1212.

107. Zhao, Y.W., Zhang, Y., Zhu, H., Hadjipianayis, G.C., Xiao, J.Q.J. (2004) Low-temperature synthesis of hexagonal (Wurtzite) ZnS nanocrystals. Am. Chem. Soc., 126, 6874–6875.

108. Falcony, C., Garcia, C., Ortiz, A., Alonso, J.C. (1992) Luminescent properties of ZnS:Mn films deposited by spray pyrolysis. J. Appl. Phys., 72, 1525.

109. Thostenson, E.T., Ren, Z.F., Chou, T.W. (2001) Advances in the science and technology of carbon nanotubes and their composites: A review. Compos Sci Technol, 61, 1899–1912.

110. Zhang, H.X., Feng, C., Zhai, Y.C., Jiang, K.L., Li, Q.Q., Fan, S.S. (2009) Crossstacked carbon nanotube sheets uniformly loaded with SnO_2 nanoparticles: A novel binder-free and high-capacity anode material for lithium-ion batteries. Adv. Mater, 21, 2299–2304.

111. Merkoci, A., Pumera, M., Llopis, X., Perez, B., del Valle, M., Alegret, S. (2005) New materials for electrochemical sensing VI: Carbon nanotubes. Trac. Trends Anal. Chem., 24, 826–838.

112. Liu, Z.L., Lin, X.H., Lee, J.Y., Zhang, W., Han, M., Gan, L.M. (2002) Preparation and characterization of platinum-based electrocatalysts on multiwalled carbon nanotubes for proton exchange membrane fuel cells. Langmuir, 18, 4054–4060.

113. Huang, L., Lau, S.P., Yang, H.Y., Leong, E.S.P., Yu, S.F., Prawer, S. (2005) Stable superhydrophobic surface via carbon nanotubes coated with a ZnO thin film. J. Phys. Chem. B, 109, 7746–7748.

114. Hasobe, T., Fukuzumi, S., Kamat, P.V. (2006) Stacked-cup carbon nanotubes for photoelectrochemical solar cells. Angew. Chem. Int. Ed., 45, 755–759.

115. Kamat, P.V. (2007) Meeting the clean energy demand: Nanostructure architectures for solar energy conversion. J. Phys. Chem. C., 111, 2834–2860.

116. Yao, Z., Wang, L., Zhang, Y., Yu, Z., Jiang, Z. (2014) Carbon nanotube modified Zn0.83Cd0.17S nanocomposite photocatalyst and its hydrogen production under visible-light. International Journal of Hydrogen Energy, 39, 1538–15386.

117. Chen, Y., Tian, G., Ren, Z., Pan, K., Shi, Y., Wang, J., Fu, H. (2014) Hierarchical core–shell carbon Nanofiber@ZnIn2S4 composites for enhanced hydrogen evolution performance. ACS Applied Materials & Interfaces, 2014, 6(16), 13841–13849.

118. Kudo, A., Miseki, Y. (2009) Heterogeneous photocatalyst materials for water splitting. Chem. Soc. Rev. 38, 253–278.

119. Kuo, Y.L., Wu, C.C., Peng, Y.H., Chang, W.S. (2012) Evaluation of the photochemical stability of zinc sulfide as protective layer on silver indium sulfide photocatalyst film. J. Chin. Chem. Soc., 59, 1323–1328.

120. Li, K., Chai, B., Peng, T.Y., Mao, J., Zan, L. (2013) Synthesis of multicomponent sulfide Ag_2ZnSnS_4 as an efficient photocatalyst for H-2 production under visible light irradiation. RSC Adv., 3, 253–258.

121. Chan, C.C., Chang, C.C., Hsu, C.H., Weng, Y.C., Chen, K.Y., Lin, H.H. et al. (2014) Efficient and stable photocatalytic hydrogen production from water splitting over $Zn_xCd_{1-x}S$ solid solutions under visible light irradiation. Int. J. Hydrogen Energy, 39, 1630–1639.

122. Meng, J.L., Yu, Z.M., Li, Y., Li, Y.D. (2014) PdS-modified CdS/NiS composite as an efficient photocatalyst for H-2 evolution in visible light. Catal Today, 225, 136–141.

123. Jia, F.Z., Yao, Z.P., Jiang, Z.H. (2012) Solvothermal synthesis ZnS–In_2S_3–Ag_2S solid solution coupled with $TiO_{2-x}S_x$ nanotubes film for photocatalytic hydrogen production. Int. J. Hydrogen Energy, 37, 3048–3055.

124. Roy, A.M., De, G.C. (2003) Immobilisation of Cds, ZnS and mixed ZnSCdS on filter paper e effect of hydrogen production from alkaline $Na_2S/Na_2S_2O_3$ solution. J. Photochem. Photobiol. A Chem., 157, 87–92.

125. Biswal, N., Das, D.P., Martha, S., Parida, K.M. (2011) Efficient hydrogen production by composite photocatalyst CdS-ZnS/Zirconiumtitanium phosphate (ZTP) under visible light illumination. Int. J. Hydrogen Energy, 36, 13452–13460.

126. Peng, S.Q., Chen, C.H., Liu, X.Y., Li, Y.X. (2013) Enhanced photocatalytic hydrogen evolution under visible light irradiation over $Cd_{0.5}Zn_{0.5}S$ solid solution by magnesium-doping. React Kinetics Mech. Catal., 110, 259–270.

127. Huang, J.D., Liu, J.Y., Han, K.L. (2012) Hybrid functionals studies of structural and electronic properties of $Zn_xCd((_{1-x}))S$ and $(Zn_xCd_{1-x})(Se_xS_{1-x})$ solid solution photocatalysts. Int. J. Hydrogen Energy, 37, 17870–17881.

128. Macias-Sanchez, S.A., Nava, R., Hernandez-Morales, V., Acosta-Silva, Y.J., Gomez-Herrera, L., Pawelec, B. et al. (2012) Cd1xZnxS solid solutions supported on ordered mesoporous silica (SBA-15): Structural features and photocatalytic activity under visible light. Int. J. Hydrogen Energy, 37, 9948–58.

129. Peng, S.Q., An, R., Li, Y.X., Lu, G.X., Li, S.B. (20120) Remarkable enhancement of photocatalytic hydrogen evolution over Cd0.5Zn0.5S by bismuth-doping. Int. J. Hydrogen Energy, 37, 1366–1374.

130. Ha, J.W., Ruberu, T.P.A., Han, R., Dong, B., Vela, J., Fang, N. (2014) Super-resolution mapping of photogenerated electron and hole separation in single metal−semiconductor nanocatalysts. J. Am. Chem. Soc., 136, 1398−1408.

131. Gratzel, M. (2001) Photoelectrochemical cells. Nature, 414, 338−344.

132. Cho, I.S., Chen, Z., Forman, A.J., Kim, D.R., Rao, P.M., Jaramillo, T.F., Zheng, X. (2011) Branched TiO_2 nanorods for photoelectrochemical hydrogen production. Nano Lett., 11, 4978−4984.

133. Chaudhari, N.S., Warule, S.S., Kale, B.B. (2014) Architecture of rose and hollow marigold-like $ZnIn_2S_4$ flowers: Structural, optical and photocatalytic study. RSC Adv., 4, 12182−12187.

134. Li, F., Chen, G.P., Luo, J.H., Huang, Q.L., Luo, Y.H., Meng, Q.B., Li, D.M. (2013) Band engineering of Cu^{2+} Doped $In_{2x}Zn_{3(1-x)}S_3$ solid solution with high photocatalytic activity for H_2 production under visible light. Catalysis Science & Technology, 3, 1993−1999.

135. Shen, S.H., Zhao, L., Guo, L.J. (2010) $Zn_mIn_2S_{3+m}$ (m = 1−5, Integer): A new series of visible-light-driven photocatalysts for splitting water to hydrogen. Int. J. Hydrogen Energy, 35, 10148−10154.

136. Chen, Z.X., Li, D.Z., Zhang, W.J., Chen, C., Li, W.J., Sun, M., He, Y.H., Fu, X.Z. (2008) Low-temperature and template-free synthesis of $ZnIn_2S_4$ microspheres. Inorg. Chem., 47, 9766−9772.

137. Shang, L., Zhou, C., Bian, T., Yu, H.J., Wu, L.Z., Tung, C.H., Zhang, T.R. (2013) Facile synthesis of hierarchical $ZnIn_2S_4$ submicrospheres composed of ultrathin mesoporous nanosheets as a highly efficient visible-light-driven photocatalyst for H_2 production. J. Mater. Chem. A, 1, 4552−4558.

138. Xu, Z.D., Li, Y.X., Peng, S.Q., Lu, G.X., Li, S.B. (2011) Composition, morphology and photocatalytic activity of Zn−In−S composite synthesized by a NaCl-assisted hydrothermal method. CrystEngComm., 13, 4770−4776.

139. Chen, Y.J., Hu, S.W., Liu, W.J., Chen, X.Y., Wu, L., Wang, X.X., Liu, P., Li, Z.H. (2011) Controlled syntheses of cubic and hexagonal $ZnIn_2S_4$ nanostructures with different visible-light photocatalytic performance. Dalton Trans., 40, 2607−2613.

140. Xu, Z.D., Li, Y.X., Peng, S.Q., Lu, G.X., Li, S.B. (2012) NaCl−assisted low temperature synthesis of layered Zn-In-S photocatalyst with high visible-light activity for hydrogen evolution. RSC Adv., 2, 3458−3466.

141. Chen, Z.X., Li, D.Z., Zhang, W.J., Shao, Y., Chen, T.W., Sun, M., Fu, X.Z. (2009) Photocatalytic degradation of dyes by $ZnIn_2S_4$ microspheres under visible light irradiation. J. Phys. Chem. C, 113, 4433−4440.

142. Lei, Z.B., You, W.S., Liu, M.Y., Zhou, G.H., Takata, T., Hara, M., Domen, K., Li, C. (2003) Photocatalytic water reduction under visible light on a novel $ZnIn_2S_4$ catalyst synthesized by hydrothermal method. Chem. Commun., 2142−2143.

143. Peng, S.J., Li, L.L., Wu, Y.Z., Jia, L., Tian, L.L., Srinivasan, M., Ramakrishna, S., Yan, Q.Y., Mhaisalkar, S.G. (2013) Size- and ShapeControlled synthesis of $ZnIn_2S_4$ nanocrystals with high photocatalytic performance. CrystEngComm., 15, 1922−1930.

144. Li, H.F., Yu, H.T., Chen, S., Zhao, H.M., Zhang, Y.B., Quan, X. (2014) Fabrication of graphene wrapped $ZnIn_2S_4$ microspheres heterojunction with enhanced interfacial contact and its improved photocatalytic performance. Dalton Trans., 43, 2888−2894.

145. Peng, S.J., Zhu, P.N., Mhaisalkar, S.G., Ramakrishna, S. (2012) Self supporting three-dimensional $ZnIn_2S_4$/PVDF–Poly(MMA-co-MAA) composite mats with hierarchical nanostructures for high photocatalytic activity. J. Phys. Chem. C, 116, 13849–13857.

146. Gao, B., Liu, L.F., Liu, J.D., Yang, F.L. (2013) Photocatalytic degradation of 2,4,6-tribromophenol over Fe-doped $ZnIn_2S_4$: Stable activity and enhanced debromination. Appl. Catal., B, 129, 89–97.

147. Fang, F., Chen, L., Chen, Y.B., Wu, L.M. (2010) Synthesis and photocatalysis of $ZnIn_2S_4$ nano/micropeony. J. Phys. Chem. C, 114, 2393–2397.

148. Xu, B., He, P.L., Liu, H.L., Wang, P.P., Zhou, G., Wang, X.A (2014) 1D/2D Helical CdS/$ZnIn_2S_4$ nano-heterostructure. Angew. Chem., Int. Ed., 53, 2339–2343.

149. Li, Y.X., Wang, J.X., Peng, S.Q., Lu, G.X., Li, S.B. (2010) Photocatalytic hydrogen generation in the presence of glucose over ZnS-coated $ZnIn_2S_4$ under visible light irradiation. Int. J. Hydrogen Energy, 35, 7116–7126.

150. Chai, B., Peng, T.Y., Zeng, P., Zhang, X.H. (2012) Preparation of a MWCNTs/$ZnIn_2S_4$ composite and its enhanced photocatalytic hydrogen production under visible-light irradiation. Dalton Trans., 41, 1179–1186.

151. Hou, J.G., Yang, C., Cheng, H.J., Wang, Z., Jiao, S.Q., Zhu, H.M. (2013) Ternary 3D architectures of CdS QDs/Graphene/$ZnIn_2S_4$ heterostructures for efficient photocatalytic H_2 production. Phys. Chem. Chem. Phys., 15, 15660–15668.

152. Wei, L., Chen, Y.J., Lin, Y.P., Wu, H.S., Yuan, R.S., Li, Z.H. (2014) MoS_2 as non-noble-metal co-catalyst for photocatalytic hydrogenevolution over hexagonal $ZnIn_2S_4$ under visible light irradiations. Appl. Catal., B, 144, 521–527.

153. Wei, L., Chen, Y., Zhao, J., Li, Z. (2013) Preparation of NiS/$ZnIn_2S_4$ as a superior photocatalyst for hydrogen evolution under visible light irradiation. Beilstein J. Nanotechnol., 4, 949–955.

154. Chen, Y.J., Ge, H., Wei, L., Li, Z.H., Yuan, R.S., Ping Liu, P., Fu, X.Z. (2013) Reduction degree of reduced graphene oxide (RGO) dependence of photocatalytic hydrogen evolution performance over RGO/$ZnIn_2S_4$ nanocomposites. Catal. Sci. Technol., 3, 1712–1717.

155. Kongkanand, A., Dominguez, R.M., Kamat, P.V. (2007) Single wall carbon nanotube scaffolds for photoelectrochemical solar cells. Capture and transport of photogenerated electrons. Nano Lett., 7, 676–680.

156. Vietmeyer, F., Seger, B., Kamat, P.V. (2007) Anchoring ZnO particles on functionalized single wall carbon nanotubes. Excited state interactions and charge collection. Adv. Mater., 19, 2935–2940.

157. Wen, Z.H., Wang, Q., Zhang, Q., Li, J.H. (2007) In situ growth of mesoporous SnO_2 on the multiwalled carbon nanotubes: A novel composite with porous-tube structure as anode for lithium batteries. Adv. Funct. Mater., 17, 2772–2778.

158. Zhou, J., Tian, G.H., Chen, Y.J., Meng, X.Y., Shi, Y.H., Cao, X.R., Pan, K., Fu, H.G. (2013) In situ controlled growth of $ZnIn_2S_4$ nanosheets on reduced graphene oxide for enhanced photocatalytic hydrogen production performance. Chem. Commun., 49, 2237–2239.

159. Ye, L., Fu, J.L., Xu, Z., Yuan, R.S., Li, Z.H. (2014) Facile one-pot solvothermal method to synthesize sheet-on-sheet reduced graphene oxide (RGO)/$ZnIn_2S_4$ nanocomposites with superior photocatalytic performance. ACS. Appl. Mater. Interfaces, 6, 3483–3490.

160. Tian, G.H., Chen, Y.J., Zhou, J., Tian, G.H., Li, R., Wang, C.J., Fu, H.G. (2014) *In situ* growth of Bi_2MoO_6 on reduced graphene oxide nanosheets for improved visible-light photocatalytic activity. CrystEngComm, 16, 842–849.

161. Chen, Y., Tian, G., Ren, Z., Pan, K., Shi, Y., Wang, J., Fu, H. (2014) Hierarchical core–shell carbon Nanofiber@$ZnIn_2S_4$ composites for enhanced hydrogen evolution performance. ACS Appl. Mater. Interfaces, 6, 13841–13849.

162. Zhu, T., Bin Wu, H.B., Wang, Y.B., Xu, R., Lou, X.W. (2012) Formation of 1D hierarchical structures composed of Ni_3S_2 nanosheets on CNTs backbone for supercapacitors and photocatalytic H_2 production. Adv. Energy Mater., 2, 1497–1502.

163. Yu, S., Feng, X., Zhang, N., Seok, J., Abruna, H. (2018) Understanding conversion-type electrodes for lithium rechargeable batteries. Acc. Chem. Res., 51, 273–281.

164. Yang, X., Zhang, J., Wang, Z., Wang, H., Zhi, C., Yu, D., Rogach, A. (2017) Carbon-supported nickel selenide hollow nanowires as advanced anode materials for sodium-ion batteries. Small, 14, 1702669.

165. Wei, Z., Wang, L., Zhuo, M., Ni, W., Wang, H., Ma, J. (2018) Layered tin sulfide and selenide anode materials for Li- and Na-ion batteries. J. Mater. Chem. A, 6, 12185–12214.

166. Xu, X., Liu, J., Ouyang, L., Hu, R., Wang, H., Yang, L., Zhu, M. (2018) A general metal-organic framework (MOF)-derived selenidation strategy for in situ carbon-encapsulated metal selenides as high-rate anodes for na-ion batteries. Adv. Funct. Mater., 28, 1707573.

167. Park, G., Kang, Y. (2018) Multiroom-structured multicomponent metal selenide–graphitic carbon–carbon nanotube hybrid microspheres as efficient anode materials for sodium-ion batteries. Nanoscale, 10, 8125–8132.

168. Wei, Y., He, J., Guo, Y., Qin, R., Li, H., Zhai, T. (2018) Healable structure triggered by thermal/electrochemical force in layered GeSe$_2$ for high performance Li-Ion batteries. Advanced Energy Materials, 8, 1703635.

169. Ali, Z., Asif, M., Huang, X., Tang, T., Hou, Y. (2018) Hierarchically porous Fe$_2$CoSe$_4$ binary-metal selenide for extraordinary rate performance and durable anode of sodium-ion batteries. Adv. Mater., 30, 1802745.

170. Han, W., Qin, X., Wu, J., Li, Q., Liu, M., Xia, Y., Du, H., Li, B., Kang, F. (2018) Electrosprayed porous Fe$_3$O$_4$/carbon microspheres as anode materials for high-performance lithium-ion batteries. Nano Res., 11, 892–904.

171. Park, J., Kang, Y. (2017) Multicomponent (Mo, Ni) metal sulfide and selenide microspheres with empty nanovoids as anode materials for Na-ion batteries. J. Mater. Chem. A 5, 8616–8623.

172. He, Y., Wang, L., Dong, C., Li, C., Ding, X., Qian, Y., Xu, L. (2019) In-situ rooting ZnSe/N-doped hollow carbon architectures as high-rate and long-life anode materials for half/full sodium-ion and potassium-ion batteries. Energy Storage Materials, 23, 35–45.

173. Shia, Z.T., Kang, W., Xua, J., Suna, Y.W., Jiang, M., Ng, T.W., Xue, H.T., Yu, D.Y.W., Zhang, W., Lee, C.S. (2016) Hierarchical nanotubes assembled from MoS$_2$-carbon monolayer sandwiched superstructure nanosheets for high-performance sodium ion batteries. Nano Energy, 22, 27–37.

174. Deng, L., Yang, Z., Tan, L., Zeng, L., Zhu, Y., Guo, L. (2018) Investigation of the prussian blue analog Co$_3$[Co(CN)$_6$]$_2$ as an anode material for nonaqueous potassium-ion batteries. Advanced Materials, 30, 1802510.

175. Wang, W., Jiang, B., Qian, C., Lv, F., Feng, J., Zhou, J., Wang, K., Yang, C., Yang, Y., Guo, S. (2018) Pistachio-shuck-like MoSe$_2$/C core/shell nanostructures for high-performance potassium-ion storage Advanced Materials, 30, 1801812.

176. Ge, P., Hou, H., Banksc, C., Fosterc, C., Lia, S., Zhang, Y., He, J., Zhang, C., Ji, C. (2018) Binding MoSe$_2$ with carbon constrained in carbonous nanosphere towards high-capacity and ultrafast Li/Na-ion storage. Energy Storage Materials, 12, 310–323.

177. Choi, J., Park, S., Kang, Y. (2018) A salt-templated strategy toward hollow iron selenides-graphitic carbon composite microspheres with interconnected multicavities as high-performance anode materials for sodium-ion batteries. Small, 15, 1803043.

178. Yang, T., Liu, Y., Yang, D., Deng, B., Huang, Z., Ling, C., Liu, H., Wang, G., Guo, Z., Zheng, R. (2019) Bimetallic metal-organic frameworks derived Ni-Co-Se@C hierarchical bundle-like nanostructures with high-rate pseudocapacitive lithium ion storage. Energy Storage Materials, 17, 374–384.

179. Yang, X., Wang, S., Denis, Y., Andrey, L. (2019) Direct conversion of metal-organic frameworks into selenium/selenide/carbon composites with high sodium storage capacity. Nano Energy, 58, 392–398.

180. Park, S., Park, G., Kang, Y. (2018) Three-dimensional porous microspheres comprising hollow Fe$_2$O$_3$ nanorods/CNT building blocks with superior electrochemical performance for lithium ion batteries. Nanoscale, 10, 11150–11157.

181. Yang, S., Park, S., Kang, Y. (2019) Mesoporous CoSe$_2$ nanoclusters threaded with nitrogen-doped carbon nanotubes for high-performance sodium-ion battery anodes. Chem. Eng. J., 370, 1008–1018.

182. Quan, L., Liu, T., Yi, M., Chen, Q., Cai, D., Zhan, H. (2018) Construction of hierarchical nickel cobalt selenide complex hollow spheres for pseudocapacitors with enhanced performance. Electrochim. Acta, 281, 109–116.

183. Wang, J., Wang, B., Liu, X., Bai, J., Wang, H., Wang, G. (2019) Prussian blue analogs (PBA) derived porous bimetal (Mn, Fe) selenide with carbon nanotubes as anode materials for sodium and potassium ion batteries. Chemical Engineering Journal, 382, 123050.

184. Hou, Y., Hu, W., Liu, L., Gui, Z., Hu, Y. (2018) In-situ synthesized CNTs/Bi$_2$Se$_3$ nanocomposites by a facile wet chemical method and its application for enhancing fire safety of epoxy resin. Com. Sci. Tech., 157, 185–194.

185. Kong, D.S., Dang, W.H., Cha, J.J., Li, H., Meister, S., Peng, H.L., Liu, Z.F., Cui, Y. (2010) Fewlayer nanoplates of Bi$_2$Se$_3$ and Bi$_2$Te$_3$ with highly tunable chemical potential. Nano Lett., 10, 2245–2250.

186. Peng, H.L., Lai, K.J., Kong, D.S., Meister, S., Chen, Y.L., Qi, X.L., Zhang, S.C., Shen, Z.X., Cui, Y. (2010) Aharonov-Bohm interference in topological insulator nanoribbons. Nat. Mater., 9, 225–229.

187. Dang, W.H., Peng, H.L., Li, H., Wang, P., Liu, Z.F. (2010) Epitaxial heterostructures of ultrathin topological insulator nanoplate and graphene. Nano Lett., 10, 2870–2876.

188. Li, H., Cao, J., Zheng, W.S., Chen, Y.L., Wu, D., Dang, W.H., Wang, K., Peng, H.L., Liu, Z.F. (2012) Controlled synthesis of topological insulator nanoplate arrays on mica. J. Am. Chem. Soc., 134, 6132–6135.

189. Chen, X., Xu, H., Xu, N., Zhao, F., Lin, W., Lin, G., Fu, Y., Huang, Z., Wang, H., Wu, M. (2003) Kinetically controlled synthesis of wurtzite ZnS nanorods through mild thermolysis of a covalent organic-inorganic network. Inorg. Chem., 42, 3100–3106.

190. Wu, C.Y., Yu, S.H., Chen, S.F., Liu, G.N., Liu, B.H. (2006) Large scale synthesis of uniform CuS nanotubes in ethylene glycol by a sacrificial templating method under mild conditions. J. Mater. Chem., 16, 3326–3331.

191. Tang, J., Alivisatos, A.P. (2006) Crystal splitting in the growth of Bi$_2$S$_3$. Nano Lett., 6, 2701–2706.

192. Xie, J., Wang, L., Anderson, J.S. (2020) Heavy chalcogenide-transition metal clusters as coordination polymer nodes. Chem. Sci., 11, 8350–8372.

193. dos Santos, V.C., Durndell, L.J., Isaacs, M.A., Parlett, C.M.A., Wilson, K., Lee, A.F. (2017) A new application for transition metal chalcogenides: WS$_2$ catalysed esterification of carboxylic acids. Catalysis Communications, 91, 16–20.

194. Liu, X., Liu, Z., Lu, J., Wu, X., Chu, W. (2014) Silver sulfide nanoparticles sensitized titanium dioxide nanotube arrays synthesized by in situ sulfurization for photocatalytic hydrogen production. J. Colloid and Inter. Sci., 413, 17–23.

195. Wang, S., Yang, S. (2000) Surfactant-assisted growth of crystalline copper sulphide nanowire arrays. Chem. Phys. Lett., 322, 567–571.

196. Blachnik, R., Müller, A. (2000) The formation of Cu$_2$S from the elements: I. Copper used in form of powders. Thermochim. Acta, 361, 31–52.

197. Jiang, X., Xie, Y., Lu, J., He, W., Zhu, L., Qian, Y. (2000) Preparation and phase transformation of nanocrystalline copper sulfides (Cu$_9$S$_8$, Cu$_7$S$_4$ and CuS) at low temperature. Journal of Materials Chemistry, 10, 2193–2196.

198. Mane, R.S., Lokhande, C.D. (2000) Chemical deposition method for metal chalcogenide thin films. Mater. Chem. Phys., 65, 1–31.

199. Nair, M.T., Nair, P.K. (1989) Chemical bath deposition of Cu$_x$S thin films and their prospective large area applications. Semiconductor Science and Technology, 4, 191.

200. Erokhina, S., Erokhin, V., Nicolini, C. (2003) Microstructure origin of the conductivity differences in aggregated CuS films of different thickness. Langmuir, 19, 766–771.

201. Chung, J., Sohn, H. (2002) Electrochemical behaviors of CuS as a cathode material for lithium secondary batteries. Journal of Power Sources, 108, 226–231.

202. Yao, Z., Zhu, X., Wu, C., Zhang, X., Xie, Y. (2007) Fabrication of micrometer-scaled hierarchical tubular structures of CuS assembled by nanoflake-built microspheres using an in situ formed Cu(I) complex as a self-sacrificed template. Cryst. Growth Des., 7, 1256–1261.

203. Gong, X., Liu, J., Baskaran, S., Voise, R.D., Young, (2000) Surfactant-assisted processing of carbon nanotube/polymer composites. Chem. Mater., 12, 1049–1052.

204. Shaffer, M.S., Windle, A.H. (1999) Fabrication and characterization of carbon nanotube/Poly(vinyl alcohol) composites. Adv. Mater., 11, 937–941.

205. Star, A., Stoddart, J.F., Steuerman, D., Boukai, A., Wong, E.W., Yang, X., Chung, S., Heath, J.R. (2001) Preparation and properties of polymer-wrapped single-walled carbon nanotubes. Angew. Chem. Int. Ed. Engl., 40 (2001) 1721–1725.

206. Olek, M., Kempa, K., Jurga, S., Giersig, M. (2005) Nanomechanical properties of silica-coated multiwall carbon nanotubes poly(methyl methacrylate) composites. Langmuir, 21, 3146–3152.

207. Reddy, K.R., Sin, B.C., Ryu, K.S., Kim, J.C., Chung, H., Lee, Y. (2009) Conducting polymer functionalized multi-walled carbon nanotubes with noble metal nanoparticles: Synthesis, morphological characteristics and electrical properties. Synth. Met., 159, 595–603.

208. Lin, J., He, C., Zhao, Y., Zhang, S. (2009) One-step synthesis of silver nanoparticles/carbon nanotubes/chitosan film and its application in glucose biosensor. Sens. Actuators B, 137, 768–773.

209. Díez-Pascual, A.M., Naffakh, M., Gómez-Fatou, M.A. (2011) Mechanical and electrical properties of novel poly(ether ether ketone)/carbon nanotube/inorganic fullerene-like WS_2 hybrid nanocomposites: Experimental measurements and theoretical predictions. Materials Chemistry and Physics, 130, 126–133.

210. Tenne, R., Margulis, L., Genut, M., Hodes, G.R. (1992) Polyhedral and cylindrical structures of tungsten disulphide. Nature, 360, 444–445.

211. Rapoport, L., Fleischer, N., Tenne, R., (2005) Applications of WS_2 (MoS_2) inorganic nanotubes and fullerene-like nanoparticles for solid lubrication and for structural nanocomposites. J. Mater. Chem., 15, 1782–1788.

212. Naffakh, M., Martín, Z., Fanegas, N., Marco, C., Gómez, M.A., Jiménez, I. (2007) Influence of inorganic fullerene-like WS_2 nanoparticles on the thermal behavior of isotactic polypropylene. J. Polym. Sci. Part B: Polym. Phys., 45, 2309–2321.

213. Naffakh, M., Marco, C., Gómez, M.A., Jiménez, I. (2008) Unique isothermal crystallization behavior of novel polyphenylene sulfide/inorganic fullerene-like WS_2 nanocomposites, J. Phys. Chem. B, 112, 14819–14828.

214. Stehlik, S., Orava, J., Kohoutek, T., Wagner, T., Frumar, M., Zima, V., Hara, T., Matsui, Y., Ueda, K., Pumera, M. (2010) Carbon nanotube—chalcogenide glass composite. J. Solid State Chem., 183, 144–149.

215. Upadhyay, A.N., Tiwari, R.S., Singh, K. (2018) Annealing effect on thermal conductivity and microhardness of carbon nanotube containing $Se_{80}Te_{16}Cu_4$ glassy composites. Mater. Res. Express 5, 025203–9.

216. Rahy, A., Choudhury, A., Kim, C., Ryu, S., Hwang, J., Hong, S.H., Yang, D.J. (2014) A simple/green process for the preparation of composite carbon nanotube fibers/yarns. RSC Adv. 4, 43235–40.

217. Kaul, P.B., Bifano, M.F.P., Prakash, V. (2012) Multifunctional carbon nanotube–epoxy composites for thermal energy management. J. Compos. Mater., 47, 77–95.

218. Dresselhaus, M.S., Jorio, A., Hofmann, M. (2010) Perspectives on carbon nanotubes and graphene raman spectroscopy. Nano Lett., 10, 751–8.

219. Andrews, R., Jacques, D., Qian, D. (2001) Purification and structural annealing of multiwalled carbon nanotubes at graphitization temperatures. Carbon, 39, 1681–7.

220. Kurochka, K.V., Melnikova, N.V. (2017) Investigation of electrical properties of glassy $AgGe_{1+x}As_{1-x}(S+CNT)_3$ (x = 0.4; 0.5; 0.6) at temperature range from 10 to 300 K. Solid State Ionics, 300, 53–59.

221. Banerjee, S., Wong (2012) Synthesis and characterization of carbon nanotube-nanocrystal heterostructures. Nano Lett., 2, 195–200.

222. Brus, L.E. (1991) Quantum crystallites and nonlinear optics. Appl. Phys. A, 53, 463–474.

223. Dabbousi, B.O., Rodriguez-Viejo, J., Mikulec, F.V., Heine, J.R., Mattoussi, H., Ober, R., Jensen, K.F., Bawendi, M.G. (1997) (CdSe)ZnS Core—shell quantum dots: Synthesis and characterization of a size series of highly luminescent nanocrystallites. J. Phys. Chem. B, 101, 9463.

224. Das, A., Wai, C.M. (2014) Ultrasound-assisted synthesis of PbS quantum dots stabilized by 1,2-benzenedimethanethiol and attachment to single-walled carbon nanotubes. Ultrason Sonochem, 21, 892–900.

225. Gendrona, D., Bubaka, G., Ceseracciu, L., Ricciardella, F., Ansaldoa, A., Ricci, D. (2016) Significant strain and force improvements of single-walled carbon nanotube actuator: A metal chalcogenides approach. Sensors and Actuators B, 230, 673–683.

226. Lee, J., Lee, W., Shrestha, N.K., Lee, D.Y., Lim, I., Kang, S.H., Nah, Y.C., Lee, S.H. Yi, W., Han, S.H. (2014) Influence of encapsulated electron active molecules of single walled-carbon nanotubes on superstrate-type Cu(In,Ga)Se$_2$ solar cells. Mater. Chem. Phy., 144, 49–54.

227. Maser, W., Benito, A.M., Muñoz, E., Martínez, M.T. (2007) Carbon nanotubes: From fundamental nano-scale Objects towards functional nanocomposites and applications. *In*: Vaseashta, A., Mihailescu, I.N. (eds), Functionalized Nanoscale Materials, Devices and Systems. Springer, Dordrecht, pp. 101–119.

228. Kurochka, K.V., Melnikova, N.V., Alikin, D.O., Kurennykh, T.E. (2019) Irradiation effect on the structural and electrical properties of the glassy Ag-Ge-As-S composite material containing carbon nanotubes. Solid State Ionics, 341, 115026–115026.

229. Zhang, R., Ren, J., Jain, H., Liu, Y., Xing, Z., Chen, G. (2014) In-situ raman spectroscopy study of photoinduced structural changes in ge-rich chalcogenide films. J. Am. Ceram. Soc., 97, 1421–1424.

230. Yang, S.M., Strelcov, E., Paranthaman, M.P., Tselev, A., Noh, T.W., Kalinin, S. (2015) Humidity effect on nanoscale electrochemistry in solid silver ion conductors and the dual nature of its locality. Nano Lett., 15, 1062–1069.

231. Sagadevan, S., Chandraseelan, E. (2014) Applications of chalcogenide glasses: An overview. Inter. J. ChemTech Research, 6, 4682–4686.

232. Ahluwalia, G. (2017) Applications of Chalcogenides: S, Se, and Te. Springer, Cham. ISBN: 978-3-319-41188-0.

233. Martinez, A., Yamashita, S. (2011) Carbon nanotube-based photonic devices: Applications in nonlinear optics: *In*: Marulanda, J.M. (ed.), Carbon Nanotubes: Applications on Electron Devices. InTech, Croatia, pp. 367–386.

234. Shanmugaratnam, S., Rasalingam, S. (2019) Transition Metal Chalcogenide (TMC) Nanocomposites for Environmental Remediation Application over Extended Solar Irradiation. InTech Book, DOI: 10.5772/intechopen.83628.

235. Juodkazis, S., Misawa, H., Louchev, O.A., Kitamura, K. (2006) Femtosecond laser ablation of chalcogenide glass: Explosive formation of nano-fibres against thermo-capillary growth of micro-spheres. Nanotechnology, 17, 4802.

236. Shportko, K., Kremers, S., Woda, M., Lencer, D., Robertson, J., Wuttig, M. (2008) Resonant bonding in crystalline phase-change materials. Nature Materials, 7, 653–658.

237. Milliron, D.J., Raoux, S., Shelby, R.M., Sweet, J.J. (2007) Solution-phase deposition and nanopatterning of GeSbSe phase-change materials. Nature Materials, 6, 352–356.

238. Harris, P.J.F. (2019) Carbon Nanotube Science: Synthesis, Properties and Applications, Cambridge University Press, ISBN 978-0-521-82895-6.

239. Saito, R., Hofmann, M., Dresselhaus, G., Jorio, A., Dresselhaus, M.S. (2012) Raman spectroscopy of graphene and carbon nanotubes. Adv. Phys., 60, 413–550.

240. Singh, A.K. (2013) SeZnSb alloy and its nano tubes, graphene composites properties. AIP Advances, 3, 042124–11.

241. Singh, A.K. (2013) Crystallization kinetics of Se–Zn–Sb nano composites chalcogenide alloys. J. Alloys Comp., 552, 166–172.

242. Singh, A.K. (2014) Optical properties of the chalcogenide-multi walled carbon nano tubes and chalcogenide-graphene composite materials. J. Nanoeng. Nanomanuf., 4, 1–9.

243. Singh, A.K. (2013) Microscopic study on the Se-Te-Ge alloy and its composite with carbon nanotubes and graphene. J. Adv. Micro. Res., 7, 1–7.

244. Singh, A.K., Kim, J.H., Park, J.T., Sangunni, K.S. (2015) Properties of the chalcogenide–carbon nano tubes and graphene composite materials. J. Alloys Comp., 627, 468–475.

245. Ganaie, M., Zulfequar, M. (2016) Structural, electrical and dielectric properties of CNT doped SeTe glassy alloys. Mater. Chem. Phy., 177, 455–462.

246. Ram, I.S., Singh, K. (2013) Thermal and mechanical properties of CNT-$Se_{90-x}Te_{10}Ag_x$ (x = 0, 5 and 10) glassy composites. Journal of Alloys and Compounds, 576, 358–362.

247. Upadhyay, A.N., Tiwari, R.S., Mehta, N., Singh, K. (2014) Enhancement of electrical, thermal and mechanical properties of carbon nanotube additive $Se_{85}Te_{10}Ag_5$ glassy composites. Mater. Lett., 136, 445–448.

248. Upadhyay, A.N., Tiwari, R.S., Singh, K. (2015) Electrical and dielectric properties of carbon nanotube containing $Se_{85}Te_{10}Ag_5$ glassy composites. Adv. Mater. Lett., 6, 1098–1103.

249. Upadhyay, A.N., Tiwari, R.S., Singh, K. (2016) Optical and electrical properties of carbon nanotube-containing $Se_{85}Te_{10}Ag_5$ glassy composites. Philos. Mag., 96, 576–583.

250. Upadhyay, A.N., Singh, K. (2016) Kinetics of phase transformation of carbon nanotubes containing $Se_{85}Te_{10}Ag_5$ glassy composites. Mater. Res. Express, 3, 125201–8.

251. Jaiswal, P., Dwivedi, D.K. (2019) Investigation of structural, electrical properties and dielectric relaxation of CNT doped Cu–Se–Ge–In chalcogenide glassy alloy. Mater. Res. Express, 6, 015202–17.

252. Banik, I. (2010) Photoconductivity in chalcogenide glasses in non-stationary regime and the barrier-cluster model. Acta Electrotechnica et Informatica, 10, 52–58.

253. I. Banik, (2009) Relationship between optical absorption and photoluminescence in non-crystalline semiconductors. J. Opto. Adv. Mater., 11, 91–103.

254. Elim, H.I., Ji, W., Ma, G.H., Lim, K.Y., Sow, C.H., Huan, C.H.A. (2004) Ultrafast absorptive and refractive nonlinearities in multiwalled carbon nanotube films. Appl. Phys. Lett., 84, 1799.

255. Vinod, E.M., Singh, A.K., Ganesan, R., Sangunni, K.S. (2012) Effect of selenium addition on the GeTe phase change memory alloys. J. Alloys Comp., 537, 127–132.

256. Lu, Y., Song, S., Gong, Y., Song, Z., Rao, F., Wu, L., Liu, B., Yao, D. (2011) Ga-Sb-Se material for low-power phase change memory. Appl. Phys. Lett., 99, 243111–3.

257. Chong, T.C., Shi, L.P., Zhao, R., Tan, P.K., Li, J.M., Lee, H.K., Miao, X.S., Du, A.Y., Tung, C.H. (2006) Phase change random access memory cell with superlattice-like structure. Appl. Phys. Lett., 88, 122114–3.

258. Ram, I.S., Singh, K. (2013) Thermal and mechanical properties of CNT-$Se_{90-x}Te_{10}Ag_x$ ($x = 0, 5$ and 10) glassy composites. J. Alloys. Comp., 576, 358–362.

259. Cardona, M., Cardona, M., Güntherodt, G. (Eds.) (1982) Resonance Phenomena in Light Scattering in Solids II, vol. 50, Springer, Berlin, 19 (Chapter-2).

260. Holubova, J., Cernosek, Z., Cernoskova, E. (2007) Sb_xSe_{100-x} system ($0 \leq x \leq 8$) studied by DSC and Raman spectroscopy. Opto. and Adv. Mater. – Rapid Comm., 1, 663.

261. Baeck, J.H., Kim, T.H., Choi, H.J., Jeong, K.H., Cho, M.H. (2011) Phase transformation through metastable structures in atomically controlled Se/Sb MultiLayers. The Journal of Physical Chemistry C, 115, 13462–13470.

262. Triches, D.M., Souza, S.M., de Lima, J.C., Grandi, T.A., Campos, C.E.M., Polian, A., Itie, J.P., Baudelet, F., Chervin, J.C. (2009) High-pressure phase transformation of nanometric ZnSb prepared by mechanical alloying. J. Appl. Phys., 106, 013509–6.

263. Kolobov, A.V., Fons, P., Tominaga, J., Ankudinov, A.L., Yannopoulos, S.N., Andrikopoulos, K.S. (2004) Crystallization-induced short-range order changes in amorphous GeTe. J. Phys. Condens. Matter, 16, S5103.

264. Andrikopoulos, K.S., Yannopoulos, S.N., Voyiatzis, G.A., Kolobov, A.V., Ribes, M., Tominaga, J. (2006) Raman scattering study of the a-GeTe structure and possible mechanism for the amorphous-to-crystal transition. J. Phys. Cond. Mater., 18 (2006) 965–979.

265. Gourvest, E., Lhostis, S., Kreisel, J., Armand, M., Maitrejean, S., Roule, A., Vallee, C. (2009) Evidence of germanium precipitation in phase-change $Ge_{1-x}Te_x$ thin films by raman scattering. Appl. Phys. Lett., 95, 031908–3.

266. Phillips, J.C., (1981) Topology of covalent non-crystalline solids II: Medium-range order in chalcogenide alloys and A-Si(Ge). J. Non-Cryst Solids, 43, 37–77.

267. Ferrari, A.C., Meyer, J.C., Scardaci, V., Casiraghi, C., Lazzeri, M., Mauri, F., Piscanec, S., Jiang, D., Novoselov, K.S., Roth, S., Geim, A.K. (2006) Raman spectrum of graphene and graphene layers. Phys. Rev. Lett., 97, 187401–4.

268. Banik, I. (2009) On photoluminescence in chalcogenide glasses based on barrier-cluster model. J. Non–Oxide Photonic Glasses, 1, 6–18.

269. O'Leary, S.K., Johnson, S.R., Lim, P.K. (1997) The relationship between the distribution of electronic states and the optical absorption spectrum of an amorphous semiconductor: An empirical analysis. J. Appl. Phys., 82, 3334–7.

270. Tintu, R., Saurav, K., Sulakshna, K., Nampoori, V.P.N., Radhakrishnan, P., Thomas, S. (2010) $Ge_{28}Se_{60}Sb_{12}$/PVA composite films for photonic applications. J. Non-Oxide Glasses, 2, 167–174.

271. Aggarwal, I.D., Sanghera, J.S. (2002) Development and applications of chalcogenide glass optical fibers at NRL. J. Opto. and Adv. Mater., 4, 665–678.

272. Seddon, A.B. (2011) A prospective for new mid-infrared medical endoscopy using chalcogenide glasses. J. Appl. Glass Sci., 2, 177–191.

273. Pamukchieva, V., Todorova, K., Mocioiu, O.C., Zaharescu, M., Szekeres, A., Gartner, M. (2012) IR studies of impurities in chalcogenide glasses and thin filmsof the Ge-Sb-S-Te system. J. Phy. Conf. Series, 356, 012047.

274. Salvetat, J.P., Bonard, J.M., Thomson, N.H., Kulik, A.J., Forro, L., Benoit, W., Zuppiroli, L. (1999) Mechanical properties of carbon nanotubes. Appl. Phys. A, 69, 255–260.

275. Ruoff, R.S., Qian, D., Liu, W.K. (2003) Mechanical properties of carbon nanotubes: Theoretical predictions and experimental measurements. Comptes Rendus Physique, 4, 993–1008.

276. Maiti. S., Shrivastava, N.K., Suin, S., Khatua, B.B. (2013) A strategy for achieving low percolation and high electrical conductivity in melt-blended polycarbonate (PC)/multiwall carbon nanotube (MWCNT) nanocomposites: Electrical and thermo-mechanical properties. Express Polym. Lett., 7, 505–518.

277. Yao, Z., Kane, C.L., Dekker, C. (2000) High-field electrical transport in single-wall carbon nanotubes. Phys. Rev. Lett., 84, 2941–4.

278. Durkop, T., Getty, S.A., Cobas, E., Fuhrer, M.S. (2004) Extraordinary mobility in semiconducting carbon nanotubes. Nano Lett., 4, 35–39.

Chapter 6

Chalcogenide Systems— Graphene Composites

INTRODUCTION

Discovery of the two-dimensional materials is extremely exciting due to their unique properties, resulting from the lowering of dimensionality. Physics of the 2D is quite rich (e.g., high temperature superconductivity, fractional quantum Hall effect etc.) and is different from its other dimensional counterparts. A 2D material acts as the bridge between bulk 3D systems and 0D quantum dots or 1D chain materials. This can well be the building block for materials with other dimensions. Graphene, an atomically thin layer, has broken the jinx of impossibility of the formation of a 2D structure at a finite temperature. The novel discovery of the graphene has been boosted the research communities a lot of interest in this material owing to its unique properties. Consequently, the number of publications on graphene has dramatically increased in the recent past. It has been recognized that graphene possesses very peculiar electrical properties such as anomalous quantum hall effect, and high electron mobility at room temperature. Graphene is also one of the stiffest and strongest materials. In addition, it has exceptional thermal conductivity. All these impassive features make them promising candidate for the various potential applications in distinct areas such as field effect devices, sensors, electrodes, solar cells, energy storage devices and nanocomposites. By the addition of a small volume per cent of graphene into polymer, the overall nanocomposite conductivity can enhance drastically, that could be enough for many electrical applications. Significantly it can also improve the strength, fracture toughness and fatigue strength for such nanocomposite materials. Due these potential abilities, graphene nanocomposites can be considered with a great potential to serve as next generation functional or structural materials. However, relatively limited research has been conducted to understand the intrinsic structureproperty relationship in graphene-based composites such as graphene-polymer nanocomposites. The mechanical property enhancement observed in graphene-polymer nanocomposites is generally attributed to the high specific surface area, excellent mechanical properties of graphene, and its capacity to deflect crack growth in a far more effectively way than one-dimensional (e.g. nanotube) and zero-dimensional (e.g. nanoparticle) fillers. Whereas the graphene sheets or thin platelets dispersed in polymer matrix may create wavy or wrinkled structures that tend to unfold rather than stretch under applied loading. Under such action it is obvious to reduce the composites stiffness due to weak adhesion at the graphene-polymer interfaces. This kind wrinkled surface texture could create mechanical interlocking and load transfer between graphene and polymer matrix, leading to improved mechanical strength. Moreover, the structural defects and stability of graphene can significantly influence the graphene-polymer interfacial behaviour. Therefore, innovative research work is required to understand the structure-property relationship in graphene and the graphene composite materials behaviour.

Since carbon nanotubes (CNTs) are tubular structures composed of curved graphene sheets with diameter up to several tens of nanometers with typical length up to several micrometers. The single and doublewall CNTs have diameters from 1.2 to 3.0 nm and are usually packed in relatively dense structures (ropes). Multiwall carbon nanotubes can contain up to tens of concentrically aligned tubules and have diameter from 3–4 to tens of nanometers. Carbon nanotubes, both single and multiwall, can have outstanding mechanical and electrical properties. Nowadays carbon-based materials are regarded as one of the key subjects for development of various nanotechnology applications – new materials, sensors, actuators, field emitters. In the last decade great effort was done in this field by many research groups, investigating structural, physical, mechanical, and electrical properties of CNTs or graphene. Therefore, every kinds of nanotubes single-wall nanotubes (SWNTs or graphene sheet) have been widely recognized as most perspective in regard of their predicted properties. Depending on chirality and diameter that may have significantly different electronic structure to reveal metallic to semiconducting properties. The innovation of the multiwall carbon nanotubes (MWNTs) in 1953 has also boosted to the area of the composite materials, now days it is to be considered one of the most common and widely used nanotubes allotrope. Usually multiwall nanotubes are composed with several concentrically aligned tubular graphene sheets, with typical diameter in range 8–30 nm. Physical and mechanical properties of MWNTs are significantly lower than that for SWNTs but still are higher than properties of commonly used construction materials and reinforcement additives.

In view of all these facts, the chalcogenide systems-graphene composites making ability motivates us to intend this chapter work by describing building blocks capacity of the chalcogenide systems including chalcogenide clusters tetrahedral building blocks, inorganic- organic frameworks of the metal chalcogenide clusters as well as properties of the open framework chalcogenides. Similarly, graphene composites making ability are described with the help their composites forms such as graphene membrane, graphene energy, graphene sensors, graphene as thermoacoustics, magnets, superconductors and graphene in biomedicine applications. The detailed descriptions on the chalcogenide systems-graphene composites, nanocrystalline chalcogenides-graphene, polycrystalline chalcogenides-graphene and amorphous chalcogenons-graphene composites are also incorporated. In this sequence composite with the chalcogenide glass with the single layer graphene and bilayer layers graphene are separately discussed in two different segments. The possible structural modifications in this novel field is also interpreted. Along with some breakthrough experimental studies examples demonstrations, such as, surface modifications in chalcogenide-graphene glassy regimes. Moreover, modifications in raman spectroscopy, variations in optical properties (such UV/Visible, PL and FTIR interpretations), mechanical, electrical and thermal properties of these composite materials are also interpreted with the help of theoretical sound concepts in view of the experimental findings.

Chalcogenides Building Blocks

The structure of metal chalcogenides frequently consist of clusters in their structural building units. In which several clusters such as tetrahedron-shaped clusters have attracted much attention as their artificial tetrahedral units can form zeolite-like nanocluster superlattices. Therefore, the collective properties of nanocluster superlattices could depend on the individual clusters under the presence of cross linking ligands spatial organization. As an example, for non-covalent linkage, the electron activation energy transfer between nanoparticles depends on the cluster–cluster distance, whereas in the case of covalent linkage the transport properties of the spacer can play a vital role [1]. Therefore, the structural chemistry of open up framework of chalcogenides can be used to describe their properties such as ion exchange, photoluminescence, optical absorption and ionic conductivity as well as potential applications of these kinds of semiconducting materials, specifically in photocatalysis and ion exchange [2, 3]. Some useful building blocks of chalcogenides for composites formations are described below:

Chalcogenide Clusters Tetrahedral Building Blocks

The simplest class of chalcogenide clusters tetrahedral building blocks is known as supertetrahedral clusters and it is expressed as Tn [2–4]. This kind of tetrahedral cluster is in a regular cubic shape, such as ZnS-type lattice, as depicted in Fig. 6.1. Specifically T1 refers to a tetrahedron having the compositional formula MX$_4$, whereas M corresponds to metal and X = S^{2-}, Se^{2-}, Te^{2-}. In the case of isolated Tn clusters compositions such as, T2, T3, T4, T5 can be correlated to M$_4$X$_{10}$, M$_{10}$X$_{20}$, M$_{20}$X$_{35}$, M$_{35}$X$_{56}$, as illustrated in Fig. 6.1. Where M represents metal cation and X corresponds to chalcogenide anion. Hence in general cations of a Tn clusters follow a simple series of rule 4, 10, 20, . . . , [$n(n + 1)(n + 2)$]/6($n \geq 2$), while the number of anions is equal to T(n + 1) clusters, their typical values are summarized in Table 6.1. Here 'n' is represents the number of metal layers in each cluster. Typically in a covalently connected clusters network, the overall stoichiometry of the framework varies depending on the pattern of connectivity. Under similar circumstances, each of the corners of a supertetrahedral cluster can share with another supertetrahedron in an infinite three-dimensional framework. Therefore, the total number of anions per supertetrahedral cluster can be reduced by factor of 2.

Table 6.1 Significate common points of supertetrahedral clusters

Tn	Stoichiometry	Key systems
T2	M$_4$X$_{10}$	Sn$_4$E$_{10}^{4-}$ (E = Se, Te), Ge$_4$X$_{10}^{4-}$ (E = S, Se), In$_4$X$_{10}^{8-}$ (X = S, Se), M$_4$(SPh)$_{10}^{2-}$ (M = Fe, Co, Cd)
T3	M$_{10}$X$_{20}$	M$_{10}$X$_{20}^{10-}$ (M = GA, In; X = S, Se), M$_{10}$X$_4$(SPh)$_{16}^{4-}$ (M = Zn, Cd; X = S, Se), Ga$_{10}$S$_{16}$L$_4^{2-}$ (L = 3, 5-dimethylpyridine)
T4	M$_{20}$X$_{35}$	M$_4$In$_{16}$S$_{35}^{14-}$ (M = Mn, Fe, Co, Zn, Cd), Zn$_4$Ga$_{16}$Se$_{35}^{14-}$
T5	M$_{35}$X$_{56}$	Cu$_5$In$_{30}$S$_{56}^{17-}$, Zn$_{13}$In$_{22}$S$_{56}^{20-}$
$\geq$ T6	M$_p$X$_q^{(a)}$	No system is recognized

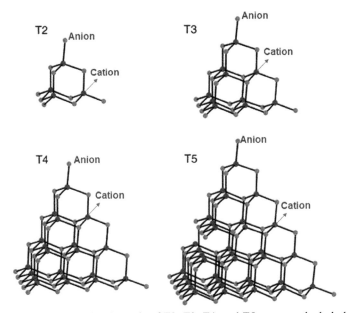

Figure 6.1 Ball and tick schematic of T2, T3, T4, and T5 supertetrahedral clusters. (Reproduced from the permission, Yang, G.Y. (2011) Modern Inorganic Synthetic Chemistry. Elsevier Book, copyright @ Elsevier)

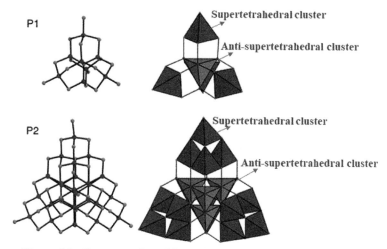

Figure 6.2   Representation of P1 and P2 pentasupertetrahedral clusters.
(Reproduced from the permission, Yang, G.Y. (2011) Modern Inorganic Synthetic Chemistry.
Elsevier Book, copyright @ Elsevier)

In a similar way the tetrahedral clusters can also form pentasupertetrahedral clusters that is usually denoted as Pn, as illustrated in Fig. 6.2 [2]. In such a formation, each Pn cluster can be conceptually constructed through the coupling of four Tn clusters onto each face of an anti-Tn cluster. In which an anti-Tn cluster can be defined as a tetrahedron-shaped cluster possessing the same geometrical feature as Tn cluster. However, they can exchange the positions of cations and anions. According to this concept, a P1 cluster can consist of one anti-T1 tetrahedron (XM$_4$) at the core and four T1 tetrahedra (MX$_4$) at corners, as a consequence composition (MX$_4$)$_4$(XM$_4$) can be formed. The typical systems formed under the pentasupertetrahedral clusters are summarized in Table 6.2.

Table 6.2   Important parameters of the pentasupertetrahedral clusters

Pn	Stoichiometry	Key systems
P1	M$_8$X$_{17}$	M$_4$Sn$_4$S$_{17}{}^{10-}$ (M = Mn, Fe, Co, Zn), M$_4$Sn$_4$Se$_{17}{}^{10-}$ (M = Mn, Zn), ECd$_8$(E'Ph)$_{16}{}^{2-}$, In$_8$S$_{16}$(SH)$^{9-}$
P2	M$_{26}$X$_{44}$	Li$_4$In$_{22}$S$_{44}{}^{18-}$, Cu$_{11}$In$_{15}$Se$_{16}$(SePh)$_{24}$(PPh$_3$)$_4$
≥ P3	M$_p$X$_q{}^{(a)}$	No system is recognized.

Moreover the tetrahedral clusters can also form capped supertetrahedral clusters that can be denoted as Cn, as depicted in Fig. 6.3 [3]. The Cn cluster consists of a regular supertetrahedral cluster (Tn) at the core, in which each face of the Tn core unit is covered by a single sheet of atoms that is called T(n + 1) sheet. While, each corner of this cluster is covered by the MX group. Usually the T(n + 1) sheet is defined as the bottom atomic sheet of a T(n + 1) cluster, however, the number of cations in each T(n + 1) sheet is [(n + 1)(n + 2)]/2. Hence, the number of anions in Cn clusters can be equal to the number of cations in the next member, the values are summarized in Table 6.3.

Table 6.3   Significant properties of the capped supertetrahedral clusters

Cn	Stoichiometry	Key systems
C1	M$_{17}$X$_{32}$	Cd$_{17}$S$_4$(SPh)$_{28}{}^{2-}$, Cd$_{17}$S$_4$(SC$_6$H$_4$Me-4)$_{28}{}^{2-}$
C2	M$_{32}$X$_{54}$	Cd$_{32}$Se$_{14}$(SePh)$_{36}$(PPh$_3$)$_4$, Cd$_{32}$S$_{14}$(SPh)$_{40}{}^{4-}$
C3	M$_{54}$X$_{84}$	Cd$_{54}$X$_{32}$(SPh)$_{48}$(H$_2$O)$_4$ (X = S, Se)$^{4-}$
≥ C6	M$_p$X$_q{}^{(a)}$	No system is recognized.

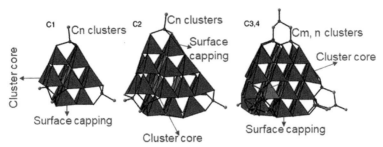

Figure 6.3   Schematic of C1, C2, and C3,4 capped supertetrahedral clusters. (Reproduced from the permission, Yang, G.Y. (2011) Modern Inorganic Synthetic Chemistry. Elsevier Book, copyright @ Elsevier).

The Cn is the regular fragment of cubic zinc-blende-type phase in which its four corners can be barrelanoid cages possessing the characteristics of the hexagonal wurtzitetype phase. The Cn clusters of each M_4X_5 barrelanoid cage at one of the four corners may independently be rotated at 60°C, this can lead to one additional class of tetrahedral clusters, denoted as Cn, m clusters. Where m is referred as the number of corners that can rotate from their original position in parent Cn clusters. Further, the addition or removal of atoms from the regular tetrahedral clusters allows the other variations in clusters such as the coreless T5 clusters, whereas the central metal site of the T5 cluster has not been occupied [5–7]. On the other hand, the T2 and T3 clusters in which an oxygen atom may exist in each adamantane cage [8].

Hence, preparing the crystalline porous chalcogenides by the directed assembly at nanosized clusters into three-dimensional superlattices along with framework topologies is a promising approach.

Inorganic–Organic Frameworks Building from Metal Chalcogenide Nanoclusters

Usually metal-chalcogenide nanoclusters link together to form the extended structures with corner-sharing chalcogens. This could limit the inflexible M–X–M angle (M = metal ions, X = chalcogens) that can place a significant constraint on the number of topological types. To overcome this and enhance the topological diversity of nanocluster superlattices, the organic ligands (or organic materials) are used as the bridge of the assembled semiconducting nanoclusters into crystallographically ordered superlattices, this could be considered as one of the promising approaches. When comparing chalcogen atoms, organic elements possessing a wide variety of shapes with rich coordination chemistry, may allow the creation of diverse nanocluster superlattices [9].

Properties of Open-Framework Chalcogenides

The open-framework solids have major advantages due to their common ion exchange properties. Specifically, metal chalcogenides with open architectures could have unique ion-exchange properties owing to their diversity in pore and channel size as well as specific affinity of the chalcogenide framework for certain cationic species. As the example, an open-framework sulfide material $K_6Sn[Zn_4Sn_4S_{17}]$ may have highly selective ion exchange properties and exchange capacities for the Cs^+ and NH_4^+ ions [10]. Surprisingly, this material has no selectivity for the Li^+ and Na^+ ions due to their large hydration sphere of the ions that prevent them from entering the framework. However, by replacing the large organic cations from small inorganic cations through the ion exchange process, the microporosity can be produced in organically templated chalcogenide frameworks. As an example, organic cations in UCR–20GaGeS–TAEA can be almost completely ion exchanged by Cs^+ at room temperature [11]. As a consequence Cs^+ exchanged material can have type I isotherm characteristic for a microporous system, possessing high Langmuir surface area and total pore volume under the presence of much heavier elements (such as Cs–Ga–Ge–S) [9]. Moreover the open-framework metal chalcogenides can also have photoluminescence with tuneable emission in typical wavelengths ranging 440 to 600 nm with varying framework compositions and structures [12]. By means of general trends, materials with heavier elements can be excited and generate luminescence at a longer wavelength. Hence, a systematic variation in spectral characteristics of open-framework chalcogenide materials may be achieved.

Usually the electronic band gaps of open-framework chalcogenides are smaller than open-framework oxides as well as many of them are in the visible range. Therefore, the optical properties of the metal chalcogenides can be tuned by varying the framework compositions. To verify the optical absorption behaviors of a series of metal selenides and tellurides with ternary anions [M_4X] [$(SnX_4)_4]^{10-}$ (M = Zn, Cd, Mn, Hg, X = Se, Te) has also been reported [13]. According to this study the optical energy band gaps decreased (0.05–0.1 eV) when it went from M = Zn through Cd and Mn to Hg. Further, changes the chalcogen atoms in the cluster from Se to Te, therefore, the band gap is decrease by 0.7–0.9 eV. Moreover in the case of open-framework metal chalcogenides, gallium selenides exhibited the optical transitions with band gaps variation between 1.4 and 1.7 eV. It was demonstrated that band gaps are smaller than those of corresponding gallium sulfides (2.6–2.8 eV). Hence, it led to the framework anions having a more significant impact on the band gap than framework cations. Additionally, the band gaps of metal chalcogenides can also be tuned by changing the extra-framework species, such as the band gap of [$(CH_3CH_2CH_2)_2NH_2]_5In_5Sb_6S_{19}$ is decreased from 2.62 to 2.38 eV after the organic cations is replaced by Cs^+ cation through ion exchange. Moreover, under the open-framework, chalcogenides can also have cluster sizes dependent on optical properties such as at quantum size chalcogenides materials generally have shown the blue shifts in their absorbance.

Open-framework chalcogenides can also have a photoelectronic effect on photoexcitation, such as SnO_2 (F-doped)/CMF-4/Nafion electrode excitation with visible light ($\lambda > 400$ nm) having a sharp photocurrent generation [14]. The photogenerated electrons in the CMF-4 can be collected through SnO_2 generated anodic current. Therefore, a steady photocurrent can be achieved in CMF-4-based photoelectrochemical cells. Their on–off cycles of illumination have confirmed the reproducibility of the transient photocurrent response of the CMF-4 film. The I-V characteristics of the SnO_2(F-doped)/CMF-4/Nafion electrode demonstrated that an increased photocurrent generation results in an increase in anodic potentials.

The electrical conductivity of chalcogenides may be due to the contributions of both electronic and ionic conductions. The open-framework construction tends to lower the electronic conductivity, and this promotes the ionic contribution. The open-framework metal chalcogenides containing organic species as the structure-directing agents, therefore, a relatively small electrical conductivity was recognized and the conductivity generally grew with increasing humidity. As an example, the alternating-current impedance analysis of the single crystal (OCF–6GaSe–TMDP) has shown a humidity dependent behavior of the electrical conductivity [15]. Under increasing humidity, the specific conductivity increases approximately two orders of magnitude (1.32×10^{-7} Ω^{-1} cm^{-1} at 84.9% relative humidity to 5.37×10^{-5} Ω^{-1} cm^{-1} at 100% relative humidity). Since direct synthesis of three-dimensional inorganic chalcogenides contains the mobile alkali and alkaline earth metal cations that leads a new class of fast-ion conductors. Such crystalline inorganic chalcogenides can integrate zeolite-like architecture with high anionic framework polarizability and high concentrations of mobile cations. Thus, the variation in their structural features can fulfill the desirable circumstances to enhance the ionic conductivity.

Graphene Composites

The strategies to use potential graphene in combination with different existing materials is called graphene composite materials. A vast number of graphene composites have been reported and research on the formation of various composites for potential applications is still ongoing. One of the promising applications of the graphene composite is with paints that can result in the formation of an entirely rust free layer [16–29]. This can be used for bricks to construct weatherproof houses. Hence the huge loss of bricks due to corrosion can be avoided [30]. Several graphene-based composites have also been utilized successfully for quality sports goods, especially for skiing, cycling and tennis. Graphene based composites materials also offer the advantage to design and fabricate lighter, stronger and safer planes by using composite aircrafts technology [31].

The synthesis of Pd–graphene composites and their use as catalytic oxidation of alcohols has also been reported [31]. In this order synthesis of graphene compounds with their sulfide containing materials to enhance their flexibility have been reported [32]. Graphene–MoS_2 composites up to 100% flexible as well as stretchable have been synthesized and used as solid-state super capacitors with a volumetric capacitance. This kind of compact graphene-based composites can be used as electrodes for stretchable electronics. Furthermore, such semisolid processing of ball-milled alloy chips with 2% graphene platelets was carried out to prepare MgLiAl base composites strengthened with graphene platelets [33]. This kind of fabricated graphene platelets composites can have higher hardness and yield stress than cast alloys. Moreover, synthesis of different shaped graphene composites were successfully used as electrodes in lithium (Li) ion batteries [34]. The surfactant-free electrodeless co-deposition technique was used to fabricate graphene oxide with copper compounds. Such composites can provide a new technical method to improve the mechanical properties of graphene composites [35]. In this order a few polymer/graphene hydrogel composites have also been used for the implantation in dorsal muscles in both *in vitro* and *in vivo* studies. It was also demonstrated that a Screen-Printed Electrode (SPE) can be modified from composite consisting Mn_3O_4 microcubes [36]. Different authors also demonstrated how thin sheets of graphene oxide can be used for amperometric determination of nitrite. Thus with these studies it is established how graphene composites can have better applications as compared to graphene materials. Useful graphene composites categories are discussed below:

Graphene Membranes

In the modern world, the availability of clean drinking water is one of the major problems in many countries. Graphene-based membranes water filtration can bring great possibilities for water purification [37]. By selecting the perfect barrier between a gas and liquid that is provided from the graphene oxide membranes, (such membranes) can separate organic solvents from water to an exceptionally high level [38]. It is a fact that every year, several thousands of people worldwide die due to diseases caused by dirty water. Graphene has the potential to provide safe drinking water to millions of people, by removal of carbon dioxide from flue gases. This has not been done before and is possible to do from graphene membranes [39].

The layered graphene oxide membranes mainly depend on lateral dimension of constituting sheets for salt rejection efficiency and water permeability [40]. Their pore offset distances are greatly influenced by the velocity and permeation time of the water molecules. Therefore, the water molecules and ions can traverse and permeate through the layered graphene oxide membranes due to their increasingly large pore offset distance and path length [41].

In this order Reddy et al., used a polytetrafuoroethylene membrane surface for the immobilization of graphene oxide for desalination through direct contact membrane distillation [42]. Later several authors showed the overall permeate flux (97 kg/m^2 h at 80°C) significantly enhanced with the graphene oxide-immobilized membrane to complete salt rejection [42–44].

Graphene can have hydrophilic and hydrophobic ability at the same time. Therefore, the graphene membrane can also be used as coatings on foods and pharmaceutical packaging, to prevent the entry of water and oxygen and keep the goods fresh for a longer time. The graphene oxide can immobilized the membrane, therefore, graphene this property can be useful for the direct contact membrane distillation process [38]. An advantage of this kind of immobilized graphene membrane, is good salt rejection with a significant enhancement in permeate flux (like 97 kg/m^2 h at 80°C). It depends on several factors including selective sorption, nanocapillary effect and the presence of polar functional groups [39]. As almost all carbon materials, gas sorption, storage and separation take place through physisorption. Their adsorbate surfaces with high specific surface areas can be very good candidates for adsorption with high polarizability. Additionally, carbon-based materials can also have low capacity of adsorption for H_2 and N_2 as well as medium capacity values for CO, CH_4, CO_2 and relatively high for H_2S, NH_3 and H_2O [37, 45–47].

Usually gas separations were conducted through polymers made of synthetic membranes or other such materials [48]. After the innovation of graphene it was recognized that a single atom thickness layer of graphene can also be used for gas separations. Since the graphene layer is the thinnest and most efficient molecular barrier for separation with high scalability. Considering their two major problems (i) how to incorporate molecular sized pores into the layer of graphene, (ii) the deficiency of the method in manufacturing large areas of mechanically robust and crack-free membranes [49]. Kumar Varoon Agrawal solved both the problems by synthesizing large-area single-layered membrane for the separation of hydrogen from methane. Such graphene membranes may contain nanopores for hydrogen gas sieving at a high surface area (about 1 mm to 2 mm with no cracks) [50, 51].

A graphene membrane is just an atom thick, and can perform the same function of filtration in a better way. The fabricated graphene membranes have also been reported for the both nanoporous graphene as well as graphene oxide membranes for gas separation purposes [52]. In this order researchers also found an efficient inorganic membrane with the magic material graphene [53]. A successful synthesis of graphene membranes with the YSZ hollow fiber ceramic and their utilization for better gas separation have been reported. Additionally, various graphene composite membranes have also been synthesized from intercalation of UiO–66–NH_2 and graphene oxide [54].

Molecular adsorption paid attention on graphene surfaces as well as graphene oxide based functionalized surfaces. In general, the adsorption of oxides and nitrogen are greater on graphene oxide surfaces owing to the presence of hydroxyl and carbonyl functional groups. Whenever these were coordinated with elements like Li and aluminum, the hydroxyl and epoxy groups of oxides acted as strong binding sites for the adsorption of CO_2, NH_3 and SO_2 [55]. To get a better performance, the multi-scale models of graphene nanoporous membranes are usually preferred compared to conventional membranes. At sub-nanometer, nanopores of graphene membranes can allow gas separation to higher levels as compared to centimeter scale graphene porous membranes [56–60].

Graphene Energy

Graphene is usually called a wonder material owing to the mysteries it holds [61]. Potentially graphene can develop lightweight, durable, high capacity energy storage batteries possessing shortened charging time. The graphene based smart phone has the ability to charge with electric power in seconds. Using this lifespan of a Li-ion battery can be enhanced; therefore, they can be charged more quickly with the advantage more holding power for a longer time [62]. Thus, graphene-based batteries can be lightweight and flexible with the stretchable property that can be easily fit into clothes of soldiers. Such light weight batteries can also recharged from body heat, this could allow soldiers to remain in the field for longer periods. Therefore, the graphene based super capacitors could not only enhance the energy of cars, planes but also could reduce the weight. Moreover, graphene potential has also been recognized in storing wind and solar power for grid applications [63].

The incorporation of either graphene or its hybrid into the battery anode in a Li-ion battery can enhance its efficiency. Using the phosphate containing lithium ion, the batteries can be made more lightweight and with a faster charging capacity [64]. Such light weight batteries can store greater amounts of energy and release energy slowly. However, conventionally capacitors can charge and discharge quickly but they are not able to store large amounts of energy such as batteries [65]. Hence, graphene-based batteries can remove the conventional difference between batteries and capacitors. Further, the graphene-enhanced Li-ion battery can work at extremely high temperatures with a long operation time [66]. According to this demonstration the miracle object graphene can bring about a revolutionary development in Li-ion batteries, although such batteries are as yet not generally available on a commercial scale. Graphene based Li-ion batteries have encouraged researchers to try to use them commercially. Such as a class of Li-ion batteries based on graphene ink anode and Li-iron phosphate cathode; optimal battery performance of 190 Wh kg^{-1} has also been reported by researchers at Kansas State University who developed a technique of using graphene electrode with the maximum capacity in a combination of silicon and graphene [67]. Moreover, a class of Li-ion rechargeable batteries based on graphene oxide has also been developed [68, 69]. Additionally, researchers have also paid great attention to develop graphene based super capacitors, considering the advantages of graphene on high surface areas that can store more electrostatic charges, therefore, very light weight and low production cost supercapacitors possessing high efficiency of charge storage can be developed [70–74].

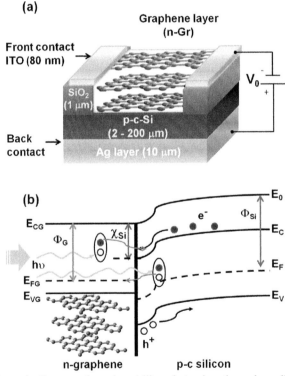

Figure 6.4 (a) Schematic diagram of graphene/silicon heterojunction solar cell. (b) Sprayed organic photovoltaic cells and mini-modules based on chemical vapor deposited graphene [67].

Graphene has also been recognized as useful for photovoltaic application. It has been reported that graphene/silicon heterojunction solar cells are working well under simulated conditions. According to this demonstration it was considered that the three-dimensional nature of graphene can form a Schottky junction, in which electrical contacts can be made along the axis to collect minority carriers, that generates on illumination [75]. The *n*-type multilayer graphene best-simulated cell power conversion efficiency at room temperature has achieved 7.62%. This efficiency of *n*-type graphene cell remained almost the same up to 40 layers. Although optimize *p*-crystalline silicon wafer can lead to efficiency up to 11.23%. However, 11.38% efficiency was achieved at 270 K as its efficiency is greatly influenced by temperature. Therefore, transparent conductance of *n*-type multilayer graphene can act as an excellent electrode [76, 77]. The doped graphene can enhance the solar panel absorption of a photon from a few electrons [78]. Graphene based solar cells absorbance spectrum of electromagnetic radiations can expand due to their low coefficient of light absorption (2.3%). Also very high tensile strength of graphene can also be useful to fabricate silicon cells on flexible as well as organic substrates. Thus, the overall performance of photovoltaic devices can be improved by the incorporation graphene into the cell matrix.

More specifically, in photovoltaic cells, graphene electrode can play diverse but positive roles in the form of an active layer or an interfacial layer [79], schematic of graphene/silicon heterojunction is given in Fig. 6.4(a, b). In order to advance in this field, different investigators have contributed their work with improved parameters as well as overall cells efficiencies and innovations [80–84].

Graphene Sensors

Graphene can also act as an efficient sensing material by exposing each atom in the environment to provide information from the surroundings [85]. To detect an individual case at a molecular level, a micrometer-size graphene sensor has been fabricated. With the help of graphene oxide sensors smart food packaging, it is possible to prevent the food wastage. These sensors are also useful to detect food decay due to changes in the environment [86]. Graphene based vital sensors can also boost the effectiveness of crops by monitoring the presence of any harmful gases. Therefore, using the graphene sensor in the field of agriculture can also discover the best surfaces for growth of specific crops [87]. They can also detect chemical warfare agents and explosives as well as are potentially advantageous for defense purposes in order to save lives [88].

Single and double layer graphene sensor attached nanoparticles (NPs) can show vibrational properties. Therefore, using graphene sensor defined boundary conditions they can be potentially be used for atomic-scale mass sensing. Hence, the nanomass can be detected easily using a graphene resonator; it can provide a highly sensitive nanomechanical element in sensor systems. With increasing temperature the vibrational frequency shifts of the graphene sensors are also increased [89]. A double layer graphene sensor can have higher sensitivity than a single layer due to high order frequency shifts. A variety of flexible and stretchable graphene sensors have been developed from different investigators to sense gases, nanomass and pressure [90–94].

Graphene based analytical device can also efficiently detect biomolecule-related elements through an appropriate transducer that generates a measurable signal; the device is called a biosensor [95]. Usually a biosensor consists of a bioreceptor interfaced with a transducer on a typical platform. To recognize a biomolecule element, the bioreceptor should be capable of recognizing it, such as, enzymes, antibodies, DNA, RNA and cells. Such biological signals can be detected by various quantities in terms of their physical, chemical, optical, thermal or electrochemical actions through the receptor that produces observable information. Clark and Lyons developed the first generation bio-sensing devices to monitor chemical components in the blood of a surgical patient [96]. They recorded the biomolecules quantitatively in blood. In the modern world, the utility of biosensors in the field of biomedical and global healthcare has become indispensable for improving human lives [97]. Graphene based bio-sensors have also been successfully used for drug delivery and food safety.

Suvarnaphaet et al., demonstrated the mechanisms of bio sensing based on their charge separations of biomolecules/nanoparticle interactions [78]. They also reviewed different synthesis and bio-sensing properties of graphene-based materials, including the pristine graphene and functionalized graphene (i.e., GO, rGO) and graphene quantum dot.

Graphene for Thermoacoustics

Graphene materials electrical properties are unique for showing the ability to transfer heat better than any other material [98]. To be more precise, some magnetic nanomaterials and carbon nanotubes have been used as audio speakers. Therefore, these materials can also produce sound without any acoustic box [99]. Hence, graphene can act as thermo-acoustic transducer from the thinnest speaker in the world (a single atom thick) due to their ability to heat up and cool down quickly [100]. Usually a small square of graphene film fixed to a piece of paper and made into a printed circuit board can serve as a thermos phone [101]. During the heating up process, it expands under an applied electrical current through silver ink. Generally a two steps process is used for the yield of graphene based thermoacoustic speaker under the relatively simple methods [102]. These vibrations can not only come from the source but also be due to thermoacoustics under rapid heating and cooling [103]. The solution of graphene oxide flakes have also been used for initial freeze-drying. As a consequence, their electrical properties have been improved for oxidize graphene. However, in the case of reduced graphene oxide aerogel can be produced that may reduce the speakers doping by embedding in walls as well as other surfaces, therefore, overall the device becomes fat and does not vibrate. It was also demonstrated that graphene-based earphone-sized, transparent and lightweight speakers can be made from graphene.

However, the demonstrated wide band frequency range is not up to the desired range. None the less, graphene earphones packaged into casings that can serve well in a wide range from 100 Hz to 50 kHz without any fluctuations, these can be used not only for humans but also to train some animals [104]. A nickel-template assisted by low voltage driven thermos acoustic speaker has also been reported. The low leakage substrate with feasible tenability and their applications in terms of acoustic ultra-sonics devices have also been recognized, therefore, at present graphene is widely used in acoustics applications [105].

Moreover, a simple template-free fabrication method can be used to produce thermoacoustic loudspeakers with N-doped three-dimensional reduced graphene oxide (N-Rgoa). This kind of speaker has an input power as high as 40 W [106]. Further, graphene-based acoustic materials in which sounds can be mixed, amplified and modulated into a single device have also been demonstrated [94]. The construction of the rich sonic pellet that possesses good control over composition as well as flow of electronic current can also be achieved. Additionally the low-voltage driven thermoacoustic speaker on three-dimensional graphene foam can also be fabricated and its microstructure may be related to the corresponding thermoacoustic performances [107].

Graphene as Magnets

Current era digital hard-disk storage and information technologies are mainly based on magnetic materials [108]. Usually for efficient encoding information, just a few layers atoms thick magnets can be used. These kinds of magnets can be used for both the cloud computing technologies as well as consumer electronics data storage. More recently, the two-dimensional magnetic insulators have also been introduced, this class of magnets can have unprecedented control over the flow of electron spin–electrons = tiny, which means that these are subatomic magnets [109]. In advancement of magnetic materials, various investigators have made efforts to create the thinnest system containing four sheets, in which each sheet is only up to a few atoms thick. Therefore, such approaches have pushed the information storage of the magnetic technologies by using the atomically thin films limit [110, 111].

Generally, at the room temperature, magnetism of materials arise from the metallic elements those possess d orbitals. These kinds of materials has been used for centuries and applied in a vast number of technologies. To improve the magnetic performance great efforts have been made with carbon magnets at room temperature that contains exclusively sp orbitals [112]. This is done by adopting a simple and controllable route from the substitution of fluorine atoms into fluoro graphene with hydroxyl groups at room temperature. It depends on the chemical composition (an F/OH ratio) and sp^3 coverage, such new graphene derivatives can have room temperature anti ferromagnetic ordering that usually does not appear in sp-based materials [110–113]. Therefore, an extraordinarily high magnetic moment is possible with these two-dimensional magnets, which usually undergo a transition to a ferromagnetic state at low temperatures due to OH functionalization [114].

Graphene as Superconductors

In fact for decades researchers have endeavored to investigate the behavior of materials for unconventional superconductors [115]. However, there is still a lack of relevant experimental techniques. More recently, a purely carbon-based two-dimensional material, twisted bilayer graphene, has been reported with tuneable property that can be used as a superconductor [116]. This opens a route for ideal material investigations with a strongly correlated phenomenon. This could lead to insights in the physics of high-critical-temperature superconductors and quantum spin liquids [117].

Such as Single-Layered Graphene (SLG) electrons can be doped on to a superconductor with a Bardeen-Cooper Schrieffer (BCS) or a non-BCS pairing symmetry. This process of doped electrons can be condensed to a superconducting state. While in pure SLG at half-filling U is B9.3 eV and $V^{\sim}$ is B5.5 eV, therefore the chiral p-wave state occurs, however with increasing U or $V^{\sim}$ the doping level can be diminished [118]. Additionally, in bilayer graphene, the possibility of spin-triplets-wave pairing may occur. Usually the intrinsic superconductivity is not observed without doping in SLG. Since SLG intercalating sheets are constructed with Ca or Li by placing SLG on a superconductor with relative atoms. Therefore, superconductivity can be enhanced theoretically with the increasing pairing potential of p- or a chiral d-wave, as the superconducting density of states in SLG. As a consequence a full transition to a superconducting state can be manifested. Their proximity effect can also enable the fabrication of devices by achieving p-wave or chiral d-wave superconductivity. In addition to this achievement of p-wave superconductivity in SLG above 4.2 K is attractive for various applications [118].

But it is a fact that hardness and transverse stiffness of atomically thin graphene is inferior to diamond, though graphene has fascinating mechanical properties. But there is a problem that there is nearly no practical or only a few demonstrations of the transformation of multilayer graphene into a diamond-like ultra hard structure [119]. The transverse stiffness and hardness compared to a diamond can be assembled from two-layer graphene on SiC (0001) at room temperature after nano-indentation tests. This has shown a reversible drop in electrical conductivity on indentation and its perforation with a diamond indenter. This could form a diamond-like film with a two-layer graphene film as theoretically suggested that their density function can produce both elastic deformations due to sp^2 to sp^3 chemical changes [120]. In the case of a buffer layer graphene film it can be thicker than three to five layers without reversible phase changes. Therefore, the conformation of the diamond-like film calculations controls over the two-layer graphene layer-stacking, as in a multilayer film hinders the phase transformation [121].

To explain the intrinsic unconventional superconductivity, a two-dimensional super lattice can be created by stacking two sheets of graphene twisted to each other at a small angle. Therefore, near zero Fermi energy, their electronic band structure of the 'twisted bilayer graphene' may possess fat bands; this could be correlated with their insulating properties. In experimental evidences, a tuneable zero-resistance state of the graphene-doped material has been observed. In which the dome-shaped regions in the temperature–carrier–density phase diagram of the twisted bilayer graphene is correlated a to superconductivity property [121]. Thus a vast number of graphene based superconductivity has been reported with the different compositions and intensive research in this area is still ongoing by making various graphene composites materials [122–128].

Graphene in Biomedicine

Graphene-based materials such as graphene sulfide, graphene flakes, graphene oxide can be used for biomedical applications. In these materials, the size of graphene molecules can be easily tuned in between millimeters to nanometers ranges for drug delivery, ultrasensitive biosensors, tissue engineering, healthcare [129].

Graphene has been used for many biomedical applications, however earlier it was used only for drug delivery, the advances in this area have been established that it can also be useful to make biosensors for tissue engineering as well as a potential candidate for the antibacterial agent [130]. The graphene-based nano/or composite materials can also be used for the fabrication of *in vitro/* or *in vivo* therapeutics cells or tissues. According to this approach, the graphene molecules can undergo complex interactions with solutes, proteins or cellular systems within the body, therefore, their interactions impact significantly on the behavior or toxicity of the molecule. The modifications in graphene or its combination with other molecules to emphasize favorable characteristics can overcome the challenges. Such directions of graphene could be useful as part of highly tailored multifunctional delivery vehicles [131].

Therefore to assist, treat, repair or replace any function in the tissue, organ or body, biomedical materials or biomaterials can be used. The carbon-based biomaterials are also known as biocompatible and they have become very common in the past decade. More frequently, the polymer containing carbon, poly-L-lactide has been used for biomedical utility. Such polymeric materials have also been used as biodegradable coronary stents and bone plates. Similar kinds of polymers, such as poly caprolactone can be used for the potential applications. As an example, contact lenses can be made from poly(methylmethacrylate) [132]. These polymers may be in the form of carbon graphene, diamond or carbon nanotubes that have been used preferably as housing for nanoparticulate as well as drug delivery.

The graphene oxides can also be used as drug delivery due to their suitable drug delivery properties by the enriched oxygen-containing groups and large planar surface areas [133]. The graphene oxides carboxylic and hydroxylic groups can also provide suitable sites for attachment of biomolecules including photo thermal therapy in tumors [134–136]. Moreover, the potential utility of specific graphene-based nanomaterials has also been recognized in various biomedical applications, such as bio-sensing, drug delivery, tissue engineering and cancer therapy [137]. It is worth noting that a few graphene-based materials can act as nanoplatforms for biomedical applications that may be useful for regenerative medicine, specifically in stem cell research. Thus, these have been intensively targeted for bio-sensing applications of graphene oxides due to their biocompatibility, in the field of biotechnology and biomedical engineering. In particular, functionalization of different oxygen-containing groups that are present in the structure can be tailored. In this case, graphene oxides can be considered as one of the most suitable materials for a wide range of applications [138–140]. To be able to detect the biological molecules of certain substances is known as a biosensor, which respond with the production of measurable signals. The graphene oxides membranes are suitable to produce biosensors since they possess high mechanical strength and thermal conductivity. Such graphene-based biosensors also involve enzymatic electrochemical sensors, which is working on immobilization of the enzymes [141]. These materials based sensors are also be useful to detect antigen–antibody complexes for disease diagnosis [142]. Additionally, the graphene-based materials high infinity can also be helpful for mammalian cells; therefore, they have been used as a scaffold for tissue engineering. The graphene-based films can also accelerate stem cell differentiation [143].

Thus graphene materials can be used in the field of biomedical applications, such as drug delivery, bio-sensing, tissue engineering and, the area of nanomedicine. Investigators have presented their innovations including synthetic strategies, functionalization and processability protocols with *in vitro* and *in vivo* applications [144, 145]. Moreover graphene materials have also been successfully used in medical electronics, tissue engineering, medical implants, medical devices, sensors, cancer therapy and biological imaging and other biomedical applications [146].

CHALCOGENIDE SYSTEMS—GRAPHENE COMPOSITES

In recent developments of the chemistry of materials, the discovery of graphene has created a great sensation due to its fascinating properties [147–149]. Graphene dimensional network of sp^2 carbon atoms has high electron mobility and ballistic conduction owing to its unusual electronic structure. Besides single-layer graphene, two-, three- and other-layered graphene were also investigated. This discovery opens the path to the investigations for other two-dimensional materials. Such as the inorganic analogues of zero-dimensional fullerenes and one-dimensional carbon nanotubes have been prepared in the past and more recently innovated graphene-like layered inorganic structures [150–152]. These materials also have interesting properties and may offer a wide scope for future study. Particularly, the layered chalcogenides like MoS_2, WS_2, $MoSe_2$, WSe_2, GaSe, $NbSe_2$, GaS, InSe, $Bi_2Sr_2CaCu_2O_x$, $CuGeO_3$, TiS_2, TiS_2,ZrS_2, HfS_2, VS_2, NbS_2, TaS_2, $TiSe_2$, $ZrSe_3$, $HfSe_3$, VSe_2, $NbSe_2$, $TaSe_2$, etc. can be used for isoelectronic with graphene [153]. Thus the layered chalcogenide and graphene basic properties are allowed to make their composites that could be impressive for potential uses, such as RGO/ MoS_2–PVP/Al thin films applicability for the flexible nonvolatile rewritable memory [154]. Thereby extensive research has been carried out on the composites of graphene with inorganic graphene analogues and fabricated a verity of their composite materials [155–158]. The most extensively investigated graphene–WS_2 composites can be made through heating the dispersed graphene oxide and few-layer WS_2 as reflux in water with the help of the sodium borohydride as a reducing agent. Such compositions can have a high specific capacity under good stability compared to the pristine WS_2. Therefore, overall the reduced graphene oxide may play an important role in enhancing the rate of WS_2-reduced graphene oxide composites [159]. Hence, different kinds of chalcogenide-graphene fabrication can be possibly used by a different route of synthesis; here some important class of chalcogenide-garphene composites will be discussed.

NANOCRYSTALLINE CHALCOGENIDES—GRAPHENE COMPOSITES

Day by day, the increasing demands of the enhancement in technologies such as energy production from renewable energy sources, fabrication of devices to store the energy in electric form (supercapacitors or ultracapacitors), batteries, etc., have boosted research in this area, considering the forefront devices for the various applications. These can deliver high output compared to conventional devices. The nanocrystalline chalcogenide–graphene composites could be one of the prospective areas to fulfill the future requirements. This can overcome the usual chalcogenide semiconducting materials drawback redox reactions among valence states of the metallic ions [160]. In order to further the enhance performance of the devices, suitable composites of various compounds with conducting materials such as graphene, carbon fiber, and carbon nanotubes have been used [161–164]. Owing to the chemical conversion by graphene can effectively prepare and form composite material for different applications with improved performances. Such formed composites can build suitable architects in a combination of nanoparticles and graphene that not only gives good electrical conductivity but also can provide robustness, stability with a high surface area [163–165]. A few researchers have reported supercapacitors based on multi-walled carbon nanotube (CNT)/Ag nanoparticle ink, metal oxide/graphene, carbon fibers coated with a metal oxide and polymer, and self-healable reduced graphene oxide (rGO) fibers [166–169]. Extensive research has been done on advanced crystalline (or nanocrystalline) chalcogenide– graphene composites [170]. Paying more focus towards the composite materials with reduced graphene with the kind cahclogenide alloys compositions [171], Ma et al., reported that the nanocrystalline MoS_2 and $MoSe_2$ supported the three-dimensional graphene foam for enhanced lithium storage [171]. Gong et al., discussed the enhancement of electrochemical and thermal transport properties for the graphene/MoS$_2$ heterostructures in terms of energy storage [172]. Wang et al., studied the relaxing volume stress and promoting active sites in vertically grown two-dimensional layered mesoporous MoS$_2$(1–x)Se$_2$x/rGO composites to enhance the capability and stability for lithium ion batteries [173]. The graphene composites with the chalcogenide halide have also been reported. According to this study HfX$_3$(X = Se and S)/graphene composites can be useful to fabricate flexible photodetectors in the range from visible to near-infrared [174].

Another most extensively explored nanocrystalline chalcogenide and reduced graphene is CdS composites that were synthesized by adopting various routes [175]. These materials have been significantly used as a buffer layer material in solar cells; it is believed that with the combination of reduced graphene composite form, an efficient electron transporter can be achieved [175]. It is also a good potential candidate of ternary composite for application in DSSCs solar cells. Similarly, the cobalt sulfide/reduced graphene oxide composite as an anode for sodium-ion batteries with superior rate capability and long cycling stability has also been reported [176]. In addition, the WS$_2$/reduced graphene oxide composite for and superior Li-ion storage and MoS$_2$/graphene counter electrodes for efficient dye-sensitized solar cells have also been demonstrated [177].

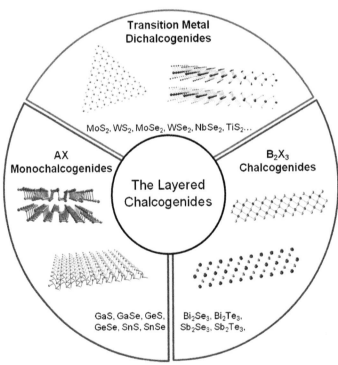

Figure 6.5   The layered chalcogenide families: the transition metal dichalcogenides, the AX mono-chalcogenides, and the B$_2$X$_3$ chalcogenides.
(Reproduced from the permission, Jeffrey, et al. (2016) Emerging opportunities in the two-dimensional chalcogenide systems and architecture. Current Opinion in Solid State and Materials Science, 20, 374, copyright @ Elsevier Ltd.)

Moreover composites with the ternary or multi-components nanocrystalline chalcogenide alloys with graphene composites have also been examined. For example, the quaternary chalcogenide Cu$_2$NiSnS$_4$ (QC) nanoparticles nanosheets can be fabricated with the reduced graphene oxide as an electrode material [178]. According to this report such composites materials can provide a narrow band gap semiconductor that is scientifically important for photovoltaics applications. Additionally, in contrast to single-phase metal sulfides, ternary or quaternary compounds can have a better performance due to their synergistic effect of two or more metallic cations. With this potential composition, several authors have also demonstrated their research outcomes by making structures such as three-dimensional flowerlike Cu_2NiSnS_4 as an anode material in Na-ion batteries and *in situ* hydrothermal synthesis of Cu_2NiSnS_4/rGO for supercapacitors application [178, 179]. In addition to this, a comprehensive review on nanocrystalline chalcogenides with reduced graphene for the removal of antibiotics from aqueous environments has been presented [180].

Considering the advances in nanocrystalline chalcogenide–graphene composite materials emerging opportunities with the two-dimensional chalcogenide systems along with their architecture have been extensively described by Cain et al. [181]. They discussed in detail different kind of architectures layered chalcogenide systems as well as their possible composites with graphene/ reduced graphene/ nanomaterials, and schematically different layered structures are represented in Fig. 6.5. Hence the field of nanocrystalline chalcogenide-graphene composites is the emergent growing area for future technologies.

POLYCRYSTALLINE CHALCOGENIDES—GRAPHENE COMPOSITES

In recent decades, metal chalcogenide semiconductors have been extensively explored as promising materials considering their potential application in various fields such as electronic and photovoltaic devices [182, 183]. More recently, the antimony trisulfide (Sb_2S_3) has been explored mainly because of its high optical absorption coefficient (1.8×10^5 cm^{-1}), and optimum optical energy band gap around 1.7 eV [184] suitable for harvesting light in the visible region, and its earth abundant and relatively environment-friendly components. Its diverse properties make it a potential candidate to utilize in various fields such as lithium and sodium ion batteries [185, 186], visible light photocatalytic [187, 188] and photovoltaic. It can also act as an absorber layer in solid state thin film and dye-sensitized solar cells [189–191]. Therefore, these kinds of materials morphologies is the crucial parameter that impacts on their physical and physiochemical properties in optoelectronic utility [192, 193]. Considering this aspect, investigators have made one-dimensional structure semiconductor materials on reduced graphene oxide (or composites) that can have two major advantages: i) tuning their diameter that increases absorption of light via enhancing aspect ratio, ii) their embedding on reduced graphene oxide (rGO), that can decrease the electron-hole recombination rate, as a consequence functionality of the desired materials [194–197]. Therefore, Sb_2S_3–rGO nanocomposite as prepared and annealed samples polycrystalline structure has been reported to have extended photocurrent performance [198]. Though this is an interesting topic, only few reports are available, therefore, serious efforts are needed to explore different kinds of chalcogenide–graphene composition composites polycrystalline structures for their better scientific and technological uses.

AMORPHOUS CHALCOGENIDES—GRAPHENE COMPOSITES

There are several inorganic solids having a tendency to intercalation or conversion reactions to build an amorphous structure, they are useful for various applications. For example, one of the most applicable areas is battery research, such as Li-ion and sodium ions anode materials are useful for battery fabrication. In recent years, to boost the area of battery research investigators have preferred to make their composites, and recognized the layered metal dichalcogenide (like SnS_2) as a potential candidate to make the amorphous composite structure. Due to the properties of layered metals chalcogenide materials they can accommodate Na^+ ions in its interlayer space and can also experience conversion reactions with Na ions [199, 200]. Considering incorporation of a layered chalcogenide can further improve the Na-ion batteries electrode performance. To explore this basic concept many attempts including nanostructure formation, carbon coating and composite formation have been made [12, 13]. Extensive research on this topic has led to the formation of two-dimensional sheet-like morphology of metal dichalcogenide composites, which may provide a highly conductive graphene two-dimensional nanosheet. Therefore it can be considered an effective approach to improve the sodium-ion electrode functionality of metal dichalcogenide through the enhancement of electrical conductivity. In this order, there are several reports on coupling of metal dichalcogenide with reduced graphene oxide [201]. It has been demonstrated that the strong self-stacking tendency of reduced graphene oxide can be prevented from closely mixing with metal dichalcogenide nanosheets. Therefore, the exfoliated metal oxide nanosheets can form homogenous colloidal mixture with reduced graphene oxide. Due to the similar nature of the metal dichalcogenides and the reduced graphene oxide surface, the crystal morphology allows that they can easily form the restacked matrix [202–204]. Hence, the intervention of a metal oxide nanosheet can optimize the pore structure of metal dichalcogenide–reduced graphene oxide nanocomposites and prevent severe self-stacking of reduced graphene oxide nanosheets [202–204]. This leads to the increase in porosity with surface expansion and possesses more reaction sites for the anchoring of Na^+ ions under overall improvement of Na^+ ion transport; therefore, it increases the Na-ion storage capacity. This can also enhance the electronic coupling of metal chalcogenide with metallic reduced graphene oxide that may also be beneficial to enhance the electrode functionality of such nanocomposite materials.

In this order, crystal structures, composite structures and porosity of the Na–SnS_2–reduced graphene oxide–titanate nanocomposites have been reported [205]. According to this study, probing of the titanate can impact additionally on physicochemical properties of the metal dichalcogenide–graphene nanocomposite. Moreover, such composite materials can be used as anode for Na-ion batteries and their electrode activity has also been verified by varying the titanate concentrations [205]. They examined the crystal structures of the Na–SnS_2–reduced graphene oxide–titanate nanocomposites by changing the titanate concentration as well as Na–SnS_2 nanocomposites and interpreted their results with the help of the X-Ray Diffraction (XRD) patterns. They found an overall crystallographic structure in amorphous nature for these materials. Additionally, the restacked Na–SnS_2 nanocomposite has a nearly identical position of the (001) peak to that of bulk SnS_2. The existed crystallographic peak (001) position they have directly related to there is no intercalation of Na^+ ions into the SnS_2 layer. While the lack of (001) reflection in the restacked Na–graphene oxide nanocomposite has related to a well-ordered intercalation structure of the Na ion into the graphene oxide layers. However, the occurrence of SnS_2, as well as graphene oxide and layered titanate phases, that is strongly correlated to homogeneous dispersion of alloying elements in nanosheets under the hybrid matrix of graphene oxide and the titanate nanosheets was not found [205].

Hence amorphous crystallographic structure of chalcogenide–graphene composite materials has established that this structure can enhance the technological performances of the devices. Though this field is still under extensive research and is not limited only to, for example, that several chalcogenide–graphene composites are disordered and can be fabricated using different synthetic method or approaches depending on their specific utility.

Chalcogenide Glass—Single Layer Graphene Composites

Considering stiffness, hardness and nearly perfect crystallographic structure of graphene it was believed that graphene could not possibly deform in low dimension chalcogenide glasses. Very recently, Lin et al., used a high performance photonic device by successive thin films deposition of $Ge_{23}Sb_7S_{70}$ on glass substrate and single layer graphene on silicon using the CVD technique during the integration process [206]. They found that the chalcogenide glass- graphene composite device overall carrier mobility of graphene remains unchanged. The capability to fabricate a ultra-broadband on-chip waveguide polarizer was also demonstrated. This is a new class multilayer waveguide platform comprising a graphene monolayer situated at the center of a symmetrically cladded strip waveguide. In which a graphene film is sandwiched between two $Ge_{23}Sb_7S_{70}$ layers of equal thickness. According to them, the fabricated waveguide can act as a polarizer, therefore, the large optical anisotropy of graphene and the polarization dependent symmetric properties of waveguide modes can be achieved.

Moreover the fabricated devices can also work as an energy-efficient photonic crystal thermo-optic switch. The waveguide property without incurring excessive optical loss was also considered. Therefore this counterintuitive concept can open up the application of graphene as a broadband transparent conductor. The embedded graphene electrode as resistive heaters to realize a thermo-optic switch with unprecedented energy efficiency was applied. Unlike traditional metal heaters the waveguide integrated graphene heater offers superior energy efficiency due to a much smaller thermal mass and large spatial overlap of the optical mode with the heating zone. They also interpreted the device utility as Mid-IR waveguide-integrated photodetector as well as broadband mid-IR waveguide modulator.

Hence with the glassy system chalcogenide-single layer composite can be fabricated through the integration approach. In which the low-temperature chalcogenide alloys can be deposited to process devices directly on two-dimensional materials without disrupting their extraordinary optoelectronic properties. Thus, this area is now open for this kind of chalcogenide–single graphene composite device fabrication using sophisticated techniques. Therefore, more research is needed with greater detail to build future prospective devices.

Chalcogenides Glass—Bilayer Layer Graphene Composites

Besides theoretical predictions, there were no deformations of bilayer or a few layers (more than two layers) graphene in low-dimensional chalcogenide glassy alloys from the direct conventional method (melt quenched); there are a few reports on the successful synthesis of chalcogenide–graphene composite materials. Singh reported the first breakthrough in his landmark research with the chalcogenide–graphene synthesis under a glassy system by adopting the conventional melt quenched technique [207].

He successfully demonstrated the formation of Se_{96}–Zn_2–Sb_2 (SZS) – 0.05% bilayer graphene (GF) composite material, by adopting the novel concept that stronger π bonds can be deformed slightly at a lower temperature than the well-defined melting temperature under the continuous heating environment, and due to excess surrounding heat, in which the '*thermonic energy tunneling effect*' process can be possible. However, graphene's strongest σ bonds may also be influenced in the continuous heating environment. Moreover, to explore the possibilities of the formation of chalcogenide composite with graphene with different elements alloys under a glassy regime, he also reported the successful synthesis of $Se_{55}Te_{25}Ge_{20}$ (GTS) with 0.025% bilayer graphene (GF) composite material [208]. To synthesize these composites materials, three step mechanisms have been adopted, as their nucleation and growth mechanisms schematic is represented in Fig. 6.6(a, b). The first step can be related to individual elements as their pure form, while the second step can be related to the formation of combined SZS and GTS alloys at quite a low temperature ($\sim 700°C$ and $950°C$). In the final third step composites can be formed at a higher temperature ($1000°C$), in which the inclusion of the SZS and GTS alloys constituents into strong organic bilayer GF configuration being visualized through breaking the weak bonding. The outcomes of these initial reports established that experimentally deformation of stiff and hard characters of graphene π and σ (partially) bonds are possible in low dimensional chalcogenide alloys under the glassy system. Hence, these landmark studies opened a new unexplored research field of composite materials associated with glassy behavior.

This is a newly established composite materials research area; therefore only very limited data is available on their different physical properties. Different properties of the chalcogenide–graphene composites materials based on their initial innovative demonstrations will be discussed.

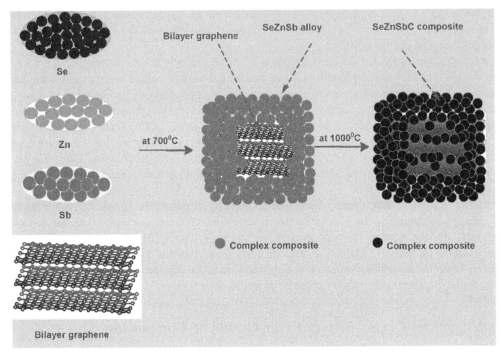

Figure 6.6(a) Illustration of the mechanism for ZSZ–GF composites, Step I: Se, Zn, Sb and bilayer graphene (GF) individual elements in the alloying proportion, Step II: Se–Zn–Sb alloy formation in which unaffected GF present, Step III: inclusion of the Se–Zn–Sb alloy constituent into the graphitic sheets.

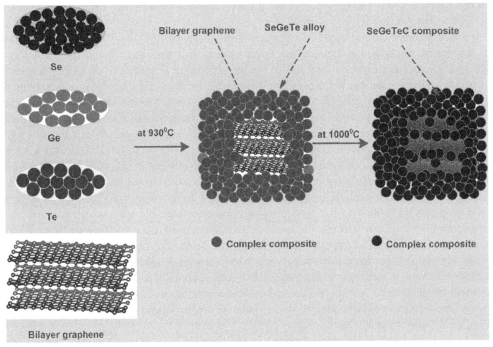

Figure 6.6(b) Illustration of the mechanism for GTS–GF composites, Step I: represents the ingredients of the alloy in proportional to their compositional amounts, Step II: represents the complex alloy formation at 930 °C, Step III: represents the Se–Ge–Te inclusion in bilayer GF through breaking the p and r (partial) symmetry.

STRUCTURAL MODIFICATIONS

The structural modifications in such materials can be interpreted with the following descriptions; since the amorphous Se structure is the mixture of two structural species, polymeric chains as well as Se_6 and Se_8 eight member ring molecules. Their bonding structural units are usually covalent, in which their inter-structural forces are the weak van der Waals type. However, the condensed form of Se possesses twofold nearest-neighbor coordination with distances d from 2.32 to 2.37 unit with the bond angles slightly above $100°$, $4s^2 4p^4$ valence electronic structure with a helical chain structure. Any variation in dihedral angle can play an important role in their twofold coordinated structural chemistry due to their bond distance and bond-angle potential functions are steeper than the dihedral angle potential. This leads to the dihedral angle being distorted more easily than other molecular parameter. Note that the coordination defects can also contribute substantially in structural modifications of Se [209, 210]. Thereby, incorporation of foreign metallic or semi-metallic elements Zn, Ge and metalloid Sb or metallic chalcogen Te affected the non-metallic Se host dihedral angle bonds with an induced structural inhomogeneity. It is expected that during thermal agitation the high energy metallic or semi metallic bonds break the low energy Se chains and rings in the presence of second high energy metalloid or metallic chalcogen Sb and Te bonds, whereas, Sb has characteristics that do not contribute to transmit the Se chain length, but under an electronegative affinity interaction it can increase the concentrations of Se–Sb electronegative bonds. On other hand, Te has the characteristic to contribute in transmission of Se chains [211]. Therefore, under cross linking whole SZS (SeZnSb) and GTS (GeTeSb) compositions alloys formed the complex configurations with a large number of cross linking bond densities. Hence, all the SZS and GTS matrices have been transformed into anion and cations containing complex alloys. As a consequence, SZS and GTS complex systems can have a large number of unsaturated hydrogen bonds accompanied by the van der Waals like bonds.

Moreover incorporation of small amounts of GF (bilayer graphene) in SZS and GTS alloys can also affect the chemical equilibrium and local density of the states by making the additional homopolar and heteropolar bonds with the inorganic and organic constituents of the systems. The incorporation GF in SZS and GTS configurations can increase the electronegativity of the materials. This could increase the number of van der Waals like bonds within complex materials [212, 213]. Subsequently, the low dimension inorganic SZS and GTS alloys constituents can interact with two-dimensional periodically structured organic bilayer GF armchair and zigzag bonds. Therefore, during the melting process the bilayer GF can be functionalized via inclusion of thermally excited SZS and GTS alloys constituents through diffusion. Under high temperature and the long diffusion process the SZS and GTS constituents could break the symmetry of the σ and π bonds of the GF. Their sudden cooling of molten composite materials could form a frozen solid solution. In which it is expected that weak zigzag bonds (correspond to π bond) of bilayer GF would be highly affected and diffused in the SZS and GTS localized states, while the high energy armchair bonds (correspond to σ bond) may lose their original periodicity.

The crystallographic structural modifications of the constituents of SZS and GTS alloys with their GF composites can be interpreted with XRD patterns, as depicted in Fig. 6.7(a, b). No appearance of any sharp peak can be directly related to the overall amorphous structures of SZS–GF, GTS–GF composites materials. This also leads the original crystallographic characteristics of GF to transform into randomized behavior in low dimension chalcogenide alloys. Hence, with these XRD patterns interpretations the randomized structures of the SZS–GF and GTS–GF composites were confirmed.

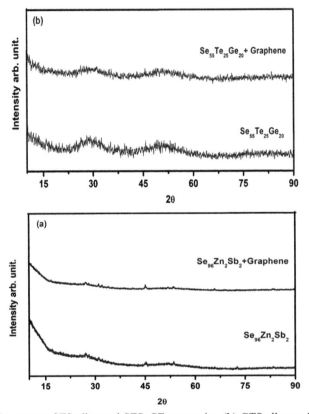

Figure 6.7   **(a)** XRD patterns; SZS alloy and SZS–GF composite, **(b)** GTS alloy and GTS–GF composite.

MODIFICATIONS IN SURFACE MORPHOLOGIES

SZS–GF Glassy Composite

Since EXD pattern of SZS–GF composite material revealed the randomized structure, the question remains about the stiffer configuration. Singh interpreted the surface morphological properties of the newly investigated SZS–GF composite with the help of Field Effect Scanning Microscopy (FESEM) [207]. The surface morphologies of SZS alloy and SZS–GF composite are illustrated in Fig. 6.8(a, b).

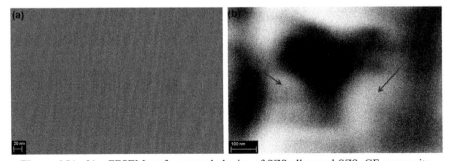

Figure 6.8(a, b)   FESEM surface morphologies of SZS alloy and SZS–GF composite.

This high resolution surface morphological image reveals a visualization of stronger GF structure diffusion within the SZS glassy system. Although GF structure was diffused in the

SZS glassy system but their back bone can be clearly identified within the configuration, this is according to the adopted concept on the partial diffusion of stronger σ bonds and the suppression of graphene layers due to complete diffusion of the π bonds. However, this observation could not report the distribution of SZS alloying elements throughout the configuration as well as their inclusion inside graphene diffused structure, therefore, Singh also performed the EDS mapping measurement for both SZS alloy and SZS–GF composite configuration, as depicted in Fig. 6.9 (a, b, c) and (d, e, f, g) [214]. This demonstration has clearly revealed the existence of the alloying elements Se, Sb, Zn within the configuration, the corresponding elemental mapping are nearly homogeneous distribution as per their compositional concentration in SZS alloy. While the alloying elemental distributions of SZS–GF were also significantly distributed in an appropriate homogenous manner throughout the complex glassy configuration. This outcome established that besides the diffuse backbone existence of GF within the SZS glassy system, the overall distributions of alloying was homogenous everywhere. However, in Chapter 5 in the case of multi-walled carbon nanotubes distributions of the alloying elements are slightly inferior within the tubes. Here the question on the presence of the percentage of elemental amounts remains unanswered; therefore, he also interpreted the EDS spectrums of both the SZS alloy and SZS–GF composite composition, as illustrated in Fig. 6.10(a, b). The EDS pattern of both SZS alloy and SZS–GF composite reported the appropriate amounts of the presence of the elements within the configuration by showing the individual elemental concentrations. Hence, the surface morphological, EDS mapping and EDS spectrum analysis revealed the diffusion of GF within the SZS glassy system and homogenous distribution of alloying elements throughout the complex configuration as well as their presence in appropriate amounts.

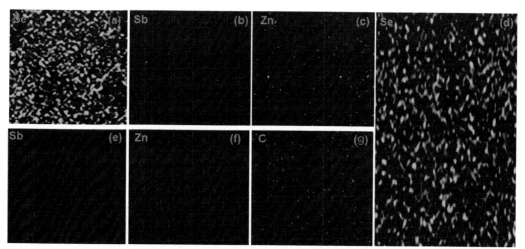

Figure 6.9 **(a, b, c)** EDS mapping: SZS alloy, **(d, e, f, g)** SZS–GF composite.

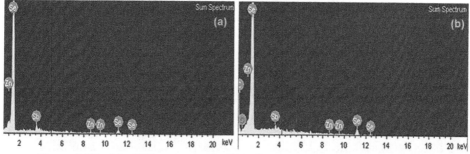

Figure 6.10(a, b) EDS spectrums for SZS alloy and SZS–GF composite material.

GTS-GF Glassy Composite

Theoretically, it has been pointed out that formation of chalcogenide–graphene glassy composite can also depend on characteristics of the alloying constituents. Considering this key point Singh also explored the GTS alloy and GTS–GF composite surface morphological images with the help of FESEM interpretation, as depicted in Fig. 6.11(a, b) [208].

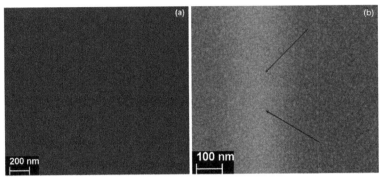

Figure 6.11(a, b)   FESEM surface morphologies of GTS alloy and GTS–GF composite.

According to this interpretation a high order diffusion of GTS alloys elements was observed, while a higher order of graphene deformation was noticed for the GTS–GF composite. This is due to the high order reactivity of alloying semi metallic Ge and metal chalcogenide Te elements with the metallic behavior graphene within the complex configuration. Therefore, almost a homogeneous surface morphology has been observed for the GTS–GF composite in which diffused and suppressed existence of graphene can be visualized. In this interpretation, the inclusion of alloying elements within the configuration as well as the graphene structure were demonstrated with the help of EDS mapping results. The GTS alloy elemental mapping outcome revealed a highly homogenous display of the elements within the glassy system, as depicted in Fig. 6.12(a, b, c). Similarly, GTS–GF composite elemental distribution throughout the system configuration was also demonstrated in a homogenous manner, as illustrated in Fig. 6.12(a, b, c, d), this is direct evidence on high order reactivity of alloying elements with graphene. In contrast to this, SZS–GF composite alloying elemental distributions, in particular elements Zn and Sb seem inferior throughout the complex glassy configuration. This could directly correlate to the reactivity of alloying elements with the graphene depending on their behavior in chalcogenide glassy composite systems. Additionally,

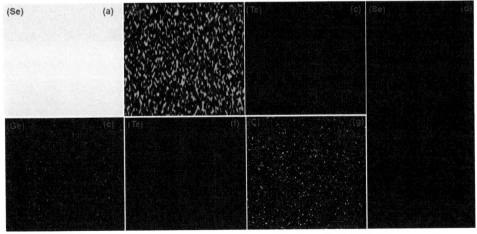

Figure 6.12   **(a, b, c)** EDS elemental mapping; GTS alloy, **(d-g)** GTS–GF composite.

to verify the existence of alloying elements in both pattern and composite glassy systems in an appropriate manner, he also demonstrated the EDS patterns, as illustrated in Fig. 6.13(a, b). The EDS patterns of the both GTS alloy and GTS–GF composite materials have been noticed according to their appropriate proportion of the compositional ratio.

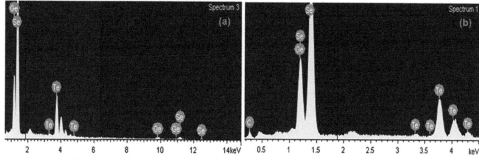

Figure 6.13(a, b) EDS Patterns of GTS alloy and GTS–GF composite.

MODIFICATIONS IN RAMAN SPECTROSCOPIC

The Raman spectroscopic tool can be useful and valuable for the materials analysis as well as their structural interpretation. Generally, Raman spectroscopic interpretation has to be considered as one of the most acceptable characterizations to determine the structure of amorphous materials. Due to its sensitive ability in signals detection even if it comes from only small amounts of crystallites. Thus, the Raman characterization tool can be significant for local ordering structures with a high precision. Raman spectral peaks of amorphous materials are very similar to crystalline material the only difference is in their peak broadening to relative peak position [215]. In order to define the SZS alloy and SZS–GF composite structural properties, Singh demonstrated the Raman spectral profile for both parent and composite material, as illustrated in Fig. 6.14(a, b) [207]. The description of SZS alloy Raman peak appearance has been given in detail in Chapter 5, therefore, here only Raman peak of SZS–GF composite will be discussed. Raman mixed Se–Se, Se–Sb and Zn–Sb peak of SZS–GF composite material has been noticed at 246, 140 and 107 cm^{-1} with the low phonon G and D mixed Raman peak at 187 cm^{-1}. The doubly degenerate highly diffused (very low intensity) D and G band peaks were recognized at 1382 and 1574 cm^{-1}, whereas the two-dimensional peak almost disappeared. This dispersion of sp^2 graphene phonon modes A, E_1 and E_2 symmetry might be due to the splitting of the doubly degenerate two-dimensional peak into a non-degenerate mode [216, 217].

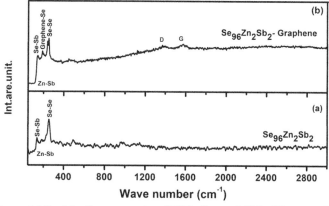

Figure 6.14(a, b) Raman spectra of SZS alloy and SZS–GF composite.

Singh et al., also interpreted the Raman spectroscopic properties GTS alloy and GTS–GF composite, as depicted in Fig. 6.15(a, b) [218]. The details of the GTS alloy Raman peaks have already been discussed in Chapter 5. Therefore, here only GTS–GF composite Raman peak are interpreted. Similar to GTS–carbon nanotubes composite, GTS–GF Raman peaks also have a weak single broad peak at 191 cm^{-1} due to the combination of homogenous GeTeSeC phase of the complex material. In which well-defined stronger graphene Raman peaks D, G and two-dimensional at 1353, 1582 and 2708 cm^{-1} disappeared. The dispersion of these sp^2 graphene phonon modes A, E_1 and E_2 symmetry could be correlated to their high order diffusion in GTS glassy configuration [216, 217].

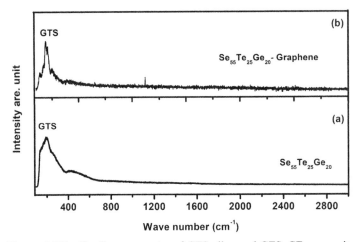

Figure 6.15(a, b)   Raman spectra of GTS alloy and GTS–GF composite.

Dispersion in G, D and two-dimensional Raman band peaks in SZS–GF and GTS–GF composite materials could be related to the change in their chirality, zigzag and armchair bonding owing to the formation of new defect states in the host Se chains and rings in the presence of other alloying elements [219]. In particular, in the bilayer graphene the Coulombic interaction depends on the metallicity/or geometry of the materials, therefore, it is quite possible that the constituents of the composites can induce more electron-hole interactions to change the bond orientations. As a consequence, composites begin to transform toward lower dimensionality under a non-periodic atomic arrangement. The appearance of considerable dispersed G, D and two-dimensional Raman bands in SZS–GF and GTS–GF indicate the composite tends toward one-dimensional structural transformation due to the suppression of the graphene layers, therefore, composite structural dispersion could be correlated to one-dimensional sp^3 hybridized (or graphene ribbon like structure) structural transformation. In such a dispersion, the stacked layers drastically suppressed and induced a strong effect to make a mixed amorphous structure of the composite [219, 220, 221].

OPTICAL PROPERTIES

UV/Visible Absorption Property

As described earlier in Chapter 5, the UV/Visible properties in amorphous semiconducting materials, the optical absorption band near its border would be smeared out and manifest itself as a tail that extends into the deep forbidden band [222]. Their corresponding profile can be interpreted in term of Urbach's formula [223], whereas the negative optical absorption correlation energy at the edge can be expressed in term of the charged defects [222–226]. Like the simulated outcomes on the arsenic chalcogenides can possess a similar configuration such as nanotubes [222–226]. Considering these structural similarities, the chalcogenide SZS alloy and their SZS–GF composite can be interpreted using this well-defined concept of amorphous semiconducting materials. Singh demonstrated the UV/Visible absorption spectrums of SZS alloy and SZS–GF composite, as depicted in Fig. 6.16(a, b), in which a large absorption tail with adjacent peaks in the range 400 to 850 nm, whereas a sharp absorption peak for the SZS–GF composite. The amount of absorption percentage is recorded higher for SZS–GF composite compared to SZS alloy. Using the well-known Tauc relationship, the optical energy band gaps for the SZS–GF composite material as well as SZS alloy have been defined at 1.37 eV and 1.39 eV, respectively [207]. The evaluated optical energy band gaps values for these materials belong to the well described range (≥ 3 eV) for chalcogenide glasses.

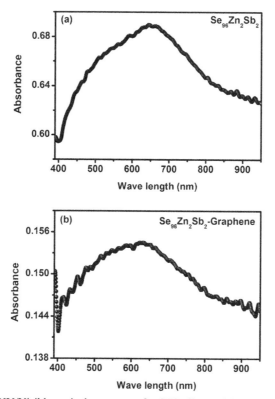

Figure 6.16(a, b) UV/Visible optical spectrums for SZS alloy and SZS–GF composite material.

Further he studied the optical properties of the GTS alloy and GTS–GF composite materials by addressing the UV/Visible absorption spectrums in the wavelength range 380–1050 nm, as illustrated in Fig. 6.17(a, b) [218]. The GTS alloys have shown a broad UV/absorption peak with a relatively sharp decline band tail edge, while the GTS–GF composite material has a relatively sharp band tail edge. Using the well-known Tauc relationship, optical energy band has been defined; the obtained optical energy band gaps of the GTS alloy and GTS–GF composite are 1.26 eV and 1.28 eV, respectively [218].

The UV/Visible absorption property in these materials could be interpreted by hybridization of the orbitals in chalcogenide materials depending on the defects creation ability of the foreign elements [227]. In which an additional impurity state π-electrons of the composite may play an important role to alter the optical properties. Therefore, a π-plasmonic resonance can occur and spans within the entire visible spectral region with the tail extending to near infrared wavelength range [221]. The π-plasmon is a collective excitation depending on surface plasmons, therefore, the optical transitions can occur between π and $\pi*$ energy bands at the same cutting line for the initial and final states. This leads to a strong absorption in between the top (π) and bottom ($\pi*$)

bands that can vanish at the wave vector k [219]. Therefore, constituents of SZS alloy are expected to create a number of defect states between the side graphene surfaces, whereas, optical absorption vanishes at the edge of the wave vector k. Therefore, these composites have exhibited better UV/Visible absorption property but its band tail could not expand longer due to the existence of strong defects armchair (corresponds to σ bond) bonds. Since GF has honeycomb structure and its optical wave vector limit is not much sharper (like carbon nanotubes). Therefore, a quadratic thermal expansion and greater suppressed ability allows a large change in chirality, as depicted in Fig. 6.18, therefore, zigzag and armchair bonds spread actively.

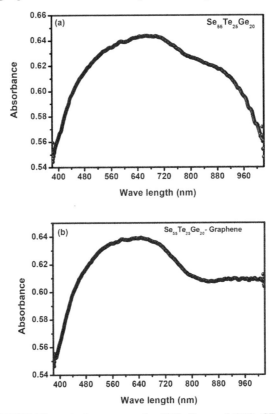

Figure 6.17(a, b) UV/Visible optical spectrums for GTS alloy and GTS–GF composite material.

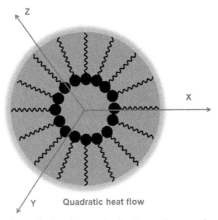

Figure 6.18 Schematic for the quadratic thermal expansion of graphene.

As a consequence, optical absorption band tail and optical energy band gaps are greater than the parent alloys. Moreover, to understand the optical property alternation in these composite materials, a schematically sub optical energy band suppression model for the chalcogenide-graphene composite glassy materials is presented here, as illustrated in Fig. 6.19.

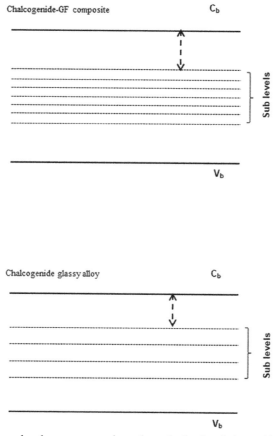

Figure 6.19 Sub-energy band gaps suppression schematic for the chalcogenide-graphene composite glassy materials.

According to this concept it is believed that newly formed composite materials bonding angle would lie between two governing constituents (Se and GF), however, there is no theoretical report on their exact bond angle modification value. Additionally, the bond angle modifications in the composite could permit the formation of additional sub energy levels within the forbidden gap, which may possibly be closer as compared to the parent chalcogenide glassy configuration. As a consequence, overall optical behavior of the chalcogenide–graphene glassy composites could be modified. This interpretation of the glassy chalcogenide–graphene composite is based on their initial results and needs further investigations on these materials.

Photoluminescence (PL) Interpretation

Generally amorphous semiconductors do not show PL property at room temperature [227], however, it could occur at low temperatures [227]. Usually PL property in such materials arise due to the optical electron transition through the tunneling effect that is acquired at an adjacent or a near region at an energy level that their excited state level just below the potential barriers [227]. Therefore, the overall wave function can be localized in a specific micro region between the adjacent potential barriers. As a consequence, under the coulombic interaction a bound state electron-hole pair can be created and forms a new energy level in the forbidden band. Such newly formed energy levels are called localized states. The potential strength of such localized states depends on specific alloying elements, and their discrete energy levels and localized states can be connected through a channel. Under suitable circumstances an optically excited electron gradually gets back to the original region of the hole. This process usually occurs in gradual tunneling and diffusion; the corresponding jumps are usually connected to an interaction with a phonon. Hence, under the non-radiant process an electron loses the energy in a localized region, subsequently a radiant optical recombination of electron-hole pair occurs, that is connected to emission of a luminescence photon [228–230]. Singh examined the PL property for theses composites and surprisingly found that PL single for the SZS–GF composite at room temperature, as depicted in Fig. 6.20. [207], however, theoretically it cannot be achieved in a chalcogenide rich glassy system at room temperature. The demonstrated PL single was in consistence with the well described behavior in chalcogenide glasses; phonon energy decline after achieving the maximum excitation. Here a schematic model to describe the PL property in such materials is presented, as depicted in Fig. 6.21; in the first step it was assumed that on initial state under the Columbic interactions atoms /or molecules formed a bound state, in the second step under a suitable light they return back to their original position, in the third and final step, while returning to their original position they transferred their energy (that is called non-radiative energy) to other atoms or molecules within the configuration, as excitation of energy acquiring atoms/molecules PL signal comes from the material.

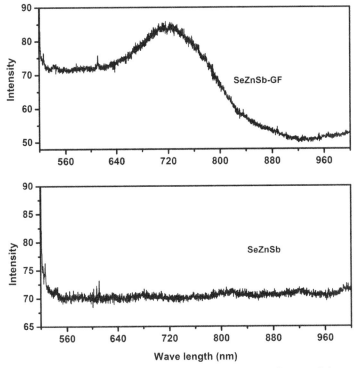

Figure 6.20   PL single for SZS–GF composite material.

FTIR Spectrum Interpretation

The usual IR properties of chalcogenide glasses were discussed in Chapter 5. As per the concept of the IR optics utility of chalcogenide glasses, it largely depends on the restrictions of the vibrational bands absorption that governs through the incorporation of the foreign impurities. Therefore, structural modifications of the chalcogenide glasses also influence IR optical property of the material. Change in IR property in these materials depends on the selection of chemical composition which is a crucial parameter to achieve higher order IR transmitting materials. Considering the key properties of these materials, Singh also demonstrated the IR properties materials in his initial report [207]. The initial recorded FTIR transmission spectrums of SZS and SZS–GF composite is shown in Fig. 6.22.

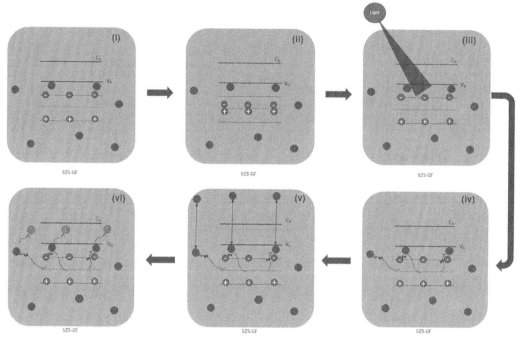

Figure 6.21   Schematic model of the PL mechanism.

This outcome has shown a drastic enhancement in IR transparency percentage in this range for the SZS–GF compared to SZS parent alloy. The recorded spectrum transparency initially is around 37% and finally reaches up to 81%. The SZS–GF composite materials IR transparency enhancement nearly doubles at lower frequency and is three times at higher frequency compared to SZS parent chalcogenide glassy alloy [207]. To confirm this huge IR transparency in different chalcogenide–graphene glassy composites, Singh also demonstrated the IR properties of GTS alloy and GTS–GF composite materials, as depicted in Fig. 6.23. A similar trend in enhancement in IR transparency was achieved for the GTS–GF composite material compared to their parent GTS glassy alloy. Such a huge enhancement in IR transparency could correlate to the change in their intrinsic molecular structure of the material caused the induced effects on stretching and bending bond vibrations. As a consequence, the internal energy can be increased due to the occurrence of vibrational energy resonance [223, 231]. Considering this approach, a three steps schematic model in Chapter 5 for IR properties of the chalcogenide–carbon nanotubes composite materials was discussed. The described schematic model approach can also extend for the chalcogenide–GF glassy composite, as depicted in Fig. 5.28. Here, it is believed that the stacked structure of graphene allowed higher order diffusion of chalcogenide parent alloys between the diffused layers compared to carbon nanotubes. Therefore, as a consequence, a large number of defect states may be created, this results in a huge IR transparency increase in graphene containing chalcogenide alloys composites.

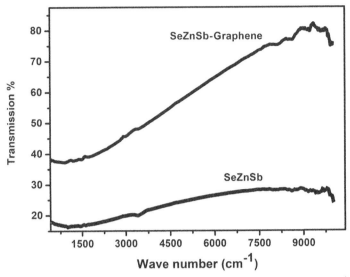

Figure 6.22 FTIR transmission spectrums of SZS and SZS–GF composite.

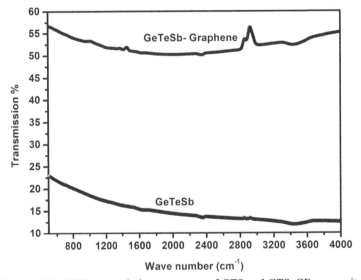

Figure 6.23 FTIR transmission spectrums of GTS and GTS–GF composite.

MECHANICAL PROPERTIES

There is no report on newly introduced chalcogenide–graphene glassy composites mechanical properties. However, some exceptional reports are available on the reinforcement of glass fiber-graphene composite materials. Ian et al., extensively reviewed composites manufactured and supported with carbon nanotubes and graphene. They also demonstrated the composites were reinforced with single-walled nanotubes, multi-walled nanotubes as well as with graphene [232]. Li et al., studied the effect of orientation of graphene based nanoplatelets on Young's modulus of composites on varying the orientation of fibers and differences in Young's modulus [233]. Meenakshi et al., demonstrated prepared flax and glass fiber reinforced polyester hybrid composite lamination using the hand layup method and tested the tensile, flexural and impact strength of laminates [234]. They concluded that the hybrid composite lamination can also possess equal mechanical strength compared to glass fiber reinforced polymers. Infantamarypriya et al., explored graphene platelet nanopowder that dispersed in epoxy resin to fabricate bi-axial Glass Fiber Reinforced Polymers (GFRP). Their result showed a large enhancement in the strength of the composite material [235]. Arun et al., investigated the mechanical properties of graphene oxide reinforced with GFRP. They studied different systems with varying concentration of graphene and found a superior strength, hardness and flexural strength for composite materials [236]. Aiyappa et al., used graphene as a nanofiller and demonstrated a considerable improvement in mechanical properties with the increased percentage of graphene [237]. Juan et al., fabricated the glass fiber with graphene oxide and modified the glass fiber reinforced plastics for high level performance [238]. Liu et al., fabricated the graphene–epoxy resin sandwiched composite and investigated their varying properties [239]. Bindusharmila et al., investigated the mechanical properties of epoxy hybrid composites with a considerable increase in tensile, impact strength and fracture toughness of composites [240]. Li et al., used graphene platelet with ultrasonicated epoxy resin to fabricate carbon fiber reinforced plastics, and found mechanical properties were enhanced when weight percentage of the graphene platelet was increased [241]. Umer et al., studied the effect of infused graphene particles with epoxy resin glass fiber composites [242]. Jyothilakshmi et al., demonstrated the mechanical properties of graphene reinforced GFRP laminates and explored the composites properties up to 5% weight [243]. Kumar et al., explored the mechanical properties of graphene-based hybrid composites reinforced with kenaf/glass fiber, they studied the different types of hybrid composites laminates with varying concentration of graphene reinforcing material with epoxy resin and found improved mechanical properties for the hybrid composites [244]. Rathinasabapathi et al., studied graphene enhanced glass fiber reinforced polymers [245]. They demonstrated that incorporation of graphene nanofillers in epoxy composites increases the tensile modulus and mechanical properties. They also commented that graphene/ epoxy composites reinforced with glass fibers can provide higher mechanical performance compared to conventional glass fiber reinforced epoxy composites.

While there was no direct report on chalcogenide–graphene glassy composites materials, the evidence described above indicates that the mechanical properties of the composites could be enhanced, therefore, it is expected that mechanical properties of such composite materials could be higher than the parent chalcogenide glassy, but a comprehensive study on mechanical properties are still needed, as these materials widely depend on their alloying continents and their concentrations.

ELECTRICAL PROPERTIES

While several studies on different kinds of chalcogenide–graphene composite materials (such as crystalline, polycrystalline and amorphous) have reported excellent electrical properties, there is no direct account on chalcogenide–graphene glassy composite. The similar crystalline–graphne, polycrystalline–graphene composites as well as chalcogenide–multi-walled carbon nanotubes composites have exhibited an impressive enhancement in their electrical properties, as published in a large number of studies. Looking closer at chalcogenide–carbon nanotubes glassy composites enhancement in electrical properties (as discussed in Chapter 5), it is expected that electrical properties could also be modified for chalcogenide–garphene composites. As compared to carbon nanotubes, graphene has a more ordered honeycomb structure and their partial structural deformation/diffusion in a chalcogenide rich glassy system may provide a more ordered channel for electrons in a complex configuration. Hence, deform and diffuse graphene in glassy composites can be considered analogous to a one dimensional flat graphene structure. According to well-defined configurations of electrical properties, an enhanced electrical conductivity can be achieved. Therefore, it is believed that electrical properties of the chalcogenide–graphene glassy composite would enhance the corresponding parent chalcogenide glassy alloy. Note that in this novel research area, investigators are actively engaged (by publishing their results) to exploring the different physical aspects, it is hoped that soon some technical reports will be able to resolve the issue in electrical properties of such composite materials. Hence, to build a concessive view on such composite materials, rigorous investigations are required owing to the fact that even in chalcogenide glasses electrical properties vary from one system to another depending on their alloying constituents. In the case of chalcogenide–graphene glassy composites the individual alloying element's reactivity with graphene cannot be ignored, furthermore, they can also impact on electrical properties of such composite materials.

THERMAL PROPERTIES

To investigate the different physical and structural properties of chalcogenide- graphene glassy composites, Singh also studied the thermal property. He performed the Differential Scanning Calorimetry (DSC) measurements in his initial report. The crystallization kinetics behavior of the parent alloy as well as the composite material to achieve the higher order crystallization kinetics characteristics along with the improved thermal stability parameters were explored by him, as well as confirming to formed composite material homogenous configuration. According to this study, the DSC experiment was performed in the temperature range 300 K to 523 K for the SZS alloy and SZS–GF composite material, as shown in Fig. 6.24 [246].

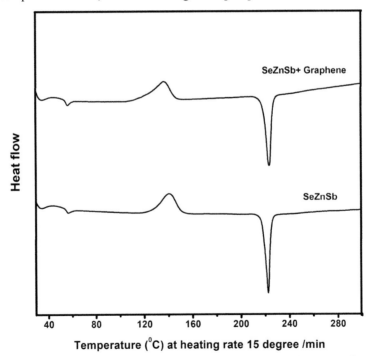

Figure 6.24   DSC thermograms for the SZS alloy and SZS–GF composite.

Both SZS alloy and SZS–GF composite material exhibited the well-defined glass transition peak (T_g), crystallization (T_c) and melting temperature peak (T_m). This is a direct confirmation about SZS alloy and SZS–GF composite material amorphous behavior with glassy nature. The defined values of T_g, T_c and T_m at different heating rates is listed in Table 6.4.

Table 6.4 Crystallization kinetic parameters such as T_g, T_c, T_p and T_m at different DSC heating rates for the SZS alloy and SZS–Graphene composite materials [246]

	Heating rates	T_g (K)	T_c (K)	T_p (K)	T_m (K)
Se–Zn–Sb	5	322	382	407	493
	10	323	387	411	494
	15	327	393	417	495
	20	328	397	433	495
Se–Zn–Sb+ Graphene	5	322	376	392	495
	10	323	384	401	496
	15	326	388	410	497
	20	327	391	414	497

By using this experimental data, he defined the kinetic properties such as glass forming ability, crystallization activation energy at T_g, T_c, and T_p (peak crystallization) using the approaches of Hruby, Ozawa relation, Augis and Bennett and Takhor, as listed in Table 6.5 [246]. The obtained T_g values of SZS–GF composite material at different heat rates are almost similar or less to SZS alloy, while the crystallization T_c and T_p values are lower than the parent alloy. On the other hand, melting temperature (T_m) is higher than parent alloys values at all four heating rates (5, 10, 15, 20°C/min). These experimental results were surprising because this study was performed with the view of achieving higher thermal parameters as predicted in various theoretical reports [246]. He further defined the thermal stability and crystallization activation energies at T_g, T_c and T_p using the listed thermal parameters. Beside the existing stiff and hard graphene back bone in glassy composite configuration, the thermal stability was found to be lower than the parent alloy. This appears to be a contradictory result from the theoretical concepts and theories. The obtained activation energies of the SZS–GF composite materials are also almost higher at T_g, T_c and T_p compared to the parent alloy, this is also a different view from the predictions on chalcogenide–graphene composite materials. These contradictory results of thermal parameters may be due to quadratic (non-linear) thermal expansion behavior of graphene, in which excess heat is generated during the synthesis process, and it is quite possible that the deformation of π bonds occur more rapidly and can also also largely affect the stronger σ bonds as well. Additionally, the individual interaction of constituents of SZS alloys could also contribute significantly. Therefore, beside the existence of the graphene backbone within the configuration, the overall thermal properties are not very impressive. Therefore, a rigorous experimental data is required to make a conclusive view on such composite materials. On the basis of initial thermal property data interpretation it could be said that this may lead to a new direction for theoretical as well as experimental investigators to rethink about the different physical properties of chalcogenide–graphene composites in the glassy form.

Table 6.5 Crystallization activation energies E_g, E_c, E_p for the SZS alloy and SZS–Graphene composite materials [246]

Compositions	Activation energy E_g (KJ/mol), at T_g	Activation energy E_c (KJ/mol), at T_c	Activation energy E_p (KJ/mol), at T_p
Se–Zn-Sb	195.49	106.21	72.06
Se–Zn-Sb + Graphene	204.77	114.86	78.43

CONCLUSIONS

Therefore a significant outcome of this study demonstrates the successful fabrication of chalcogenide–graphene glassy composites and their physical properties interpretations. By providing the descriptions of the basic concept of building blocks in chalcogenide including clusters tetrahedral building blocks, inorganic–organic frameworks built from metal chalcogenide nanoclusters, properties of open-framework in chalcogenides, led to different forms of graphene composites formation ability for various kinds of utility such as graphene membranes, graphene energy, graphene sensors, graphene thermoacoustics, graphene magnets, graphene superconductors, graphene in biomedicine. Considering the analogous structural similarities between chalcogenides and graphene, an explanation on their potential composites formation has also been provided. Additionally, a brief discussion was presented on different forms of chalcogenide–graphene composite materials including nanocrystalline, polycrystalline and amorphous chalcogenides with graphene. Considering the importance of the emerging chalcogenide glass-graphene composites a short description has also been addressed on the chalcogenide-single layer graphene composites. Moreover, this study paid great attention on the recently investigated novel potential field in chalcogenide–bilayers graphene glassy composites considering their outstanding and unexpected physical properties, along with their structural modifications, surface morphological modifications and modification in Raman spectrums. Additionally, descriptions on their optical properties such as alternation in UV/Visible properties considering the possible band structure modifications, unexpected room temperature PL property in some composites. More significantly, a huge enhancement in their IR transparency compared to parent chalcogenide glassy alloys. The large enhancement in IR transparency has been correlated with a conceptual schematic model. A brief overview on mechanical and electrical properties of the analogues materials was also discussed. This initial innovative research work has also demonstrated the major contradiction with the theoretical predictions for such composite materials, therefore, thermal properties have also been shown. According to theoretical predictions incorporation of graphene in the chalcogenide glassy system could drastically enhance thermal properties, such as thermal stability, etc. But the actual research findings were opposite to the theoretically described predictions by showing their lower thermal stability and other crystallization kinetics parameters such as lower T_g, T_c, T_p with higher values of activation energies compared to the parent chalcogenide glassy alloy, however, the increase in their melting temperatures is consistent with the theoretical predictions. Thus, intensive theoretical and experimental research is required to resolve the issue for such composite materials.

■ References

1. Torma, V., Vidoni, O., Simon, U., Schimid, G. (2003) Charge-transfer mechanisms between gold clusters. Eur. J. Inorg. Chem., 6, 1121–1127.

2. Bu, X., Zheng, N., Feng, P. (2004) Tetrahedral chalcogenide clusters and open frameworks. Chem. Eur. J., 10, 3356–3362.

3. Feng, P., Bu, X., Zheng, N. (2005) The interface chemistry between chalcogenide clusters and open framework chalcogenides. Acc. Chem. Res., 38, 293–303.

4. Li, H., Laine, A., O'Keefe, M., Yaghi, O.M. (1999) Supertetrahedral sulfide crystals with giant cavities and channels. Science, 283, 1145–1147.

5. Wang, C., Bu, X., Zheng, N., Feng, P. (2002) Nanocluster with one missing core atom: A three-dimensional hybrid superlattice built from dual-sized supertetrahedral clusters. J. Am. Chem. Soc., 124, 10268–10269.

6. Su, W., Huang, X., Li, J., Fu, H. (2002) Crystal of semiconducting quantum dots built on covalently bonded T5 $[In_{28}Cd_6S_{54}]^{-12}$: The largest supertetrahedral cluster in solid state. Journal of the American Chemical Society, 124, 12944–12945.

7. Eichhofer, A., Deglmann, P. (2004) Mercury–chalcogenide clusters: Synthesis and structure of $[Hg_{10}Te_4(SePh)_{12}(PPhnPr_2)_4]$, $[Hg_{10}Te_4(TePh)_{12}(PPhnPr_2)_4]$ and $[Hg_{34}Te_{16}(SePh)_{36}(PPhnPr_2)_4]$. Eur. J. Inorg. Chem., 2, 349.

8. Jiang, T., Ozin, G.A. (1998) New directions in tin sulfide materials chemistry. Journal of Materials Chemistry, 8, 1099–1108.

9. Lin, Z., Feng, P., Bu, X. (2011) Nano/Microporous materials: Crystalline metal-chalcogenide superlattices. *In*: Encyclopedia of Inorganic and Bioinorganic Chemistry. John Wiley & Sons, Ltd., Hoboken, NJ, United States. DOI: 10.1002/9781119951438.eibc0332.

10. Manos, M.J., Chrissafis, K., Kanatzidis, M.G. (2006) Unique pore selectivity for Cs+ and exceptionally high NH_4+ exchange capacity of the chalcogenide material $K_6Sn[Zn_4Sn_4S_{17}]$. J. Am. Chem. Soc., 128, 8875–8883.

11. Zheng, N., Bu, X., Wang, B., Feng, P. (2002) Microporous and photoluminescent chalcogenide zeolite analogs. Science, 298, 2366–2369.

12. Wang, C., Bu, X., Zheng, N., Feng, P. (2002) A 3D open-framework indium telluride and its selenide and sulfide analogues. Angewandte Chemie International Edition in English, 41, 1959–61.

13. Ruzin, E., Fuchs, A., Dehnen, S. (2006) Fine-tuning of optical properties with salts of discrete or polymeric, heterobimetallic telluride anions $[M_4(\mu_4\text{–}Te)(SnTe_4)_4]^{10-}$(M = Mn, Zn, Cd, Hg) and $^3_\infty\{[Hg_4(\mu_4\text{–}Te)(SnTe_4)_3]^{6-}\}$. Chemical Communications, 46, 4796–4798.

14. Zhang, Q.C., Liu, Y., Bu, X., Wu, T., Feng, P. (2008) A rare (3, 4)-connected chalcogenide superlattice and its photoelectric effect. Angewandte Chemie International Edition in English, 47, 113–116.

15. Bu, X., Zheng, N., Wang, X., Wang, B., Feng, P. (2004) Three-dimensional frameworks of gallium selenide supertetrahedral clusters. Angewandte Chemie International Edition in English, 43, 1502–5.

16. Cui, M., Ren, S., Qin, S., Xue, Q., Zhao, H., Wang, L (2018) Process able poly (2-butylaniline)/hexagonal boron nitride nanohybrids for synergetic anticorrosive reinforcement of epoxy coating. Corros. Sci., 131, 187–198.

17. Zhang, Q., Qin, Z., Luo, Q., Wu, Z., Liu, L., Shen, B., Hu, W (2017) Microstructure and nanoindentation behavior of Cu composites reinforced with graphene nanoplatelets by electroless co-deposition technique. Sci. Rep., 7, 1338.

18. Park, S., Ruof, R.S. (2009) Chemical methods for the production of graphenes. Nature Nanotechnology, 4, 217–224.

19. Wang, S., Wang, R., Chang, J., Hu, N., Xu, C. (2018) Self-supporting Co_3O_4/graphene hybrid films as binder-free anode materials for lithium ion batteries. Sci. Rep., 8, 3182.

20. Muthumariappan, A., Govindasamy, M., Chen, S.M. (2017) Screen-printed electrode modifed with a composite prepared from graphene oxide nanosheets and Mn_3O_4 micro-cubes for ultrasensitive determination of nitrite. Microchim. Acta, 184, 3625–3634.

21. Du, Y., Li, D., Liu, L., Gai, G. (2018) Recent achievements of self-healing graphene/polymer composites. Polymers, 10, 114.

22. Li, J.R., Kuppler, R.J., Zhou, H.C. (2009) Selective gas adsorption and separation in metal-organic frameworks. Chem. Soc. Rev., 38, 1477–1504.

23. Bhadra, M., Roy, S., Mitra, S (2016) Desalination across a graphene oxide membrane via direct contact membrane distillation. Desalination, 378, 37–43.

24. Liu, G., Han, K., Ye, H., Zhu, C., Gao, Y., Liu, Y., Zhou, Y. (2017) Graphene oxide/triethanolamine modifed titanate nanowires as photocatalytic membrane for water treatment. Chem. Eng. J., 320, 74–80.

25. Hassan, F., Mahendra, A., Klaus-Viktor, K. (2017) PeinemannPebax®1657/Graphene oxide composite membranes for improved water vapor separation. J. Memb. Sci., 525, 187–194.

26. Wang, Y., Krishna, Z., Gupta, M., Shi, Q., Lu, R. (2017) Molecular dynamics study on water desalination through functionalized nanoporous graphene author links open overlay panel. Carbon, 116, 120–127.

27. Yang, R.T. (2003) Adsorbents: Fundamentals and Applications. Wiley, Hoboken.

28. Zuttel, A., Sudan, P., Mauron, P., Wenger, P. (2004) Model for the hydrogen adsorption on carbon nanostructures. Appl. Phys. A, 78, 941–946.

29. Lee, C.K. (2014) Monatomic chemical-vapor-deposited graphene membranes. ACS Nano, 8, 2336–2344.

30. Zhou, D., Cheng, Q.Y., Cui, Y., Wang, T., Li, X., Han, B.H. (2014) Graphene–terpyridine complex hybrid porous material for carbon dioxide adsorption. Carbon, 66, 592–598.

31. Lin, Y.C. (2011) Clean transfer of graphene for isolation and suspension. ACS Nano, 5, 2362–2368.

32. Dutkiewicz, J., Rogal, Ł., Fima, P., Ozga, P. (2018) Composites strengthened with graphene platelets and formed in semisolid state based on α and α/β MgLiAl alloys. J Mater. Eng. Perform, 27, 2205–2215.

33. Ruiz-Garcia, C., Jimenez, R., Perez-Carvajal, J., Berenguer-Murcia, A., Darder, M., Aranda, P. (2014) Graphene-clay based nanomaterials for clean energy storage. Sci. Adv. Mater., 6, 151–158.

34. Liu, S., Sun, L., Xu, F., Zhang, J., Jiao, C., Li, F. (2013) Nanosized Cu-MOFs induced by graphene oxide and enhanced gas storage capacity. Energy Environ. Sci., 6, 818–823.

35. Huang, W., Zhou, X., Xia, Q., Peng, J., Wang, H., Li, Z. (2014) Preparation and adsorption performance of GrO@Cu–BTC for separation of CO_2/CH_4. Ind. Eng. Chem. Res., 53, 1176–1184.

36. Boutilier, M.S.H., Jang, D., Idrobo, J.C. (2017) Molecular sieving across centimeter-scale single-layer nanoporous graphene membranes. ACS Nano, 11, 5726–5736.

37. Huang, L., Zhang, M., Li, C., Shi, G. (2015) Graphene-based membranes for molecular separation. J. Phys. Chem. Lett., 6, 2806–2815.

38. Sun, C., Wen, B., Bai, B. (2015) Application of nanoporous graphene membranes in natural gas processing: Molecular simulations of CH_4/CO_2, CH_4/H_2S and CH_4/N_2 separation. Chem. Eng. Sci., 138, 616–621.

39. Zhu, J., Meng, X., Zhao, J., Jin, Y., Yanga, N., Zhang, S., Sunarso, J. (2017) Facile hydrogen/nitrogen separation through graphene oxide membranes supported on YSZ ceramic hollow fibers. Journal of Membrane Science, 535, 143–150.

40. Jia, M., Feng, Y., Liu, S., Qiu, J., Yao, J. (2017) Graphene oxide gas separation membranes intercalated by UiO-66-NH_2 with enhanced hydrogen separation performance. Journal of Membrane Science, 539, 172–177.

41. Azamat, J., Khataee, A., Joob, S.W. (2015) Molecular dynamics simulation of trihalomethanes separation from water by functionalized nanoporous graphene under induced pressure. Chem. Eng. Sci., 127, 285–292.

42. Anki Reddy, K., Gogoi, A., Raidongia, K., Konch, T. (2018) Data for: Water and salt dynamics in multilayer graphene oxide (GO) membrane: Role of lateral sheet dimensions. Mendeley Data. https://doi.org/10.17632/sx38wb8w3n

43. Wang, L. (2017) Fundamental transport mechanisms, fabrication and potential applications of nanoporous atomically thin membranes. Nature Nanotechnology, 12, 509–522.

44. Huang, S., Dakhchoune, M., Luo, W., Oveisi, E., He, G., Rezaei, M., Zhao, J., Alexander, D.T.L., Züttel, A., Strano, M.S., Agrawal, K.V. (2018) Single-layer graphene membranes by crack-free transfer for gas mixture. Nat. Commun. 9, 2632.

45. Gogoi, A., Konch, T.J., Raidongia, K., Reddy, K.A. (2018) Water and salt dynamics in multilayer graphene oxide (GO) membrane: Role of lateral sheet dimensions. Journal of Membrane Science, 563, 785–793.

46. Cai, X., Lai, L., Shen, Z., Lin, J. (2017) Graphene and graphene-based composites as Li-ion battery electrode materials and their application in full cells. J. Mater. Chem. A, 5, 15423–15446.

47. Bonaccorso, F., Colombo, L., Yu, G., Stoller, M., Tozzini, V., Ferrari, A.C., Ruof, R.S., Pellegrini, V. (2015) Graphene, related two-dimensional crystals, and hybrid systems for energy conversion and storage. Science, 347, 6217.

48. https://nanographene.net/graphene-based-battery-and-market-prospect/

49. Huang. Y., Liu, J., Huang, Q., Zheng, Z., Hiralal, P., Zheng, F., Ozgit, D., Su, S., Chen, S., Tan, P.H., Zhang, S., Zhou, H. (2018) Flexible high energy density zinc-ion batteries enabled by binder-free MnO_2/ reduced graphene oxide electrode. npj Flexible Electronics, 2, 21.

50. Zhang, R. (2017) Lithiophilic sites in doped graphene guide uniform lithium nucleation for dendrite-free lithium metal anodes. Angew. Chem. Int. Ed., 56, 7764–7768.

51. Zhang, Z., Sun, J., Lai, C., Wang, Q., Hu, C. (2017) High-yield ball-milling synthesis of extremely concentrated and highly conductive graphene nanoplatelet inks for rapid surface coating of diverse substrates. Carbon, 120, 411–418.

52. Xia, C., Guo, J., Li, Y., Liang, H., Zhao, C., Alshareef, H.N. (2017) Rechargeable aqueous zinc-ion battery based on porous framework zinc pyrovanadate intercalation cathode. Adv. Mater., 30, 1705580.

53. Chang, J., Adhikari, S., Lee, T.H., Li, B., Yao, F., Pham, D.T., Le, V.T., Lee, Y.H. (2015) Leaf vein-inspired nano channeled graphene flm for highly efcient micro-super capacitors. Adv. Energy. Mater., 5, 1500003.

54. Zhang, N. (2017) Rechargeable aqueous zinc-manganese dioxide batteries with high energy and power densities. Nat. Commun., 8, 405.

55. He, P. (2018) Sodium ion stabilized vanadium oxide nanowire cathode for high-performance zincion batteries. Adv. Energy. Mater., 8, 1702463.

56. Hiura, H., Lee, M.V., Tyurnina, A.V., Tsukagoshi, K. (2012) Liquid phase growth of graphene on silicon carbide. Carbon, 50, 5076–5084.

57. Bernardi, M., Palummo, M., Grossman, J.C. (2013) Extraordinary sunlight absorption and one nanometer thick photovoltaics using two-dimensional monolayer materials. Nano. Lett., 13, 3664–3670.

58. Wu, B., Zhang, G., Yan, M., Xiong, T., He, P. (2018) Graphene scroll-coated α-MnO$_2$ nanowires as highperformance cathode materials for aqueous Zn-ion battery. Small, 14, 1703850.

59. Pan, H., Zhang, G., Liao, X., Mai, X. (2016) Reversible aqueous zinc/manganese oxide energy storage from conversion reactions. Nat. Energy, 1, 16039.

60. Lee, S.W., Lee, C.W., Yoon, S.B., Kim, M.S., Nam, K.W., Jeong, J.H. (2016) Superior electrochemical properties of manganese dioxide/reduced graphene oxide nanocomposites as anode materials for highperformance lithium ion batteries. J. Power Sources, 312, 207–215.

61. Chang, J., Jin, M., Yao, F., Kim, T., Le, V.T., Yue, H., Gunes, F., Li, B., Ghosh, A., Xie, S., Lee, Y.H. (2013) Asymmetric supercapacitors based on graphene/MnO$_2$, nano-spheres and graphene/ MoO$_3$ nanosheets with high energy density. Adv. Funct. Mater., 23, 5074–5083.

62. Huang, Y., Liu, H., Gong, L., Hou, Y., Li, Q. (2017) A simple route to improve rate performance of LiFePO$_4$/reduced graphene oxide composite cathode by adding Mg^{2+} via mechanical mixing. J. Power Sources, 347, 29–36.

63. Ogata, C., Kurogi, R., Hatakeyama, K., Taniguchi, T., Koinuma, M., Matsumoto, Y. (2016) All-graphene oxide device with tunable supercapacitor and battery behavior by the working voltage. Chem. Commun., 52, 3919–3922.

64. Shia, E., Lib, H., Xua, W., Wua, S., Weic, J., Fang, F., Cao, A. (2015) Improvement of graphene-Si solar cells by embroidering grapheme with a carbon nanotube spider-web. Nano Energy, 15, 379–405.

65. Zhang, Y., Zhang, L., Zhou, C. (2013) Review of chemical vapor deposition of graphene and related applications. Acc. Chem. Res., 46, 2329–2339.

66. Notte, L.L., Bianco, G.V., Palma, A.L., Carlo, A., Bruno, G., Reale, A. (2018) Sprayed organic photovoltaic cells and mini-modules based on chemical vapor deposited graphene as transparent conductive electrode, panel. Carbon, 129, 878–883.

67. Patel, K., Tyagi, P.K. (2015) Multilayer graphene as a transparent conducting electrode in silicon heterojunction solar cells. AIP Adv., 5, 077165.

68. Zhang, Q., Wan, X., Xing, F., Huang, L., Long, G., Yi, N., Ni, W., Liu, Z., Tian, J., Chen, Y. (2013) Solutionprocess able graphene mesh transparent electrodes for organic solar cells. Nano Lett., 6, 478–484.

69. Yan, H., Wang, J., Feng, B., Duan, K., Weng, J. (2015) Graphene and Ag nanowires co-modifed photoanodes for high-efciency dye-sensitized solar cells. Sol. Energy, 122, 966–975.

70. Wang, J.T., Ball, J.M.E., Barea, M., Abate, A., Webber, J.A.A., Huang, J., Saliba, M., Bisquert, J., Snaith, H.J., Nicholas, R.J. (2014) Low-temperature processed electron collection layers of graphene/ TiO$_2$ nanocomposites in thin film perovskite solar cells. Nano Lett., 14, 724–730.

71. Chung, M.G., Kim, D.H., Seo, D.K., Kim, T., UkIm, H., Lee, H.M., Yoo, J.B., Hong, S.H., Kang, T.J., Kim, Y.H. (2012) Flexible hydrogen sensors using graphene with palladium nanoparticle decoration. Sensors and Actuators B: Chemical, 169, 387–392.

72. Cao, Y., Fatemi, V., Fang, S., Watanabe, K., Taniguchi, T., Kaxiras, E., Herrero, P.J. (2018) Unconventional superconductivity in magic-angle graphene super lattices. Nature, 556, 50.

73. Song, T., Cai, X., Yuan, M.W., Zang, X. (2018) Giant tunnelling magnetoresistance in spin-flter van der Waals hetero structures. Science, 360, 1214–1218.

74. Sattar, T. (2019) Current review on synthesis, composites and multifunctional properties of graphene. Topics in Curr. Chem., 377, 1–45.

75. Natsuki, T., Yiwada, A., Natsuki, J. (2017) Infuence of temperature on vibrational frequency of graphene sheet used as nano-scale sensing. C: Journal of Carbon Research, 3, 4. https://doi.org/10.3390/c3010004

76. Shin, S.H., Ji, S., Choi, S., Hee, L.B., Wan, A., Park, J., Kim, J., Lee, K.S. (2017) Integrated arrays of air-dielectric graphene transistors as transparent active-matrix pressure sensors for wide pressure ranges. Nat. Commun, 8, 14950.

77. Jo, G., Choe, M., Lee, S., Park, W., Kahng, Y.H., Lee, T. (2012) The application of graphene as electrodes in electrical and optical devices. Nanotechnology, 23, 112001–19.

78. Suvarnaphaet, P., Pechprasarn, S. (2017) Graphene-based materials for biosensors: A review. Sensors, 17(10), E2161. https://doi.org/10.3390/s17102161

79. Brown, J.J., Moore, N.C., Supekar, O.D., Gertsch, J.C., Bright, V.M. (2016) Ultrathin thermoacoustic nanobridge loudspeakers from ald on polyimide. Nanotechnology, 27, 475504.

80. Tian, H. (2012) Single-layer graphene sound-emitting devices: Experiments and modelling. Nanoscale, 4, 2272–2277.

81. Kaverzin, A.A., Mayorov, A.S., Shytov, A., Horsell, D.W. (2012) Impurities as a source of 1/f noise in graphene. Phys. Rev. B, 85, 075435.

82. Kim, C.S., Lee, K.E., Lee, J.M., Kim, S.O., Cho, B.J., Choi, J.W. (2016) Application of n-doped threedimensional reduced graphene oxide aerogel to thin flm loudspeaker. ACS Appl. Mater. Interfaces, 8, 22295–22300.

83. Suk, J.W., Kirk, K., Hao, Y., Hall. N.A., Ruof, R.S. (2012) Thermoacoustic sound generation from monolayer graphene for transparent and fexible sound sources. Adv. Mater., 24, 6342–6347.

84. Tian, H., Yang, Y., Xie, D., Ge, J., Ren, T.L. (2013) A reduced graphene oxide sound-emitting device: A new use for joule heating. RSC Adv., 3, 17672–17676.

85. Wang, X. et al. (2015) A spectrally tunable all-graphene-based flexible field-effect light-emitting device. Nat. Commun., 6, 7767.

86. Rengel, R., Pascual, E., Martín, M.J. (2014) Infuence of the substrate on the difusion coefcient and the momentum relaxation in graphene: The role of surface polar phonons. Appl. Phys. Lett., 104, 233107.

87. Yan, Z., Nika, D. (2015) Thermal properties of graphene and few-layer graphene: Applications in electronics. IET Circuits Devices Syst., 9, 4–12.

88. Tian, H. (2014) Graphene earphones: entertainment for both humans and animals. ACS Nano, 8, 5883–5890.

89. Shi, X., Lin, P.V., Sasagawa, T., Dobrosavljević, V., Popović, D. (2014) Two-stage magnetic-field-tuned superconductor-insulator transition in underdoped $La_{2-x}Sr_xCuO_4$. Nature Physics, 10, 437–443.

90. Ferrand, H.L., Bolisetty, S., Demirörs, A., Libanori, F.R., Studart, A.R., Mezzenga, R. (2016) Magnetic assembly of transparent and conducting graphene-based functional composites. Nat. Commun., 7, 12078.

91. Li, D., Kaner, R.B. (2008) Graphene-based materials. Science, 320, 1170–1171.

92. Xu, Z., Gao, C. (2014) Graphene in macroscopic order: Liquid crystals and wet-spun fibres. Acc. Chem. Res., 47, 1267.

93. Perreault, F., Faria, A., Elimelech, M. (2015) Environmental applications of graphene-based nanomaterials. Chem. Soc. Rev., 44, 5861–5896.

94. Heath, M.S. (2017) Multi-frequency sound production and mixing in graphene. Sci. Rep., 7, 1363.

95. Li, J., Kim, J.K. (2007) Percolation threshold of conducting polymer composites containing 3D randomly distributed graphite nanoplatelets. Compos. Sci. Technol., 67, 2114–2120.

96. Clark, L.C. Jr., Lyons, C. (1962) Electrode systems for continuous monitoring in cardiovascular surgery. Ann. NY. Acad. Sci., 102, 29–45.

97. Bornhoeft,L.R.,Castillo,A.C.,Smalley,P.,Kittrell,R.C.,James,D.K.,Brinson,B.E.,Rybolt,T.R.,Johnson, B.R., Cherukuri, T.K., Cherukuri, P. (2016) Teslaphoresis of carbon nanotubes. ACS Nano, 10, 4873.

98. Kravchyk, K.V., Wang, S., Piveteau, L., Kovalenko, M.V. (2017) Efficient aluminum chloride-natural graphite battery. Chem. Mater., 29, 4484–4492.

99. Parvez, K., Wu, Z.S., Li, R.J., Liu, X.J., Graf, R., Feng, X.L., Müllen, K. (2014) Exfoliation of graphite into graphene in aqueous solutions of inorganic salts. J. Am. Chem. Soc., 136, 6083.

100. Benardo, A.D., Millo, O., Barbon, M., Alpern, H., Kalceim, Y., Sassi, U. (2017) p-wave triggered superconductivity in single-layer graphene on an electron-doped oxide superconductor. Nat. Commun., 8, 14024.

101. Lin, F., Zhu, Z., Zhou, X. (2017) Graphene fakes: Orientation control of graphene fakes by magnetic field: broad device applications of macroscopically aligned graphene. Adv Mater., 29. https://doi.org/10.1002/adma.201770005

102. Chaste, J., Eichler, A., Moser, J., Ceballos, G., Rurali, R., Bachtold, A. (2012) A nanomechanical mass sensor with yoctogram resolution. Nat. Nanotechnol., 7, 301–304.

103. Natsuki, T. (2015) Theoretical analysis of vibration frequency of graphene sheet used as nanomechanical mass sensor. Electronics, 4, 723–738.

104. Zhang, Z., Xu, X., Qiu, L., Wang, S., Wu, T., Ding, F., Peng, H., Liu, K. (2017) The way towards ultrafast growth of single-crystal graphene on copper. Adv. Sci., (Weinh) 4, 1700087.

105. Heidarizadeh, M., Doustkhah, E. (2017) Dithiocarbamate to modify magnetic graphene oxide nanocomposite (Fe_3O_4–GO): A new strategy for covalent enzyme (lipase) immobilization to fabrication a new nanobiocatalyst for enzymatic hydrolysis of PNPD. Int. J. Biol. Macromol., 101, 696–702.

106. Fei, W., Zhou, J., Guo, W. (2015) Low-voltage driven graphene foam thermoacoustic speaker. Small, 11, 2252–2256.

107. Saito, Y., Kasahara, Y., Ye, J., Iwasa, Y., Nojima, T. (2015) Metallic ground state in an ion-gated two-dimensional superconductor. Science, 350, 409–413.

108. Tsen, A.W., Hunt, B., Kim, Y.D., Yuan, Z.J., Jia, S., Cava, R.J., Hone, J., Kim, P., Dean, C.R., Pasupathy, A.N. (2016) Nature of the quantum metal in a two-dimensional crystalline superconductor. Nat. Phys., 12, 208–212.

109. Xing, S. (2015) Quantum grifths singularity of superconductor-metal transition in Ga thin films. Science, 350, 542–545.

110. Biscaras, J., Bergeal, N., Hurand, S., Feuillet-Palma, C., Rastogi, A., Budhani, R.C., Grilli, M., Caprara, S., Lesueur, J. (2013) Multiple quantum criticality in a two-dimensional superconductor. Nat. Mater., 12, 542–548.

111. Cao, Y., Fatemi, V., Fang, S., Watanabe, K., Taniguchi, T., Kaxiras, E., Jarillo-Herrero, P. (2018) Unconventional superconductivity in magic-angle graphene superlattices. Nature, 556, 43–50.

112. Ichinokura, S., Sugawara, K., Takayama, A., Takahashi, T., Hasegawa, S. (2016) Superconducting calcium-intercalated bilayer graphene. ACS Nano, 10, 2761–2765.

113. Dai, J., Wang, D., Zhang, M., Niu, T., Li, A., Ye, M., et al. (2016) How graphene islands are unidirectionally aligned on the Ge (110) surface. Nano Lett., 16, 3160–3165.

114. Allain, A., Han, Z., Bouchiat, V. (2012) Electrical control of the superconducting-to-insulating transition in graphene-metal hybrids. Nat. Mater., 11, 590–594.

115. Han, X., Chen, Y., Zhu, H., Preston, C., Wan, J., Fang, Z., Hu, L. (2013) Scalable, printable, surfactant-free graphene ink directly from graphite. Nanotechnology, 24, 205304.

116. Afroja, S., Karima, M.N., Abdelkaderb, A., Cassonc, A., Yeatesa, S. (2018) Inkjet printing of graphene inks for wearable electronic applications. Nanomaterials, 8, 602.

117. Li, J. (2013) Efcient inkjet printing of graphene. Adv Mater., 25, 3985–3992.

118. Higginbotham, A.L., Kosynkin, D.V., Sinitskii, A., Sun, Z., Tour, J.M. (2010) Lower-defect graphene oxide nanoribbons from multi walled carbon nanotubes. ACS Nano, 4, 2059–2069.

119. Novoselov, K.S., Geim, A.K., Morozov, S.V., Jiang, D., Zhang, Y., Dubonos, S.V., Grigorieva, I.V., Firsov, A.A. (2004) Electric field effect in atomically thin carbon films. Science, 306, 666–669.

120. Kamyshny, A., Magdassi, S. (2014) Conductive nanomaterials for printed electronics. Small, 10, 3515–3535.

121. Ma, R. (2011) A drastic reduction in silver concentration of metallic ink by the use of single-walled carbon nanotubes decorated with silver nanoparticles. J. Mater. Chem., 21, 7070–7073.

122. Ichinokura, S., Sugawara, K., Takayama, A., Takahashi, T., Hasegawa, S. (2016) Superconducting calcium-intercalated bilayer graphene. ACS Nano, 10, 2761–2765.

123. Paton, K.R., Varrla, E., Backes, C., Smith, R.J., Khan, U., O'Neil, A., Boland, C., Lotya, M., Istrate, O.M., King, P., Higgins, T., Barwich, S. (2014) Scalable production of large quantities of defect-free few-layer graphene by shear exfoliation in liquids. Nat. Mater., 13, 624–630.

124. Zhao, J. (2010) Efcient preparation of large-area graphene oxide sheets for transparent conductive films. ACS Nano, 4, 5245–5252.

125. Liang, Y.T., Hersam, M.C. (2010) Highly concentrated graphene solutions via polymer enhanced solvent exfoliation and iterative solvent exchange. J. Am. Chem. Soc., 132, 17661–17663.

126. Gao, Y. (2014) Inkjet printing patterns of highly conductive pristine graphene on flexible substrates. Ind. Eng. Chem. Res., 53, 16777–16784.

127. Secor, E.B., Ahn, B.Y., Gao, T.Z., Lewis, J.A., Hersam, M.C. (2015) Rapid and versatile photonic annealing of graphene inks for flexible printed electronics. Adv. Mater., 27, 6683–6688. https://doi.org/10.1002/adma.201502866

128. Hu, G., Kang, J., Ng, L.W.T., Zhu, X., Howe, R.C.T., Jones, C.G., Hersam, M.C., Hasan, T. (2018) Functional inks and printing of two-dimensional materials. Chem. Soc. Rev., 47, 3265–3300.

129. Kitching, K.J., Pan, V., Ratner, B.D. (2004) Biomedical applications of plasma-deposited thin films. *In*: Biederman, H. (ed.), Plasma Polymer Films. Imperial College Press, London, 325–377.

130. Lai, J.Y., Ma, D.H.K., Lai, M.H., Li, Y.T., Chang, R.J., Chen, L.M. (2013) Characterization of cross-linked porous gelatin carriers and their interaction with corneal endothelium: Biopolymer concentration effect. PLoS ONE, 8, 54058.

131. Lee, O.J., Ju, H.W., Khang, G., Sun, P.P., Rivera, J., Cho, J.H., Park, S.J., Eden, J.G., Park, C.H. (2015) An experimental burn wound-healing study of non-thermal atmospheric pressure micro plasma jet arrays. J. Tissue Eng. Regen. Med., 10, 348–357.

132. Li, Y.S., Liao, J.L., Wang, S.Y., Chiang, W.H. (2016) Intercalation-assisted longitudinal unzipping of carbon nanotubes for green and scalable synthesis of graphene nanoribbons. Sci. Rep., 6, 22755.

133. Lim, H.L., Hwang, Y., Kar, M., Varghese, S. (2014) Smart hydrogels as functional biomimetic systems. Biomater. Sci., 2, 603–618.

134. University of Manchester: Biomedical Applications of Graphene. http://www.graphene.manchester.ac.uk/explore/the-applications/biomedical/

135. Shin, S.R. (2016) Graphene-based materials for tissue engineering. Adv. Drug. Deliv. Rev., 105, 255–274.

136. Nanda, S.S., Yi, D.K., Kim, K. (2016) Study of antibacterial mechanism of graphene oxide using raman spectroscopy. Nat. Sci. Rep., 6, 28443.

137. Nezakati, T., Cousins, B.G., Seifalian, A.M. (2014) Toxicology of chemically modifed graphene-based materials for medical application. Arch. Toxicol., 88, 1987–2012.

138. Nguyen, B.H., Nguyen, V.H. (2016) Promising applications of graphene and graphene-based nanostructures. Adv. Nat. Sci. Nanosci. Nanotechnol., 7, 023002.

139. Shinde, U.P., Yeon, B., Jeong, B. (2013) Recent progress of in situ formed gels for biomedical applications. Prog. Polym. Sci., 38, 672–701.

140. Tibbitt, M.W., Rodell, C.B., Burdick, J.A., Anseth, K.S. (2015) Progress in material design for biomedical applications. Proc. Natl. Acad. Sci. USA, 112, 14444–14451.

141. Wang, C., Li, Y.S., Jiang, J., Chiang, W.H. (2015) Controllable tailoring graphene nanoribbons with tunable surface functionalities: An efective strategy toward high-performance lithium-ion batteries. ACS Appl. Mater. Int., 7, 17441–17449.

142. Whang, K., Healy, K., Elenz, D., Nam, E., Tsai, D., Thomas, C., Nuber, G., Glorieux, F., Travers, R., Sprague, S. (1999) Engineering bone regeneration with bio absorbable scafolds with novel micro-architecture. Tissue Eng., 5, 35–51.

143. Lee, J., Kim, J., Kim, S., Min, D.H. (2016) Biosensors based on graphene oxide and its biomedical application. Adv. Drug. Deliv. Rev., 105, 275–287.

144. Xie, H., Yin, F., Yu, T., Wang, J.T., Liang, C. (2014) Mechanism for direct graphite-to-diamond phase transition. Sci. Rep., 4, 5930.

145. Gao, Y., Cao, T., Cellini, F., Berger, C., Heer, W.A., Tosatti, E., Riedo, E., Bongiorno, A. (2018) Ultrahard carbon flm from epitaxial two-layer graphene. Nat. Nanotechnol., 13, 133–138.

146. Rifai, A., Pirogova, E., Fox, K. (2018) Diamond, carbon nanotubes and graphene for biomedical applications. *In*: Reference Module in Biomedical Sciences: Encyclopedia of Biomedical Engineering, vol. 1, Elsevier, pp. 97–107. https://doi.org/10.1016/b978-0-12-801238-3.99874-X.

147. Novoselov, K.S., Geim, A.K., Morozov, S.V., Jiang, D., Zhang, Y., Dubonos, S.V., Grigorieva, I.V., Firsov, A.A. (2004) Electric field effect in atomically thin carbon films. Science, 306, 666–669.

148. Rao, C.N.R., Sood, A.K., Subrahmanyam, K.S., Govindaraj, A. (2009) Graphene: The new two-dimensional nanomaterial. Angew. Chem., 121, 7890–7916.

149. Rao, C.N.R., Matte, H.S.S.R., Subrahmanyam, K.S. (2013) Synthesis and selected properties of graphene and graphene mimics. Acc. Chem. Res., 46, 149–159.

150. Novoselov, K.S., Jiang, D., Schedin, F., Booth, T.J., Khotkevich, V.V., Morozov, S.V., Geim, A.K. (2005) Two-dimensional atomic crystals. Proc. Natl. Acad. Sci. USA, 102, 10451–10453.

151. Raidongia, K., Gomathi, A., Rao, C.N.R. (2010) Synthesis and characterization of nanoparticles, nanotubes, nanopans, and graphene-like structures of boron nitride. Israel J. Chem., 50, 399–404.

152. Song, X., Hu, J., Zeng, H. (2013) Two-dimensional semiconductors: Recent progress and future perspectives. J. Mater. Chem. C, 1, 2952–2969.

153. Rao, C.N.R., Ramakrishna Matte, H.S.S., Maitra, U. (2013) Graphene analogues of inorganic layered materials. Angew. Chem. Int. Ed., 52, 13162–13185.

154. Liu, J., Zeng, Z., Cao, X., Lu, G., Wang, L.H., Fan, Q.L., Huang, W., Zhang, H. (2012) Preparation of MoS_2-polyvinylpyrrolidone nanocomposites for flexible nonvolatile rewritable memory devices with reduced graphene oxide electrodes. Small, 8, 3517–3522.

155. Chang, K., Chen, W. (2011) *In situ* synthesis of MoS_2/graphenenanosheet composites with extraordinarily high electrochemical performance for lithium ion batteries. Chem. Commun., 47, 4252–4254.

156. Das, S.K., Mallavajula, R., Jayaprakash, N., Archer, L.A. (2012) Self-assembled MoS_2-carbon nanostructures: Influence of nanostructuring and carbon on lithium battery performance. J. Mater. Chem., 22, 12988–12992.

157. Chang, K., Chen, W., Ma, L., Li, H., Li, H., Huang, F., Xu, Z., Zhang, Q., Lee, J.Y. (2011) Graphene-like MoS_2/amorphous carbon composites with high capacity and excellent stability as anode materials for lithium ion batteries. J. Mater. Chem., 21, 6251–6257.

158. Chang, K., Chen, W. (2011) l-Cysteine-assisted synthesis of layered MoS_2/Graphene composites with excellent electrochemical performances for lithium ion batteries. ACS Nano, 5, 4720–4728.

159. Shiva, K., Matte, H.S.S.R., Rajendra, H.B., Bhattacharyya, A.J., Rao, C.N.R. (2013) Employing synergistic interactions between few-layer WS_2 and reduced graphene oxide to improve lithium storage, cyclability and rate capability of Li-ion batteries. Nano Energy, 2(5), 787–793. DOI: 10.1016/j.nanoen.2013.02.001

160. Hou, D., Zhou, W., Liu, X., Zhou, K., Xie, J., Li, G., Chen, S. (2015) Pt Nanoparticles/MoS_2 nanosheets/carbon fibers as efficient catalyst for the hydrogen evolution reaction. Electrochim. Acta, 166, 26–31.

161. Candelaria, S.L., Shao, Y., Zhou, W., Li, X., Xiao, J., Zhang, J.G., Wang, Y., Liu, J., Li, J., Cao, G. (2012) Nanostructured carbon for energy storage and conversion. Nano Energy, 1, 195–220.

162. Cheng, X., Zhang, J., Ren, J., Liu, N., Chen, P., Zhang, Y., Deng, J., Wang, Y., Peng, H. (2016) Design of a hierarchical ternary hybrid for a fiber-shaped asymmetric supercapacitor with high volumetric energy density. J. Phys. Chem. C, 120, 9685–9691.

163. Chi, K., Zhang, Z., Xi, J., Huang, Y., Xiao, F., Wang, S., Liu, Y. (2014) Freestanding graphene paper supported three-dimensional porous graphene-polyaniline nanocomposite synthesized by inkjet printing and in flexible all-solid-state supercapacitor. ACS Appl. Mater. Interfaces, 6, 16312–16319.

164. Zhang, Z., Wang, Q., Zhao, C., Min, S., Qian, X. (2015) One-step hydrothermal synthesis of 3D petal-like Co_9S_8/RGO/Ni_3S_2 composite on nickel foam for high-performance supercapacitors. ACS Appl. Mater. Interfaces, 7, 4861–4868.

165. Ma, L., Shen, X., Ji, Z., Wang, S., Zhou, H., Zhu, G. (2014) Carbon coated nickel sulfide/reduced graphene oxide nanocomposites: Facile synthesis and excellent supercapacitor performance. Electrochim. Acta, 146, 525–532.

166. Wang, S., Liu, N., Tao, J., Yang, C., Liu, W., Shi, Y., Wang, Y., Su, J., Li, L., Gao, Y. (2015) Inkjet printing of conductive patterns and supercapacitors using a multi-walled carbon nanotube/Ag nanoparticle based ink. J. Mater. Chem. A, 3, 2407–2413.

167. Yang, C., Shi, Y., Liu, N., Tao, J., Wang, S., Liu, W., Wang, Y., Su, J., Li, L., Yang, C., Gao, Y. (2015) Freestanding and flexible graphene wrapped MnO_2/MoO_3 nanoparticle based asymmetric supercapacitors for high energy density and output voltage. RSC Adv., 5, 45129–45135.

168. Liu, W., Liu, N., Shi, Y., Chen, Y., Yang, C., Tao, J., Wang, S., Wang, Y., Su, J., Li, L., Gao, Y. (2015) A wire-shaped flexible asymmetric supercapacitor based on carbonfiber coated with a metal oxide and a polymer. J. Mater. Chem. A, 3, 13461−13467.

169. Wang, S., Liu, N., Su, J., Li, L., Long, F., Zou, Z., Jiang, X., Gao, Y. (2017) Highly stretchable and self-healable supercapacitor with reduced graphene oxide based fiber springs. ACS Nano, 11, 2066−2074.

170. Askari, M.B., Salarizadeh, P., Rozati, S.M., Seifi, M. (2019) Two-dimensional transition metal chalcogenide composite/reduced graphene oxide hybrid materials for hydrogen evolution application. Polyhedron, 162, 201–206.

171. Ma, L., Wang, X., Zhao, B., Yang, J., Zhang, X., Zhou, Y., Chen, J. (2018) Novel metal chalcogenide supported on three-dimensional graphene foam for enhanced lithium storage. J. Alloys and Comp., 762, 149–156.

172. Gong, F., Ding, Z., Fang, Y., Tong, C.J., Xia, D., Lv, Y., Wang, B., Papavassiliou, D.V., Liao, J., Wu, M. (2018) Enhanced electrochemical and thermal transport properties of graphene/MoS_2 heterostructures for energy storage: Insights from multiscale modeling. ACS Appl. Mater. Interfaces, 10, 14614−14621.

173. Wang, S., Liu, B., Zhi, G., Gong, X., Gao, Z., Zhang, J. (2018) Relaxing volume stress and promoting active sites in vertically grown 2D layered mesoporous $MoS_{2(1-x)}Se_{2x}$/rGO composites with enhanced capability and stability for lithium ion batteries. Electrochimica Acta, 268, 424–434.

174. Fan, L., Tao, Y., Wu, X., Wu, Z., Wu, J. (2017) HfX_3(X = Se and S)/graphene composites for flexible photodetectors from visible to near-infrared. Mater. Res. Bull., 93, 21–27.

175. Nosheen, E., Shah, S.M., Iqbal, Z. (2017) Ru-dye grafted CdS and reduced graphene oxide Ru/CdS/rGO composite: An efficient and photo tuneable electrode material for solid state dye sensitized polymer solar cells. J. Photochem. Photobiol. B: Biology, 167, 117–127.

176. Peng, S., Han, X., Li, L., Zhu, Z., Cheng, F., Srinivansan, M., Adams, S., Ramakrishna, S. (2016) Unique cobalt sulfide/reduced graphene oxide composite as an anode for sodium-ion batteries with superior rate capability and long cycling stability. Small, 12, 1359–1368.

177. Zhou, L., Yan, S., Lin, Z., Shi, Y. (2016) *In situ* reduction of WS_2 nanosheets for WS_2/reduced graphene oxide composite with superior Li-ion storage. Mater. Chem. Phy., 171, 16–21.

178. Sarkar, S., Howli, P., Das, B., Das, N.S., Samanta, M., Das, G.C., Chattopadhyay, K.K. (2017) Novel quaternary chalcogenide/reduced graphene oxide-based asymmetric supercapacitor with high energy density. ACS Appl. Mater. Interfaces, 9, 22652−22664.

179. Yuan, S., Wang, S., Li, L., Zhu, Y. H., Zhang, X. B., Yan, J. M. (2016) Integrating 3D flower-like hierarchical Cu_2NiSnS_4 with reduced graphene oxide as advanced anode materials for na-ion batteries. ACS Appl. Mater. Interfaces, 8, 9178−9184.

180. Wang, X., Yin, R., Zeng, L., Zhu, M. (2019) A review of graphene-based nanomaterials for removal of antibiotics from aqueous environments. Environ. Pollut., 253, 100–110.

181. Cain, J.D., Hanson, E.D., Shi, F., Dravid, V.P. (2016) Emerging opportunities in the two-dimensional chalcogenide systems and architecture. Cur. Opin. Solid State and Mater. Sci., 20, 374–387.

182. Carey, J.J., Allen, J.P., Scanlon, D.O., Watson, G.W. (2014) The electronic structure of the antimony chalcogenide series: Prospects for optoelectronic applications. J. Solid State Chem., 213, 116–125.

183. Pawbake, A.S., Pawar, M.S., Jadkar, S.R., Late, D.J. (2016) Large area chemical vapor deposition of monolayer transition metal dichalcogenides and their temperature dependent raman spectroscopy studies. Nanoscale, 8, 3008–3018.

184. Chang, J.A., Rhee, J.H., Im, S.H., Lee, Y.H., Kim, H.J., Seok, S.I., Nazeeruddin, M.K., Gratzel, M. (2010) High-performance nanostructured inorganic-organic heterojunction solar cells. Nano Lett., 10, 2609–2612.

185. Hou, H., Jing, M., Huang, Z., Yang, Y., Zhang, Y., Chen, J., Wu, Z., Ji, X. (2015) One-dimensional rod-like Sb_2S_3-based anode for high-performance sodium-ion batteries. ACS Appl. Mater. Interfaces, 7, 19362–19369.

186. Yu, D.Y., Prikhodchenko, P.V., Mason, C.W., Batabyal, S.K., Gun, J.S., ladkevich, S., Medvedev, A.G., Lev, O. (2013) High-capacity antimony sulphide nanoparticle-decorated graphene composite as anode for sodium-ion batteries. Nat. Commun., 4, 2922–2928.

187. Guo, H., Hou, W., Liang, B., Zhang, H. (2017) Fabrication and photocatalytic performance of Sb_2S_3 film/ITO combination. Catal. Lett., 147, 2592–2599.

188. Zhang, H., Hu, C., Ding, Y., Lin, Y. (2015) Synthesis of 1D Sb_2S_3 nanostructures and its application in visible-light-driven photodegradation for MO. J. Alloy. Compd., 625, 90–94.

189. Kamruzzaman, M., Chaoping, L., Yishu, F., Farid Ul Islam, A.K.M., Zapien, J.A. (2016) Atmospheric annealing effect on TiO_2/Sb_2S_3/P3HT heterojunction hybrid solar cell performance. RSC Adv., 6, 99282–99290.

190. Chen, X., Li, Z., Zhu, H., Wang, Y., Liang, B., Chen, J., Xu, Y., Mai, Y. (2017) CdS/ Sb_2S_3 heterojunction thin film solar cells with a thermally evaporated absorber. J. Mater. Chem. C, 5, 9421–9428.

191. Wang, X., Li, J., Liu, W., Yang, S., Zhu, C., Chen, T. (2017) A fast chemical approach towards Sb_2S_3 film with a large grain size for high-performance planar heterojunction solar cells. Nanoscale, 9, 3386–3390.

192. Kazeminezhad, I., Hekmat, N., Kiasat, A. (2014) Effect of growth parameters on structural and optical properties of CdS nanowires prepared by polymer controlled solvothermal route, Fiber Polym., 15, 672–679.

193. Farbod, M., Ghaffari, N.M., Kazeminezhad, I. (2012) Effect of growth parameters on photocatalytic properties of CuO nanowires fabricated by direct oxidation. Mater. Lett., 81, 258–260.

194. Shiravizadeh, A.G., Elahi, S.M., Sebt, S.A., Yousefi, R. (2018) High performance of visible NIR broad spectral photocurrent application of monodisperse PbSe nanocubes decorated on rGO sheets. J. Appl. Phys., 123, 083102.

195. Yousefi, R., Azimi, H.R., Mahmoudian, M.R., Basirun, W.J. (2018) The effect of defect emissions on enhancement photocatalytic performance of ZnSe QDs and ZnSe/Rgo nanocomposites. Appl. Surf. Sci., 435, 886–893.

196. Yousefi, R., Azimi, H.R., Beheshtian, J., Seyed-Talebi, S.M., Azimi, H.R., Jamali-Sheini, F. (2018) Experimental and theoretical study of enhanced photocatalytic activity of Mg-doped ZnO NPs and ZnO/rGO nanocomposites. Chem. Asian J., 13, 194–203.

197. Nouri, M., Saray, A.M., Azimi, H.R., Yousefi, R. (2017) High solar-light photocatalytic activity of using Cu_3Se_2/rGO nanocomposites synthesized by a green co-precipitation method. Solid State Sci., 73, 7–12.

198. Molaeia, P., Kazeminezhad, I. (2018) Extended photocurrent performance of antimony trisulfide/ reduced graphene oxide composite prepared via a facile hot-injection route. Ceramics Int., 44, 13191–13196.

199. Yabuuchi, N., Kucota, K., Dahbi, M., Komaba, S. (2014) Research development on sodium-ion batteries. Chem. Rev., 114, 11636–11682.

200. Choi, J., Jin, J., Jung, I.G., Kim, J., Kim, H.J., Son, S.U. (2011) $SnSe_2$ nanoplate-graphene composites as anode materials for lithium ion batteries. Chem. Commun., 47, 5241–5243.

201. Zhang, Y., Zhu, P., Huang, L., Xie, J., Zhang, S., Cao, G., Zhao, X. (2015) Few-layered SnS_2 on few-layered reduced graphene oxide as na-ion battery anode with ultralong cycle life and superior rate capability. Adv. Funct. Mater., 25, 481–489.

202. Gu, T.H., Gunjakar, J.L., Kim, I.Y., Patil, S.B., Lee, J.M., Jin, X., Lee, N.S., Hwang, S.J. (2015) Porous hybrid network of graphene and metal oxide nanosheets as useful matrix for improving the electrode performance of layered double hydroxides. Small, 11, 3921–3931.

203. Jin, X., Adpakpang, K., Kim, I.Y., Oh, S.M., Lee, N.S., Hwang, S.J. (2015) An effective way to optimize the functionality of graphene-based nanocomposite: Use of the colloidal mixture of graphene and inorganic nanosheets. Sci. Rep., 5, 11057.

204. Shin, S.I., Go, A., Kim, I.Y., Lee, J.M., Lee, Y., Hwang, S.J. (2013) Solar CO_2 reduction using H_2O by a semiconductor/metal-complex hybrid photocatalyst: Enhanced efficiency and demonstration of a wireless system using $SrTiO_3$ photoanodes. Energy Environ. Sci., 6, 608–617.

205. Park, B., Oh, S.M., Jin, X., Adpakpang, K., Lee, N.S., Hwang, S.J. (2017) A 2D metal oxide nanosheet as an efficient additive for improving na-ion electrode activity of graphene-based nanocomposites. Chem. Eur. J., 23, 6544–6551.

206. Lin, H., Song, Y., Huang, Y., Kita, D., Deckoff-Jones, S., Wang, K., Li, L. et. al. (2017) Chalcogenide glass-on-graphene photonics. Nature Photon, 11, 798–805. https://arxiv.org/ftp/arxiv/papers/1703/1703.01666.pdf

207. Singh, A.K. (2013) SeZnSb alloy and its nano tubes, graphene composites properties. AIP Advances, 3, 042124–11.

208. Singh, A.K. (2013) Microscopic study on the Se–Te–Ge alloy and its composite with carbon nanotubes and graphene. J. Adv. Microsc. Res., 7, 1–7.

209. Lucovsky, G. (1969) Comments on the structure of chalcogenide glasses from infrared spectroscopy. Mater. Res. Bull., 4(8), 505–514.

210. Hohl, D., Jones, R.O. (1991) First-principles molecular-dynamics simulation of liquid and amorphous selenium. Phy. Rev. B, 43, 3856.

211. Vazquejz, J., Wagnerc, C., Villares, P. (1998) Glass transition and crystallization kinetics in $Sb_{0.18}As_{0.34}Se_{0.48}$ glassy alloy by using non-isothermal techniques. J. Non-Cryst. Solids, 235, 548–553.

212. Eder, D. (2010) Carbon Nanotube–Inorganic Hybrids. Chem. Rev., 110, 1348–1385.

213. Abergel, D.S.L., Apalkov, V., Berashevich, J., Ziegler, K., Chakraborty, T. (2010) Properties of graphene: A theoretical perspective. Adv. Phys., 59, 261–482.

214. Singh, A.K. (2014) Optical properties of the chalcogenide-multi walled carbon nano tubes and chalcogenide-graphene composite materials. J. Nanoeng. Nanomanuf., 4, 1–9.

215. Lu, Y., Song, S., Gong, Y., Song, Z., Rao, F., Wu, L., Liu, B., Yao, D. (2011) Ga–Sb–Se material for low-power phase change memory. Appl. Phys. Lett., 99, 243111–3.

216. Triches, D.M., Souza, S.M., de Lima, J.C., Grandi, T.A., Campos, C.E.M., Polian, A., Itie, J.P., Baudelet, F., Chervin, J.C. (2009) High-pressure phase transformation of nanometric ZnSb prepared by mechanical alloying. J. Appl. Phys., 106, 013509–6.

217. Ferrari, A.C., Meyer, J.C., Scardaci, V., Casiraghi, C., Lazzeri, M., Mauri, F., Piscanec, S., Jiang, D., Novoselov, K.S., Roth, S., Geim, A.K. (2006) Raman spectrum of graphene and graphene layers. Phys. Rev. Lett., 97, 187401–4.

218. Singh, A.K., Kim, J.H., Park, J.T., Sangunni, K.S. (2015) Properties of the chalcogenide–carbon nano tubes and graphene composite materials. J. Alloys and Comp., 627, 468–475.

219. Saito, R., Hofmann, M., Dresselhaus, G., Jorio, A., Dresselhaus, M.S. (2012) Raman spectroscopy of graphene and carbon nanotubes. Adv. Phys., 60, 413–550.

220. Abergela, D.S.L., Apalkovb, V., Berashevicha, J., Zieglerc, K., Chakrabortya, T. (2010) Properties of graphene: A theoretical perspective. Adv. Phy., 59, 261–482.

221. Elim, H.I., Ji, W., Ma, G.H., Lim, K.Y., Sow, C.H., Huan, C.H.A. (2004) Ultrafast absorptive and refractive nonlinearities in multiwalled carbon nanotube films. Appl. Phy. Lett., 84, 1799.

222. Banik, I. (2009) On photoluminescence in chalcogenide glasses based on barrier-cluster model. Journal Non-Oxide and Photonic Glasses, 1, 6–18.

223. Banik, I. (2010) Photoconductivity in chalcogenide glasses in non-stationary regime and the barrier-cluster model. Acta Electrotechnica et Informatica, 10, 52–58.

224. Anderson, P.W. (1975) Model for the electronic structure of amorphous semiconductors. Phys. Rev. Lett., 34, 953–955.

225. Street, R.A., Mott, N.F. (1975) States in the gap in glassy semiconductors. Phys. Rev. Lett., 35, 1293–1296.

226. Kastner, M., Adler, D., Fritzsche, H. (1976) Valence-alternation model for localized gap states in lone-pair semiconductors. Phys. Rev. Lett., 37, 1504–1507.

227. Adler, D., Yoffa, E.J. (1977) Localized electronic states in amorphous semiconductors. Can. J. Chern., 95, 1920–1929.

228. Banik, I. (2009) Relationship between optical absorption and photoluminescence in non-crystalline semiconductors. J. Opto. Adv. Mater., 11, 91–103.

229. Bishop, S.G., Mitchell, D.L. (1973) Photoluminescence excitation spectra in chalcogenide glasses. Phy. Rev. B, 8, 5696–5703.

230. Li, Z.Q., Henriksen, E.A., Jiang, Z., Hao, Z., Martin, M.C., Kim, P., Stormer, H.L., Basov, D.N. (2009) Band structure asymmetry of bilayer graphene revealed by infrared spectroscopy. Phy. Rev. Lett., 102, 037403.

231. Kuzmenko, A.B., Heumen, E.V., Marel, D.V., Lerch, P. (2009) Infrared spectroscopy of electronic bands in bilayer graphene. Phys. Rev. B, 79, 115441 (2009).

232. Kinloch, I.A., Suhr, J., Lou, J., Young, R.J., Ajayan, P.M. (2018) Composites with carbon nanotubes and graphene: An outlook. Compos. Mater., 362, 547–553.

233. Li, Z., Young, R.J., Wilson, N.R., Kinloch, I.A., Vallees, C., Li, Z. (2016) Effect of the orientation of graphene-based nanoplatelets upon the Young's modulus of nanocomposites. Compos. Sci. Technol., 123, 125–133.

234. Meenakshi, C.M., Krishnamoorthy, A. (2018) Preparation and mechanical characterization of flax and glass fiber reinforced polyester hybrid composite laminate by hand lay-up method. Materials Today: Proceedings, 5, 26934–26940.

235. Priya, I.I.M., Vinayagam, B.K. (2015) Enhancement of bi-axial glass fibre reinforced polymer composite with graphene platelet nanopowder modifies epoxy resin. Adv. Mech. Eng., 10, 1.

236. Arun, G.K. et al. (2018) Investigation on mechanical properties of graphene oxide reinforced GFRP. Mater. Sci. Eng., 310, 012158.

237. Aiyappa, M.R.P., Babu, S.L (2016) Studies on mechanical properties of graphene based GFRP laminates. International Journal of Engineering Research & Technology, 5, 411–415.

238. Juan, C., Zhao, D., Jin, X., Wang, C., Wang, D. Ge, H. (2014) Modifying glass fibers with graphene oxide: Towards high-performance polymer composites. Compos. Sci. Technol., 97, 41–45.

239. Liu, D., Liu, Y., Sui, G. (2016) Synthesis and properties of sandwiched films of epoxy resin and graphene/cellulose nanowhiskers paper. Composites: Part A, 84, 87–95.

240. Bindu Sharmila, T.K., Antony, J.V., Jayakrishnand, M.P., Sabura Beegum, P.M., Thomas Thachil, E. (2016) Mechanical, thermal and dielectric properties of hybrid composites of epoxy and reduced graphene oxide/iron oxide. Mater. Des., 90, 66–75.

241. Li, Y., Zhang, H., Huang, Z., Bilotti, E., Peijs, T. (2017) Graphite nanoplatelet modified epoxy resin for carbon fibre reinforced plastics with enhanced properties. J. Nanomater., 5194872, 10 p.

242. Umer, R., Li, Y., Dong, Y., Haroosh, H.J., Liao, K. (2015) The effect of graphene oxide (GO) nanoparticles on the processing of epoxy/glass fiber composites using resin infusion. Int. J. Adv. Manuf. Technol., 81, 2183–2192.

243. Jyothilakshmi, R., SunithBabu, L., Sridhar, B.S., Balasubramanya, H.S. (2017) Mechanical characterization and analysis of GFRP laminates with graphene reinforcement. Research & Reviews: Journal of Engineering and Technology, 6, 26–35.

244. Ramesh Kumar, S.C., Shivanand, H.K., Vidayasagar, H.N., Nagabhushan, V. (2018) Studies on mechanical properties of graphene based hybrid composites reinforced with kenaf/glass fiber. AIP Conference Proceedings, 1943, 020115.

245. Rathinasabapathi, G., Krishnamoorthy, A. (2019) Reinforcement effect of graphene enhanced glass fibre reinforced polymers: A prominence on graphene content. Digest J. Nano. Bio., 14, 641–653.

246. Singh, A.K. (2013) Crystallization kinetics of Se–Zn–Sb nano composites chalcogenide alloys. J. Alloys Compd., 552, 166–172.

Chapter 7

Applications of the Chalcogenide Systems– CNTs Nanocomposites

INTRODUCTION

Nanocomposites play a significant role in one of the most promising technologies, which is called as nanotechnology. Nanocarbon polymers can provide enormous possibilities to human civilization. The activated carbons have been used since the prehistoric age and they are playing major roles in various applications. Nanocomposites usually consists polymeric matrix materials and nanofillers, which held scientific, industrial and academic significance due to their improved properties. Even low filler contents as compared with the conventional micro and macro or neat counterparts, they exhibit superior property enhancements. Like thermoplastic composites can have a superior to conventional microscale composites and they can be synthesized using simple and inexpensive techniques. Similarly, polymer nanocomposite markets have also become attractive due to increasing application in the automotive and packaging sectors, aviation and many more. Therefore, the advantage of nanocarbons meant useful to large industries in tires, cars, printing, pencils, labtops etc. By controlling the two common things; storing hydrogen gas phase and electrochemical adsorption due to their cylindrical and hollow geometry at nanometer-scale diameters. Regarding to CNTs it has been predicted that they can store a liquid or a gas in the inner cores through a capillary effect. As a threshold for economical storage has set storage about 6.5% weight is minimum level for hydrogen fuel cells. Past several experimental findings have been demonstrated that single walled carbon nanotubes (SWCNTs) are able to meet and sometimes exceed this level by using gas phase adsorption (physisorption). However, most experimental findings with the high storage capacities are rather controversial, therefore, so it is debatable to assess the storage of the electrochemical storage. The main limitation is that a detailed understanding of the hydrogen storage mechanism and the effect of materials processing and its mechanism. These are a few major issues to the decades with nanocarbon process, to bring new technology/challenges in the area of hydrogen storage.

CNTs composite materials are only in demands since decades due discussed applications but they can be useful in many more areas due to their significant potentials. The stiffness of CNTs are also make them ideal candidates for structural applications. As an intense, they can be used as reinforcements in high-strength, low-weight and high-performance composites. Because, theoretically SWCNTs could have a Young's modulus of 1TPa, while the multi walled carbon nano tubes (MWCNTs) could be less strong. Owing to individual cylinders slide with respect to each other. In a similar way the ropes of SWCNTs can also less strong. Due to individual tubes may pull out by shearing and finally the whole rope could be break. This could results stresses far below the tensile strength of individual nanotubes. Moreover, nanotubes can also sustain large strains in tension without showing signs of fracture. Nanotubes are also highly flexible that is one of the most significant property of the nanotubes to be use as reinforcements for the composite materials. Although it has also pointed out in various studies nanotubes are not better fillers than the traditionally used carbon fibers. The key issue is to create a good interface between the nanotubes and the polymer matrix, because nanotubes are possessing very smooth and small diameter, that is nearly the same to the polymer chain. Second key issue is the nanotube common aggregation, this behave differently toward loads than individual nanotubes. These are the key limiting factors to a good load transfer for the cylinders in MWCNTs and shearing of tubes in SWCNT ropes. Therefore, to resolve these problems the aggregates need to be broken up and dispersed or cross-linked to prevent slippage. Thus, the main advantage to use nanotubes for structural polymeric composites is that the nanotube reinforcements could increase the toughness of the composites by absorbing energy during their highly flexible elastic behavior. Such composite materials are also show a large increase in conductivity, with only a little loss in photoluminescence and electroluminescence yields. Additionally, these composites are more robust than the pure materials. Nanotube–polymeric chalcogenide composites can be also used in other areas. Like one of the fastest-developing areas is in the biochemical field. As the membranes for molecular separations for osteointegration (growth of bone cells). Though these areas are less well explored. The most important thing about nanotubes for efficient use of them as reinforcing fibers is how to maneuver the surfaces chemically to enhance interfacial behavior between the individual nanotubes and the parent material matrix configuration. This could play a role in the development of CNTs composite based devices performances.

Similarly, another carbon allotrope i.e. graphene has also revolutionized after its discovery. It has completely redefined the modern-day technology with its remarkable properties. The research has exponentially grown by numerous universities, R&D establishments, and many more private and governmental bodies around the world. The versatile research on graphene can be found in every discipline, as of today. Therefore, prospects of graphene composites are carried out around the globe and several technical and scientific works presented every day. Due to graphene's exceptional thermal, mechanical and electronic properties, it stands out as the most promising candidate to be a major filling agent for composite applications. Graphene nanocomposites at very low loading can substantial enhancements in their multifunctional aspects, compared to conventional composites and their materials. This not only makes the material lighter with simple processing, but also makes it stronger for various multifunctional utility. Graphene properties can also able to improve the physicochemical qualities of the host matrix upon distribution. This could help the strengthening and increasing the interfacial bonds between the layers of graphene and the host matrix. This kind of bonding could dictate the emergence of the cumulative properties of graphene in reinforced nanocomposites. Graphene nano composites can have a great number of applications encompassing engineering, electronics, medicine, energy, industrial, household design, and many more. Majority of the work dealt with electronic/sensor-oriented applications, to generalize the broad applications of graphene and graphene-based nanocomposite into their respective disciplines. Graphene composites have also been extensively examined in the distinct areas such as, biomedical applications including drug delivery, gene delivery, cancer therapy, biosensing and bioimaging, graphene oxide based antibacterial materials, and scaffolds for tissue/cell culturing. Similarly, graphene composites are also applicable in the field of the field of memory devices for the high performances' electronics, ranging from electrochemical sensors to instrumentation. Therefore, graphene-based nanocomposites have a promising growth in technology and applications. There are several studies have directed toward its good dispersion ability with its derivatives in different kinds of the host matrices. But there are certain challenges remain to be resolved in obtaining large-scale, defect-free exfoliation of graphene with high quality and good properties. The high quality and quantity research work on graphene has attracted worldwide attention in them. It is expected that upto end of the year 2020, the graphene market could rise around 60%. Such remarkable growth in the coming years predicts an exponential boom in graphene research and development worldwide.

Considering the remarkable potentials in applicability of the novel carbon nanotubes-chalcogenide and graphene-chalcogenide systems composites under the glassy regime are taken-up in chapter. This work predominately focuses toward to provide a comprehensive overview on applications of the chalcogenide systems-CNTs nanocomposites including applications based on the nanocrystalline chalcogenide systems-CNTs composites as well as applications based on the polycrystalline chalcogenide systems-CNTs composites. An extensive detail has also been provided on applications based on the amorphous chalcogenide systems-CNTs composites. Moreover, a detail overview section on applications based on the amorphous chalcogenide systems-GF composites has been also accommodated. At the end of this chapter some futuristic emerging applications based on carbon nanotubes -chalcogenide glassy systems and graphene-chalcogenide glassy systems are also discussed.

OVERVIEW ON APPLICATIONS OF
THE CHALCOGENIDE SYSTEMS—CNTS NANOCOMPOSITES

Nanocomposites are novel words in modern nanotechnology, but their existence in functional devices and structures of nanometer dimensions is old; for example, Roman glass makers formulated glasses that incorporated nanosized metals in as early as the 4th century. Nanocomposite materials have the potential to overcome or improve on the disadvantages that currently impact microcomposite materials. Taking this into the account, preliminary testing of various groups of materials for nanocomposite potential in the 21st century revealed unconventional composite properties [1–3]. However, nanocomposite assembly also poses many preparation challenges associated with arrangement and stoichiometry in the nanocluster phase. It was also stated that nanocomposites could have high efficiency with rare property and arrangements [4–6]. As such, novel composites have clear beneficial advantages in many areas ranging from packaging to biomedical applications.

Nanocomposites with rare earth element properties ascend due to the materials' small size, large surface area and molecular interface phase relationship. These can be used to develop better performance of drugs, catalysts, biomaterials and other high value added materials. Changes in particle properties can also be manifested when the particle size is below a certain threshold–typically defined as 'critical size' [7, 8]. It is a well established concept that whenever their dimensions reach the nanometer level, the interactions at phase interfaces are mostly enhanced. Therefore, inclusion of carbon nanotubes as well as their successive use in terms of composites can have unique mechanical, thermal and electrical properties in an interesting dimensional area. At present, nanocomposites offer new technology and business opportunities for nearly every kind of industry, in particular those involved with environmentally friendly behavior [9–11].

Hence, nanocomposites offer an exceptionally extensive range of prospective applications from electronics, optical communications and biological systems to new materials development. Many possible nanocomposites based applications have been explored by fabricating various devices. Several potential applications and their new devices are being proposed. It is not possible here to summarize all the devices and applications that have been studied in the past. However, it is significant to note that the applications of nanocomposites in diverse fields have clearly different demands, and thus face different challenges, therefore, this field of innovations requires a distinct approach to conventional methods. [12–14]. As such composite reinforcements can refract the crack to deliver connecting elements that further deters opening of the crack. Moreover their integrated phase may experience phase transition in conjunction with the volume increase due to the stress field of propagating the crack that contributes in toughness and strengthening [12–14].

Studies examining properties of composites have indicated that nanocomposites are the potential structure material for ultra-lightweight spacecrafts used in space mission projects. Such composites materials can be useful to fabricate spacecraft devices with mobile mechanical parts such as gyroscopes, gears, solar arrays, antennae, drives, sunshields, rovers, radars, solar concentrators and reflector arrays. The spacecraft parts can be manufactured from suitable materials that allow the parts to be folded or packaged into small volumes, as spacecrafts require ultra-lightweight parts that can be deployed mechanically into a large ultra-lightweight functioning. Likewise, rocket propellants can also be prepared from a polymer-Al/Al$_2$O$_3$ nanocomposite to improve ballistic performance [15–18].

Carbon nanotubes are also considered due to their superior properties such as rigidity, strength, elasticity, electric conductivity and field emission. Carbon nanotubes carbon allotrope can form from cylindrical carbon molecules with novel properties which makes them potentially useful in a wide variety of applications such as nanotechnology, electronics, optics, etc. Among these, carbon nanotubes may have many different structures, differing in length, thickness and in the type of helicity and number of layers. All of these were formed from essentially the same graphite sheet and their electrical characteristics differ depending on these variations, acting either as metals or as semiconductors [19–24]. Carbon nanotubes are also highly conductive both to electricity and heat, with an electrical conductivity as high as copper and a thermal conductivity such as diamond. Their extraordinary mechanical properties are 100 times stronger than steel that can be used in military, aerospace and medical applications such as lubricants, coatings, catalysts and electro-optical devices. The carbon nanotubes composites materials stiffness, strength, and tenacity compared to other fibers and their thermal and electrical conductivity can also be very high, compared to other high conductive materials. Thus, the carbon nanotubes unique property as reinforcement filler materials makes them applicable in the following key areas:

- Drug delivery systems
- UV protection gels
- Lubricants and scratch free paints
- New fire retardant materials
- New scratch/abrasion resistant materials
- Superior strength fibers and films

Such nanocomposite materials also have numerous automotive and general/industrial applications such as mirror housings for various types of vehicles, mobile phones, pagers, films, food packaging, environmental protection, etc. [25]. Recently, investigators paid attention towards carbon nanotubes composites biological and biomedical applications considering their covalent and non-covalent functionalizations, which makes carbon nanotube composites a potential candidate for applications in the field of biosensors, tissue engineering, as well as biomedical devices [26–32].

Considering the physical properties described above and their analogous properties of the chalcogenide based nanocomposites, these have been fabricated for many specific potential applications, such as nanoelectronics, sensors, etc. Such composites can also be useful for the substantial global upsurge in the depletion of fossil fuels from the rapid growth of global economy. In recent years, a consensus was made about the exhaustion of existing fossil fuel reserves and to prevent the increase in greenhouse gas emissions to save from environmental pollution. Therefore, it is desirable to develop and commercialize sustainable environment friendly energy sources and their related technologies are being developed globally as a matter of urgency [33–36]. These technologies include, but are not limited to as for example, electrochemical supercapacitors that can have overriding importance because of their exceptional power density and storage properties compared to other contemporary energy storage devices. Composites based supercapacitors may have a number of great advantages including long life cycle, high power density, high efficiency, high specific capacitance, flexible operating temperature and be environmentally friendly. Additionally, they can be quickly charged with fast power delivery and are capable to bridge the gap between batteries and conventional capacitors [37].

Due to industrial and scientific importance, the metal chalcogenides (S, Se, and Te) have drawn great attention in the past two decades specifically because of their anisotropic property. Usually, the transition elements of groups IV to VII B can combine with VI A group elements, such as S, Se, and Te to form binary stable layered crystalline structures [38]. In general, such layered transition metal chalcogenides are represented by the simple formula MX_2, here M is the transition element in groups IV B (Ti, Zr, Hf), V B (V, Nb, Ta), VI B (Mo, W), or VII B (Tc, Re) and X is a chalcogenide atom in the VI A group (S, Se, Te). The structure and properties of most of the transition metal chalcogenides comparatively resemble semimetal pristine graphene, except for the band gap [39], that is nearly zero in pristine graphene whereas in transition metal chalcogenides this would be non-zero. The value depends on the elemental combination as well as the number of layers and the presence or lack of adopting atoms. This directly indicates their band gap values lie between 0 and 2 eV. Such composites materials variation in band gap for different transition metal chalcogenides structures can be tuned; therefore, they might become industrially important materials [40]. They have also gained considerable attention due to their high specific power as well as long stability and life cycle, which offer better safety tolerance relative to batteries for their wide range applications in consumer electronics, electric tools, buffer powers, hybrid electronic vehicles and so forth [39]. Metal chalcogenides can also be applied in the fields of fuel cells, solar cells, light-emitting diodes, sensors, lithium-ion batteries, electrocatalysts, thermoelectric devices and memory devices due to their excellent properties. Additionally their advantages also include (i) improved life cycle; (ii) flexibility; (iii) provide additional reactive sites as well as catalytic activity; (iv) improved conductivity with the reduction of inner resistance and ohmic loss; (v) short path lengths to electron transport; and (vi) quantum-sized effects. There are a number of binary, ternary compositions have been extensively used for device fabrications such as Ni_3S_2, CuS, Co_3S_4, Bi_2S_3, La_2S_3, WS_2, $CuSe$, $MoSe_2$, $CoSe_2$, $NiCo_2S_4$, MoS_2, $NiSe\text{-}MoSe_2$, $MoSe_2@CN$, etc. [37, 41]. Therefore chalcogenide–carbon nanotubes are the new potential field and can can be applied in different areas, and not only limited to the fields described above.

APPLICATIONS BASED ON THE NANOCRYSTALLINE CHALCOGENIDE SYSTEMS—CNTs COMPOSITES

The emerging nanocrystalline chalcogenide–carbon nanotubes have also been recognized as the field for a wide range of applications, such as carbonous metallic framework of multi-walled carbon nanotubes/Bi_2S_3 nanorods as heterostructure composite films for efficient quasi solid state dye sensitized solar cells (QDSSCs). According to reports, bismuth sulfide (Bi_2S_3) is a layered semiconductor with a direct band gap of 1.3 eV with a high absorption coefficient, superior photocatalytic activity and reasonable conversion efficiency. In addition, Bi_2S_3 can be easily synthesized by consuming low energy and cost effective methods. MWCNTs can provide fast electron transport due to the coexistence of a tubular morphology as well as diffusive transport and good mechanical properties [42, 43]. These exceptional properties and immense suitability in electronic composite materials enable MWCNTs to be assembled with a well-known transition metal sulfide, bismuth sulfide [44]. Such heterostructured materials have two different energy level systems that play an important role to achieve electron hole separation. Additionally, the coupling of MWCNTs with Bi_2S_3 can also reduce the electron-hole recombination and enhance the light absorption ability. The typical schematic of the composites based DSSC is shown in Fig. 7.1. To fabricate photo anodes, double-holed FTO glasses can be used, the TiO_2 layer can then be deposited on FTO glass and the photo anodes dried. The annealed photo anodes are immersed into a dye solution for a desired time. The DSSCs can be fabricated by sandwiching the dye-immersed photo anodes and composites material-coated counter electrodes with an appropriate thickness surlyn as a sealant and cell spacer. Additionally, the desired amount of gel electrolyte is also filled into the cell through a hole drilled into the photo anodes that is sealed by surlyn and a cover glass [44].

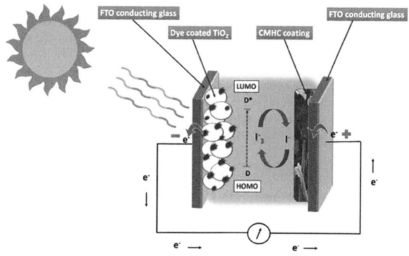

Figure 7.1 Illustration of DSSC cell.
(Reproduced from the permission, Memon, et al., 2018. Carbonous metallic framework of multi-walled carbon Nanotubes/Bi$_2$S$_3$ nanorods as heterostructure composite films for efficient quasisolid state DSSCs. Electrochimica Acta 283, 997–1005. copyright © 2018, Elsevier Ltd. [44])

Moreover, a verity of applications of the nanocrystalline chalcogenides–carbon nanotubes composite materials have been reported, He et al., demonstrated that such composite materials have potential application for the ZnSe/N-doped hollow carbon architectures as high-rate and long-life anode materials for half/full sodium-ion and potassium-ion batteries [45]. Xu et al., studied the Ni$_x$Co$_{1-x}$Se$_2$ nanosheets composite for superior lithium storage capability [46]. Etogo et al., presented the enhanced potassium-ion storage in Co$_{0.85}$Se@C nanoboxes film [47]. Similarly Wei et al., demonstrated the applicability of the zinc-cobalt binary metal sulfide @ N-doped carbon with the enhanced lithium-ion storage capability [48].

The nanocrystalline chalcogenide–carbon nanotubes composites can also be applicable as biomaterials with improved parameters, Huang et al., presented the results on the tungsten-cobalt sulfide-based heteroatom doped porous carbon (WS$_2$/Co$_{1-x}$S@N, S co-doped porous carbon) nanocomposites utilizing the *in situ* synthesized PTA@ZIF-67 as precursors, in which the PTA was used to an abundant tungsten source, while the metal ion in ZIF-67 acted as a cobalt source for the organic linker in ZIF-67 [49]. These results have shown that such composites are useful for an efficient bifunctional electrocatalysts [49].

Moreover incorporation of the MWCNTs in HgS can form composite materials for a high performance superconductor. The composite of MWCNTs with HgS as metal sulfide can enhance the electrochemical activity towards the fabrication of supercapacitive electrode. Pande et al., explored the possibility in their study using the facile chemical route for multi-walled carbon nanotube/mercury sulfide nanocomposite to produce high performance supercapacitive electrode [50]. They demonstrated that the nanocrystalline HgS–MWCNTs composite material have a high specific capacitance (946.43 F/g) with synergetic energy density of 42.97 Wh/kg along with the enhanced cycling stability of 93% over 4000 cycles for the single electrode system. Such excellent electrochemical performance was related to the fast reversible redox reactions in nanoregime, which enabled short diffusion paths of electrolyte ions that was supplied by composite material. Thus, the nanostructured composite can be applied in high performance supercapacitor application. Their complete solid-state device fabrication can replace the liquid electrolyte with polymer gel to overcome some major limitations of aqueous route issues such as improper sealing and solvent evaporation. Hence, such composites solid-state device have great potential as high performance power sources for flexible, lightweight and portable electronics in the future [50].

In a similar way, potential composites material applications have been also demonstrated with the chalcocgenides and Single Walled Carbon Tubes (SWCNTs) composites materials [51, 52]. Such composite materials physical properties can be useful for the fabrication of high quality advanced fibers and wires. Since carbon nanotubes display very interesting fluidic properties, they are useful for high quality membrane fabrication for water desalination. The chalcogenide–SWCNTs composites are also suitable for the fabrication of conducting thin films for optoelectronics as well as transistors applications. Such composite materials are also a potential candidate to make the thermal interface for electronics packaging. Additionally, chalcogenide–SWCNTs can also be useful for the fabrication of high performance energy storage materials. Hence, the applicability of these composites materials can have a broader range and not only in the areas described above [51, 52].

APPLICATIONS BASED ON THE POLYCRYSTALLINE CHALCOGENIDE SYSTEMS—CNTs COMPOSITES

Polycrystalline chalcogenides–CNTs composites are also a useful class of material that can be used for various applications [53]. Recent investigations in this field have shown the polycrystalline chalcogenide–CNTs configuration alloys can possess enhanced electrical properties. Singh et al., demonstrated the enhanced electrical conductivity and lower activation energy as well as other improved electrical parameters for the polycrystalline $(Se_{80}Te_{20})_{100-x}Ag_x$ $(0 \leq x \leq 4)$–CNTs composite systems [54]. According to this study, despite the low concentration addition of CNTs in $(Se_{80}Te_{20})_{100-x}Ag_x$ $(0 \leq x \leq 4)$ system, there is a rapid increment in electrical conductivity due to the increase in silver ion transport. Upadhyay et al., also found similar trends in enhancement electrical properties in their systematic studies with the $[(Se_{85}Te_{10}Ag_5)_{100-X}(CNT)_X]$ systems [55–58]. According to these studies on polycrystalline chalcogenide–CNTs composites materials; the low frequency ac conductivity plateau and high-frequency dispersion system can be directly responsible for the onset of the conductivity relaxation (switch over frequency) that shifts to higher frequencies from the increasing CNT content in pure $Se_{85}Te_{10}Ag_5$ alloy, this leads to a better CNT–CNT connectivity due to incorporation of CNT. A significant increment in dc conductivity value can also be achieved with the increasing CNT concentration, their conductivity enhancement may be in the order of 6 to 9 in magnitude. The incorporation of small amounts of the CNTs in these systems can also widely impact dielectric constant values with the increasing amount of CNTs within the chalcogenide rich configurations. This could directly relate to more charge creation within the configuration due to the incorporation of a small amount of CNTs. The enhancement in thermal conductivity has also been noticed in such systems [55–58]. Considering the rapid enhancement in electrical and thermal conductivities in such composite materials, it could be noted that they would be a potential candidate for the fabrication of future high performing optoelectronics devices. More recently investigations with the polycrystalline chalcogenide–CNT have also shown that these composites materials can be useful to fabricate hybrid counter electrode and flexible electrodes for efficient dye-sensitized solar cells as well as photocatalyst for the hydrogen production under visible-light [59–62].

Hybrid materials can deliver better properties compared with their individual counterparts, in which inorganic material can play several roles: enhancing the mechanical, thermal stability, the refractive index with an accessible and interconnected network for sensing/or catalysts or to contribute in magnetic, electronic, redox, electrochemical or chemical properties. Additionally, organic materials have the ability to modify the mechanical properties for the production of films and fibers, and various geometric structures for integrated optics under the controlled network connectivity for a specific physical or chemical property including electrical or optical characteristics, electrochemical behavior, chemical or biochemical reactivity, etc. [63].

APPLICATIONS BASED ON THE AMORPHOUS CHALCOGENIDE SYSTEMS—CNTs COMPOSITES

Amorphous chalcogenide–CNTs composites are another class of materials that could fulfill various drawbacks of other forms of materials, such as bulk metal selenides tolerate dramatic volume variation and particles agglomeration during the repeated charge-discharge process in lithium-ion batteries application, this causes severe polarization and rapid capacity fading problems [64–66]. According to different reports, the zinc selenide (ZnSe) can have a high theoretical capacity based on the alloying and conversion reactions [67, 68]. ZnSe has also been widely used in various laser devices [69], semiconductors [70] and solar cells [71], this may help develop their commercial use for energy storage. Several researchers have investigated the application of ZnSe-based electrodes for lithium-ion batteries. As an example, ZnSe composite with hollow carbon could deliver a high reversible discharge capacity of 1134 mAh g^{-1} after 500 cycles at 0.6 A g^{-1} [72]. This rising capacity could be connected to the generation and activation of Se during the electrochemical process [73]. The microspheres ZnSe–CNT composite has also been used as an anode for the sodium-ion batteries, this result has indicated the possibility of delivering capacity of 387 mAh g^{-1} for 180 cycles [74]. But ZnSe composites still have aggregation problems to some extent, therefore, this causes the overall reduction in cycle-life performance of electrodes. Therefore, it is essential to engineer a more robust electrode that can accommodate the structural changes in the ZnSe during the insertion/desertion of ions. Thus, the novel hierarchical hybrid nanocomposite of ultrafine ZnSe nanoparticles directly grow on the outer surface and in the inner cavity of amorphous hollow carbon nanospheres (ZnSe@CNTs) via a facile hydrothermal process could resolve this issue. This can prevent the severe aggregation and ensure the structural integrity of ZnSe@CNTs electrodes during cycling compared to the aggregated ZnSe microspheres, numerous ultrafine ZnSe nanoparticles are firmly anchored onto the conductive CNTs to get a more accessible surface to carry out high ratio surface and near-surface redox reactions. This is to ensure the ZnSe@CNTs electrode long-term cycle stability and high-rate performance of more than 1000 cycles at 1.0 A g^{-1}, with a stable reversible discharge capacity of 361.9 mAh g^{-1} and capacity retention of 87.0% under the Coulombic efficiency above 99.9% [75]. The typical $Na_3V_2(PO_4)_3$ (NVP)//ZnSe@CNTs electrode schematic is illustrated in Fig. 7.2. Under such a

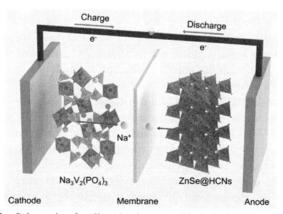

Figure 7.2 Schematic of sodium-ion battery with ZnSe@CNTs//NVP couple.

(Reproduced from the permission, Lu et al., 2019. Construction of ultrafne ZnSe nanoparticles on/in amorphous carbon hollow nanospheres with high-power-density sodium storage. Nano Energy, 59, 762–772. copyright © 2019 Elsevier Ltd. [75])

tailored structure, the ultrafine ZnSe precursor particles both on/in their surface can significantly inhibit the volume expansion during the repeated sodiation and desodiation process. Specifically, for the sodium-ion batteries anodes in the ether-based electrolyte the ZnSe@CNTs (amorphous) can have a high reversible capacity of 361.9 and 285.9 mAh g^{-1} at 1 and 10 A g^{-1} more than 1000 cycles [75].

In addition to these key applications based on the amorphous form of the chalcogenide–CNTs composites, innovations have also been achieved from the chalcogenide alloys–CNTs composites for various applications such as memory elements, cryogenic microelectronics, ion batteries, solar cells and sensing. For example, enhanced electrical properties for the glassy $AgGe_{1+x}As_{1-x}(S+CNT)_3$ at temperatures ranging from 10 to 300 K [76]. Moreover, the improved thermal and mechanical properties have also been reported for the $CNT–Se_{90-x}Te_{10}Ag_x$ glassy composites [77]. The report of innovations to explore different specific parameters is still ongoing in this novel research field for high efficient future optoelectronics devices by exploring distinct physical properties of chalcogenide–CNTs glassy composites.

OVERVIEW AND APPLICATIONS BASED ON THE AMORPHOUS CHALCOGENIDE SYSTEMS—GF COMPOSITES

In recent years, specifically transition metal chalcogenides, such as the sulfides and selenides have received growing research interest as potential electrode materials for energy storage and conversion due to its tuneable stoichiometric compositions, unique crystal structures and rich redox sites as well as relatively higher electrical conductivity in comparison to their transition metal oxide counterparts [78–80]. As an example, compared to routine anode materials (graphite) in lithium-ion batteries based on insertion/deinsertion mechanism, the use of transition metal chalcogenides can provide a higher theoretical special capacity [78, 80]. Their conversional mechanism is mainly governed by the MS $(Se)_n + 2nLi^+ + 2ne^- \leftrightarrow nLi_2S(Se) + M$, whereas M (M = Sn, In, Sb , Bi, Mo, W, V, etc.) [79, 81–85]. The nanostructured metal chalcogenides could act as polar hosts with a strong affinity that is useful for the long cycling life of sulfur utilization [86]. Metal chalcogenides can also be practical for fabricating supercapacitors that can have excellent energy density [81]. They are also prominent candidates to be used for the fabrication of Dye-Sensitized Solar Cells (DSSCs) to deliver superior energy conversion efficiency of the cells. Their water splitting ability that derives from the electro- and photoelectron-chemistry can also be considered a promising pathway for hydrogen production. Moreover, metal chalcogenides can also be applied in metal-air batteries and various kinds of photo-/electro-catalyst to catalyze water dissociation, due to their special band structure as well as unique electronic configuration [87]. Besides several advantages, metal chalcogenides are also free from the drawbacks such as low special surface area, inferior reactivity, low electron/ion transfer rate and rapid recombination rate of electrons and holes [88]. Such drawbacks of the chalcogenide materials can be overcome by making their composite materials [81].

Graphene is a promising electron-shuttling material with excellent transport behavior with sp^2 network [89]. It has an unusually high charge-carrier mobility of ca. 200 000 $cm^2V^{-1}s^{-1}$ at room temperature [90–94]. Therefore, graphene–semiconductor composite can shuttle the photogenerated electrons as well as reduce charge-recombination losses [95]. Additionally, graphene high electrical conductivity, mechanical strength and flexible charge carrier mobility has provided a huge potential in electrode materials for next-generation energy storage and conversion [81, 96–98]. Graphene two-dimensional thin sheet structure makes it ideal to support inorganic nanomaterials to build nanostructured devices [99–102].

Considering the application of the various forms of the existing metal chalcogenides–graphene composites materials, it is expected that newly established chalcogenides–graphene glassy field can also have a great potential for administration in the area of optoelectronics, medicine, etc. Yet the direct implemention of this class of materials has not been extensively explored. Considering, the large change in their physical properties in composites form, as demonstrated from their initial studies (shown in Chapter 6). Therefore, a scientific quest is desired to explore this class of materials for the future potential application in different fields.

FUTURE APPLICATIONS BASED ON THE AMORPHOUS CHALCOGENIDE SYSTEMS—CNTs AND CHALCOGENIDE—GF COMPOSITES

The potential chalcogenide–CNTs and chalcogenide–GF glassy composites can have a wide range of future prospective application in marked areas including optoelectronics, biomedical, etc. For instance, Singh's systematically initial research work on such composites materials have shown an impassive IR transparency percentage increments compared to parent glassy configurations. On that basis, a prospective sensing application of chalcogenide–MWCNTs and chalcogenide–GF composites materials have been discussed here.

A schematic representation is illustrated in Fig. 7.3, for the IR naval sensing application. From this, it was demonstrated that if a ship at a particular distance from the sensing system spreads over a low IR region which cannot be easily detected from the usual chalcogenide glass based device but can be noticed from such composites devices. Additionally, chalcogenide–MWCNTs and chalcogenide–GF based sensing devices can also provide a broader range of detection strong singles beyond the conventional IR sensing devices.

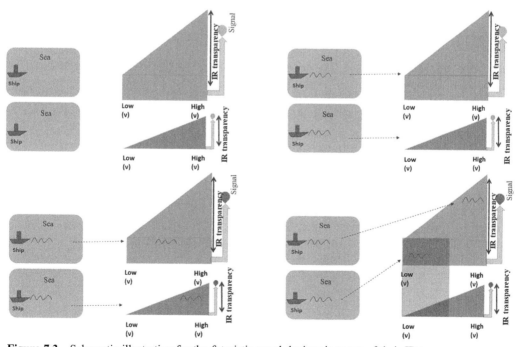

Figure 7.3 Schematic illustration for the futuristic naval devices in terms of their IR transparency range.

Similarly another possible sensing application based on chalcogenide–MWCNTs and chalcogenide–GF glassy composites can also be useful for a biomedical utility, as shown in Fig. 7.4. From this, it can be assumed that a low energy species I was detected from both conversional as well as composite based sensing devices, a high energy species II then cannot be identified from conventional material based device but can be discovered from the chalcogenide–MWCNTs and chalcogenide–GF devices due to their initial high and broader range IR transparency region.

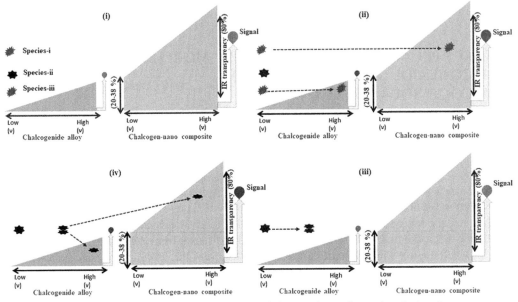

Figure 7.4   Schematic illustration for the futuristic optoelectronic sensing devices in terms of their IR transparency range.

Additionally, if we assume a metabolism having moderate energy and initially it breaks into parts having both low and high energies segments, in this case only low energy metabolism segments can be detected from the conventional material based device, while chalcogenide–MWCNTs and chalcogenide–GF based device can easily detect both parts of the metabolism. Hence, with these examples, efforts were made to show the superiority of chalcogenide–MWCNTs and chalcogenide–GF composites materials devices over the conventional chalcogenide glassy devices for future applications. Therefore, such novel composites materials prospective applications are not limited to these representations and can be in numerous new distinct areas.

CONCLUSIONS

The key technical applications of the chalcognide composite materials including their distinct compositions in the form of nanocrystalline chalcogenide systems-CNTs, polycrystalline chalcogenide systems-CNTs and amorphous chalcogenide systems-CNTs composites have been reviewed here. Such composites can be applicable for different purposes like optoelctronics, photovoltaic, nanoelectronics, with improved performances. Moreover, possible applications of the distinct graphene composites including amorphous chalcogenide–graphene composite materials by providing a brief overview on their existing applicability have also been discussed. In this order some possible futuristic applications based on the recently introduced new class of amorphous chalcogenide–MWCNTs and chalcogenide–graphene composite materials physical properties have also been described. These futuristic applications are mainly based on their unique IR transparency property, and are not limited to that, their applicability range can be wider in various technological areas. Hence, these new class composite materials are futuristic applications still open for innovations, to fabricate high performance devices which may fulfill future demands.

References

1. Ounaies, Z., Park, C., Wise, K.E. et al. (2003) Electrical properties of single wall carbon nanotube reinforced polyimide composites. Comp. Sci. Tech., 63, 1637–1646.

2. Schmidt, D., Shah, D., Giannelis, E.P. (2002) New advances in polymer/layered silicate nanocomposites. Cur. Opin. Solid State & Mater. Sci., 6, 205–212.

3. Gleiter, H. (1992) Materials with ultrafine microstructures: Retrospectives and perspectives. Nanostr. Mater., 1, 1–19.

4. Kamigaito, O. (1991) What can be improved by nanometer composites? J. Japan Soc. Powder Metal., 38, 315–321.

5. Iijima, S. (1991) Helical microtubes of graphitic carbon. Nature, 354, 56–58.

6. Braun, T., Schubert, A., Sindelys, Z. (1997) Nanoscience and nanotechnology on the balance. Scientometrics, 38, 321–325.

7. Choa, Y.H., Yang, J.K., Kim, B.H. et al. (2003) Preparation and characterization of metal: Ceramic nanoporous nanocomposite powders. J. Magnet. Magnetic Mater., 266, 12–19.

8. Wypych, F., Seefeld, N., Denicolo, I. (1997) Preparation of nanocomposites based on the encapsulation of conducting polymers into $2H-MoS_2$ and $1T-TiS_2$. Quimica Nova, 20, 356–360.

9. Aruna, S.T., Rajam, K.S. (2003) Synthesis, characterisation and properties of Ni/PSZ and Ni/YSZ nanocomposites. Scripta Materialia, 48, 507–512.

10. Giannelis, E.P. (1996) Polymer layered silicate nanocomposites. Adv. Mater., 8, 29–35.

11. Sternitzke, M. (1997) Review: Structural ceramic nanocomposites. J. Eur. Ceram. Soc., 17, 1061–1082.

12. Peigney, A., Laurent, C.H., Flahaut, E. et al. (2000) Carbon nanotubes in novel ceramic matrix nano-composites. Ceramic Inter., 26, 677–683.

13. Alexandre, M., Dubois, P. (2000) Polymer-layered silicate nanocomposites: Preparation, properties and uses of a new class of materials. Mater. Sci. & Eng., 28, 1–63.

14. Gangopadhyay, R., Amitabha, D. (2000) Conducting polymer nanocomposites: A brief overview. Chem. Mater., 12, 608–622.

15. Jordan, J., Jacob, K.I., Tannenbaum, R. et al. (2005) Experimental trends in polymer nanocomposites: A review. Mater. Sci. Eng. A., 393, 1–11.

16. Choi, S.M., Awaji, H. (2005) Nanocomposites: A new material design concept. Sci. Tech. Adv. Mater., 6, 2–10.

17. Xie, X.L., Mai, Y.W., Zhou, X.P. (2005) Dispersion and alignment of carbon nanotubes in polymer matrix: A review. Mater. Sci. Eng. R., 49, 89–112.

18. Din, S.H., Shah, M.A., Sheikh, N.A., Butt, M.M. (2018), Nano-composites and their applications: A Review. Characterization and Application of Nanomaterials, 2, 9 p. DOI: 10.24294/can.v2i1.875.

19. Iijima, S., Ichihashi, T. (1993) Single-shell carbon nanotubes of 1-nm diameter. Nature, 363, 603–605.

20. Salvetat, J.P., Bonard, J.M., Thomson, N.H., Kulik, A.J., Forró, L., Benoit, W., Zuppiroli, L. (1999) Mechanical properties of carbon nanotubes. Appl. Phys. A., 69, 255–260.

21. Khan, U., Ryan, K., Blau, W.J., Coleman, J.N. (2007) The effect of solvent choice on the mechanical properties of carbon nanotubepolymer composites. Compos. Sci. Technol., 67, 3158–3167.

22. Kearns, J.C., Shambaugh, R.L. (2002) Polypropylene fibers reinforced with carbon nanotubes. J. Appl. Polym. Sci., 86, 2079–2084.

23. Moniruzzaman, M., Winey, K.I. (2006) Polymer nanocomposites containing carbon nanotubes. Macromolecules, 39, 5194–5205.

24. Khare, R., Bose, S. 2005. Carbon nanotube based composites–A review. J. Minerals & Mater. Chara. & Eng., 4, 31–46.

25. Baksi, S., Biswas, S. (2014) Nanocomposites–An overview. The Scitech Journal, 1, 22–30.

26. Hirsch, A. (2002) Functionalization of single-walled carbon nanotubes. Angew Chem. Int. Ed. Engl., 241, 1853–9.

27. Chen, R.J., Zhang, Y., Wang, D., Dai, H. (2001) Noncovalent sidewall functionalization of single-walled carbon nanotubes for protein immobilization. J. Am. Chem. Soc., 123, 3838–9.

28. Besteman, K., Lee, J.O., Wiertz, F.G.M., Heering H.A., Dekker C. (2003) Enzyme-coated carbon nanotubes as single-molecule biosensors. Nano Lett., 3, 727–30.

29. Chen, R.J., Bangsaruntip, S., Drouvalakis, K.A., Kam, N.W.S., Shim, M., Li, Y.M. et al. (2003) Noncovalent functionalization of carbon nanotubes for highly specific electronic biosensors. Proc. Natl. Acad. Sci. USA, 100, 4984–9.

30. Chen, X., Tam, U.C., Czlapinski, J.L., Lee, G.S., Rabuka, D., Zettl, A. et al. (2006) Interfacing carbon nanotubes with living cells. J. Am. Chem. Soc., 128, 6292–3.

31. Kam, N.W.S., O'Connell, M., Wisdom, J.A., Dai, H. (2005) Carbon nanotubes as multifunctional biological transporters and near-infrared agents for selective cancer cell destruction. Proc. Natl. Acad. Sci. USA, 102, 11600–5.

32. Kang, Y., Liu, Y.C., Wang, Q., Shen, J.W., Wu, T., Guan, W.J. (2009) On the spontaneous encapsulation of proteins in carbon nanotubes. Biomaterials, 30, 2807–2815.

33. Theerthagiri, J., Senthil, R., Senthilkumar, B., Polu, A.R., Madhavan, J., Ashokkumar, M. (2017) Recent advances in MoS_2 nanostructured materials for energy and environmental applications—A review. J. Solid State Chem., 252, 43–71.

34. Thiagarajan, K., Theerthagiri, J., Senthil, R., Arunachalam, P., Madhavan, J., Ghanem, M.A. (2017) Synthesis of $Ni_3V_2O_8$@graphene oxide nanocomposite as an efficient electrode material for supercapacitor applications. J. Solid State Electrochem., 22, 527–536.

35. Arunachalam, P., Ghanem, M.A., Al-Mayouf, A.M., Al-shalwi, M. (2017) Enhanced electrocatalytic performance of mesoporous nickel-cobalt oxide electrode for methanol oxidation in alkaline solution. Mater. Lett., 196, 365–368.

36. Theerthagiri, J., Sudha, R., Premnath, K., Arunachalam, P., Madhavan, J., Al-Mayouf, A.M. (2017) Growth of iron diselenide nanorods on graphene oxide nanosheets as advanced electrocatalyst for hydrogen evolution reaction. Int. J. Hydrogen Energy, 42, 13020–13030.

37. Theerthagiri, J., Karuppasamy, K., Durai, G., Rana, A.H.S., Arunachalam, P., Sangeetha, K. Kuppusami, P., Kim, H.S. (2018) Recent advances in metal chalcogenides (MX; $X = S$, Se) nanostructures for electrochemical supercapacitor applications: A brief review. Nanomaterials, 8, 256–28.

38. Simon, P., Gogotsi, Y. (2008) Materials for electrochemical capacitors. Nat. Mater., 7, 845–854.

39. Wang, G., Zhang, L., Zhang, J. (2012) A review of electrode materials for electrochemical supercapacitors. Chem. Soc. Rev., 41, 797–828.

40. Kötz, R., Carlen, M. (2000) Principles and applications of electrochemical capacitors. Electrochim. Acta, 45, 2483–2498.

41. Ojha, M., Deepa, M. (2019) Molybdenum selenide nanotubes decorated carbon net for a high performance supercapacitor. Chem. Eng. J., 368, 772–783.

42. Li, G.R., Wang, F., Jiang, Q.W., Gao, X.P., Shen, P.W. (2010) Carbon nanotubes with titanium nitride as a low-cost counter-electrode material for dye-sensitized solar cells. Angew. Chem. Int. Ed., 49, 3653–3656.

43. Luo, Y., Gong, Z., He, M., Wang, X., Tang, Z., Chen, H. (2012) Fabrication of high-quality carbon nanotube fibers for optoelectronic applications. Sol. Energy Mater. Sol. Cell., 97, 78–82.

44. Memon, A.A., Patil, S.A., Sun, K.C., Mengal, N., Arbab, A.A., Sahito, I.A., Jeong, S.H., Kim, H.S. (2018) Carbonous metallic framework of multi-walled carbon Nanotubes/Bi_2S_3 nanorods as heterostructure composite films for efficient quasisolid state DSSCs. Electrochimica Acta 283, 997–1005.

45. He, Y., Wang, L., Dong, C., Li, C., Ding, X., Qian, Y., Xu, L. (2019) In-situ rooting ZnSe/N-doped hollow carbon architectures as high-rate and long-life anode materials for half/full sodium-ion and potassium-ion batteries. Energy Storage Materials, 23, 35–45.

46. Xu, H., Wu, R. (2019) Porous hollow composites assembled by $NixCo_{1-x}Se_2$ nanosheets rooted on carbon polyhedra for superior lithium storage capability. J. Collo. Inter. Sci., 536, 673–680.

47. Etogo, C.A., Huang, H., Hong, H., Liu, G., Zhang, L. (2020) Metal–organic-frameworks-engaged formation of $Co_{0.85}Se$@C nanoboxes embedded in carbon nanofibers film for enhanced potassium-ion storage. Energy Storage Materials, 24, 167–176.

48. Wei, X., Zhang, Y., Zhang, B., Lin, Z., Wang, X., Hu, P., Li, S., Tan, X., Cai, X., Yang, W., Mai, L. (2019) Yolk-shell-structured zinc-cobalt binary metal sulfide @ N-doped carbon for enhanced lithium-ion storage. Nano Energy, 64, 103899–8.

49. Huang, Z., Yang, Z., Hussain, M.Z., Chen, B., Jia, Q., Zhu, Y., Xia, Y. (2020) Polyoxometallates@ zeolitic-imidazolate-framework derived bimetallic tungsten-cobalt sulfide/porous carbon nanocomposites as efficient bifunctional electrocatalysts for hydrogen and oxygen evolution. Electrochimica Acta, 330, 135335–13.

50. Pande, S.A., Pandit, B., Sankapal, B.R. (2018) Facile chemical route for multiwalled carbon nanotube/mercury sulfide nanocomposite: High performance supercapacitive electrode. J. Colloid Inter. Sci., 514, 740–749.

51. Kuganathan, N., Chroneos, A. (2019) Encapsulation of cadmium telluride nanocrystals within single walled carbon nanotubes. Inorg. Chimica Acta, 488, 246–254.

52. Rao, R., Pint, C.L., Islam, A.E. et al. (2018) Carbon nanotubes and related nanomaterials: Critical advances and challenges for synthesis toward mainstream commercial applications. ACS Nano, 12, 11756–11784.

53. Camargo, P.H.C., Satyanarayana, K.G., Wypych, F. (2009) Nanocomposites: Synthesis, structure, properties and new application opportunities. Mater. Res., 12, 1–39.

54. Singh, D., Kumar, S., Thangaraj, R. (2014) Electrical properties of MWCNT-composite $(Se_{80}Te_{20})_{100-x}Ag_x$ ($0 \leq x \leq 4$) chalcogenide glasses. Phase Transitions: A Multinational Journal, 87(2), 148–156.

55. Upadhyay, A.N., Tiwari, R.S., Singh, K. (2015) Effect of carbon nanotube additive on the structural and thermal properties of $Se_{85}Te_{10}Ag_5$ glassy alloy. J. Therm. Ana. Calor., 122, 547–552.

56. Upadhyay, A.N., Tiwari, R.S., Singh, K. (2015) Electrical and dielectric properties of carbon nanotube containing $Se_{85}Te_{10}Ag_5$ glassy composites. Adv. Mater. Lett., 6, 1098–1103.

57. Upadhyay, A.N., Tiwari, R.S., Mehta, N., Singh, K. (2014) Enhancement of electrical, thermal and mechanical properties of carbon nanotube additive $Se_{85}Te_{10}Ag_5$ glassy composites. Mater. Lett., 136, 445–448.

58. Upadhyay, A.N., Tiwari, R.S., Singh, K. (2018) Annealing effect on thermal conductivity and microhardness of carbon nanotube containing $Se_{80}Te_{16}Cu_4$ glassy composites. Mater. Res. Express, 5, 025203–9.

59. Liu, X., Gao, L., Yue, G., Zheng, H., Zhang, W. (2017) Efficient dye-sensitized solar cells incorporating hybrid counter electrode of $CuMnSnS_4$ microspers/carbon nanotubes. Solar Energy, 158, 952–959.

60. Gopi, C.V.V.M., Ravi, S., Rao, S.S., Reddy, A.E., Kim, H.J. (2017) Carbon nanotube/metal-sulfde composite flexible electrodes for high-performance quantum dot-sensitized solar cells and supercapacitors. Sci. Repo., 7, 46519–12.

61. Yao, Z., Wang, L., Zhang, Y., Yu, Z., Jiang, Z. (2014) Carbon nanotube modified $Zn_{0.83}Cd_{0.17}S$ nanocomposite photocatalyst and its hydrogen production under visible-light. Int. J. Hydrog. Energy, 39, 15380–15386.

62. Lv, J., Li, D., Dai, K., Liang, C., Jiang, D., Lu, L., Zhu, G. (2017) Multi-walled carbon nanotube supported CdS-DETA nanocomposite for efficient visible light photocatalysis. Mater. Chem. Phy., 186, 372–381.

63. Mir, S.H., Nagahara, L.A., Thundat, T., Tabari, P.M., Furukawa, H., Khosla, A. (2018) Review–organic-inorganic hybrid functional materials: An integrated platform for applied technologies. J. Electro. Society, 165, B3137–B3156.

64. Ou, X., Li, J., Zheng, F.H., Wu, P., Pan, Q.C., Xiong, X.H., Yang, C.H., Liu, M.L. (2017) In situ X-ray diffraction characterization of $NiSe_2$ as a promising anode material for sodium ion batteries. J. Power Sources, 343, 483–491.

65. Park, S.K., Kim, J.K., Chan Kang, Y. (2017) Metal–organic framework-derived $CoSe_2$/(NiCo)Se_2 box-in-box hollow nanocubes with enhanced electrochemical properties for sodium-ion storage and hydrogen evolution. J. Mater. Chem. A, 5, 18823–18830.

66. Zhang, F., Xia, C., Zhu, J.J., Ahmed, B., Liang, H.F., Velusamy, D.B., Schwingenschlögl, U., Alshareef, H.N. (2016) $SnSe_2$ 2D Anodes for advanced sodium ion batteries. Adv. Energy Mater., 6, 1601188.

67. Dong, S.H., Li, C.X., Ge, X.L., Li, Z.Q., Miao, X.G., Yin, L.W. (2017) ZnS-Sb$_2$S$_3$@C Core-double shell polyhedron structure derived from metal–Organic framework as anodes for high performance sodium ion batteries. ACS Nano, 11, 6474–6482.

68. Fang, G.Z., Wu, Z.X., Zhou, J., Zhu, C.Y., Cao, X.X., Lin, T.Q., Chen, Y.M., Wang, C., Pan, A.Q., Liang, S.Q. (2018) Sodium-ion batteries: Observation of pseudocapacitive effect and fast ion diffusion in bimetallic sulfides as an advanced sodium-ion battery anode. Adv. Energy Mater., 8, 1703155.

69. Vasilyev, S., Moskalev, I., Mirov, M., Mirov, S., Gapontsev, V. (2016) Multi-Watt mid-IR femtosecond polycrystalline Cr^{2+}:ZnS and Cr^{2+}:ZnSe laser amplifiers with the spectrum spanning 2.0–2.6 μm. Optic Express, 24, 1616–1623.

70. Kim, S., Marshall, A.R., Kroupa, D.M., Miller, E.M., Luther, J.M., Jeong, S., Beard, M.C. (2015) Air-stable and efficient PbSe quantum-dot solar cells based upon ZnSe to PbSe cation-exchanged quantum dots. ACS Nano, 9, 8157–8164.

71. Yang, W.J., Liu, B.D., Yang, B., Wang, J.Y., Sekiguchi, T., Thorsten, S., Jiang, X. (2015) Pseudobinary Solid-solution: An alternative way for the bandgap engineering of semiconductor nanowires in the case of GaP–ZnSe. Adv. Funct. Mater., 25, 2543–2551.

72. Chen, Z.L., Wu, R.B., Wang, H., Zhang, K.H.L., Song, Y., Wu, F.L., Fang, F., Sun, D.L. (2017) Embedding ZnSe nanodots in nitrogen-doped hollow carbon architectures for superior lithium storage. Nano Res., 11, 966–978.

73. Xu, Y.H., Liang, J.W., Zhang, K.L., Zhu, Y.C., Wei, D.H., Qian, Y.T. (2016) Origin of additional capacities in selenium-based ZnSe@C nanocomposite Li-ion battery electrodes. Electrochem. Commun., 65, 44–47.

74. Tang, C.J., Wei, X.J., Cai, X.Y., An, Q.Y., Hu, P., Sheng, J.Z., Zhu, J.X., Chou, S.L., Wu, L.M., Mai, L.Q. (2018) Hollow cobalt selenide microspheres: Synthesis and application as anode materials for Na-Ion batteries. ACS Appl. Mater. Interfaces, 10, 19626–19632.

75. Lu, S., Zhu, T., Wu, H., Wang, Y., Li, J., Abdelkader, A., Xi, K., Wang, W., Li, Y., Ding, S., Gao, G., Kumar, R.V. (2019) Construction of ultrafne ZnSe nanoparticles on/in amorphous carbon hollow nanospheres with high-power-density sodium storage. Nano Energy, 59, 762–772.

76. Kurochka, K.V., Melnikova, N.V. (2017) Investigation of electrical properties of glassy AgGe$_{1+x}$As$_{1-x}$(S+CNT)$_3$(x = 0.4; 0.5; 0.6) at temperature range from 10 to 300 K. Solid State Ionics, 300, 53–59.

77. Ram, I.S., Singh, K. (2013) Thermal and mechanical properties of CNT–Se$_{90-x}$Te$_{10}$Ag$_x$ (x = 0, 5 and 10) glassy composites. J. Alloys and Comp., 576, 358–362.

78. Yu, X.Y., Yu, L., Lou, X.W.D. (2016) Metal sulfide hollow nanostructures for electrochemical energy storage. Adv. Energy Mater., 6, 1501333.

79. Rui, X., Tan, H., Yan, Q. (2014) Nanostructured metal sulfides for energy storage. Nanoscale, 6, 9889–9924.

80. Li, H., Su, Y., Sun, W. et al., (2016) Carbon nanotubes rooted in porous ternary metal sulfide@N/S-doped carbon dodecahedron: Bimetal-organic-frameworks derivation and electrochemical application for high-capacity and long-life lithium-ion batteries. Adv. Funct. Mater., 26, 8345–8353.

81. Zhang, K., Park, M., Zhou, L. et al., (2016) Urchin-like CoSe$_2$ as a high-performance anode material for sodium-ion batteries. Adv. Funct. Mater., 26, 6728–6735.

82. Zhang, F., Xia, C., Zhu, J. et al., (2016) SnSe$_2$ 2D anodes for advanced sodium ion batteries. Adv. Energy Mater., 6, 1601188.

83. Xu, X., Liu, W., Kim, Y. et al., (2014) Nanostructured transition metal sulfides for lithium ion batteries: Progress and challenges. Nano Today, 9, 604–630.

84. Li, T., Li, X., Wang, Z. et al., (2017) A short process for the efficient utilization of transition-metal chlorides in lithium-ion batteries: A case of Ni$_{0.8}$Co$_{0.1}$Mn$_{0.1}$O$_{1.1}$ and LiNi$_{0.8}$Co$_{0.1}$Mn$_{0.1}$O$_2$. J. Power Sources. 342, 495–503.

85. Yuan, H., Kong, L., Li, T. et al. (2017) A review of transition metal chalcogenide/graphene nanocomposites for energy storage and conversion. Chinese Chemical Lett., 28, 2180–2194.

86. Liu, X., Huang, J.Q., Zhang, Q. et al. (2017) Nanostructured metal oxides and sulfides for lithium–sulfur batteries. Adv. Mater., 29, 1601759.

87. Zou, X., Zhang, Y. (2015) Noble metal-free hydrogen evolution catalysts for water splitting. Chem. Soc. Rev., 44, 5148–5180.

88. Kagkoura, A., Skaltsas, T., Tagmatarchis, N. (2017) Transition-metal chalcogenide/graphene ensembles for light-induced energy applications. J. Chem. Eur., 23, 12967–12979.

89. Gdmez-Navarro, C., Weitz, R.T., Bittner, A.M., Scolari, M., Mews, A., Burghard, M., Kern, K. (2007) Electronic transport properties of individual chemically reduced graphene oxide sheets. Nano Lett., 7, 3499–3503.

90. Farrow, B., Kamat, P.V. (2009) CdSe Quantum dot sensitized solar cells. Shuttling electrons through stacked carbon nanocups. J. Am. Chem. Soc., 2009, 131, 11124.

91. Allen, M.J., Tung, V.C., Kaner, R.B. (2010) Honeycomb carbon: A review of graphene. Chem. Rev., 110, 132–145.

92. Iwase, A., Ng, Y.H., Ishiguro, Y., Kudo, A., Amal, R. (2011) Reduced graphene oxide as a solid-state electron mediator in Z-scheme photocatalytic water splitting under visible light. J. Am. Chem. Soc., 133, 11054.

93. Lightcap, I.V., Kosel, T.H., Kamat, P.V. (2010) Anchoring semiconductor and metal nanoparticles on a two-dimensional catalyst mat. storing and shuttling electrons with reduced graphene oxide. Nano Lett., 10, 577–583.

94. Balandin, A.A., Ghosh, S., Bao, W., Calizo, I., Teweldebrhan, D., Miao, F., Lau, C.N. (2008) Superior thermal conductivity of single-layer graphene. Nano Lett., 2008, 8, 902–907.

95. Cheng, G., Akhtar, M.S., Yang, O.B., Stadler, F.J. (2013) Novel preparation of anatase TiO_2@reduced graphene oxide hybrids for high-performance dye-sensitized solar cells. ACS Appl. Mater. Interfaces, 5, 6635–6642.

96. Park, B., Oh, S.M., Jin, X., Adpakpang, K., Lee, N.S., Hwang, S.J. (2017) A 2D metal oxide nanosheet as an efficient additive for improving Na-ion electrode activity of graphene-based nanocomposites. Chem. Eur. J., 23, 6544–6551.

97. Fan, L., Tao, Y., Wu, X., Wu, Z., Wu, J. (2017) $HfX_3(X = Se$ and $S)$/graphene composites for flexible photodetectors from visible to near-infrared. Mater. Res. Bull., 93, 21–27.

98. Ma, L., Wang, X., Zhao, B., Yang, J., Zhang, X., Zhou, Y., Chen, J. (2018) Novel metal chalcogenide supported on three-dimensional graphene foam for enhanced lithium storage. J. Alloys Compd., 762, 149–156.

99. Wang, Y., Wang, F., He, J. (2013) Controlled fabrication and photocatalytic properties of a three-dimensional ZnO nanowire/reduced graphene oxide/CdS heterostructure on carbon cloth. Nanoscale, 5, 11291.

100. Jaeger, V., Wilson, W., Subramanian, V. (2011) Photodegradation of methyl orange and 2,3-butanedione on titanium-dioxide nanotube arrays efficiently synthesized on titanium coils. Appl. Catal., B: Environmental, 110, 6–13.

101. Mukherjee, B., Smith, Y.R., Subramanian, V. (2012) CdSe Nanocrystal assemblies on anodized TiO_2 nanotubes: Optical, surface, and photoelectrochemical properties. J. Phys. Chem. C, 116, 15175–15184.

102. Mukherjee, B., Gupta, S., Peterson, A., Imahori, H., Manivannan, A., Subramanian, V. (2014) A unique architecture based on 1D semiconductor, reduced graphene oxide, and chalcogenide with multifunctional properties. Chem. Eur. J., 20, 10456–10465.

Index

Printed and bound by CPI Group (UK) Ltd, Croydon, CR0 4YY

24/10/2024

01778294-0005